125 **Topics in Current Chemistry**

Fortschritte der Chemischen Forschung

Managing Editor: F. L. Boschke

Stereochemistry

Editors: F. Vögtle, E. Weber

With Contributions by
W. H. Laarhoven, J. Martens, W. M. Müller,
K. Naemura, M. Nakazaki, W. J. C. Prinsen,
K. Schlögl, F. Vögtle, W. H. Watson,
K. Yamamoto

With 61 Figures and 37 Tables

Springer-Verlag
Berlin Heidelberg GmbH 1984

This series presents critical reviews of the present position and future trends in modern chemical research. It is addressed to all research and industrial chemists who wish to keep abreast of advances in their subject.

As a rule, contributions are specially commissioned. The editors and publishers will, however, always be pleased to receive suggestions and supplementary information. Papers are accepted for "Topics in Current Chemistry" in English.

ISBN 978-3-662-15271-3 ISBN 978-3-540-39015-2 (eBook)
DOI 1.1007/978-3-540-39015-2

Library of Congress Cataloging in Publication Data. Main entry under title: Stereochemistry. (Topics in current chemistry = Fortschritte der chemischen Forschung; 125)
Bibliography: p.
Includes index.
Contents: Stereochemistry of twisted double bond systems/M. Nakazaki, K. Yamamoto, K. Naemura — Planarchial molecular structures/K. Schlögel — Carbohelicenes and heterohelicenes/W. H. Laarhoven, W. J. C. Prinsen — [etc.]
1. Stereochemistry—Addresses, essays, lectures.
I. Vögtle, F. (Fritz), 1939– . II. Weber, E. III. Laarhoven, W. H. IV. Series: Topics in current chemistry; 125.
QD1.F58 vol. 125 [QD481] 540s [541.2'23] 84-5638

Originally published by Springer-Verlag Berlin Heidelberg New York in 1984.
Softcover reprint of the hardcover 1st edition 1984

2152/3020-543210

Table of Contents

Stereochemistry of Twisted Double Bond Systems

Masao Nakazaki, Koji Yamamoto, and Koichiro Naemura

Department of Chemistry, Faculty of Engineering Science, Osaka University, Toyonaka, Osaka 560, Japan

Table of Contents

I Introduction

Principally, the double bond in any chiral molecule is dissymmetrically deformed even if this unsaturated center is situated far from its chiral center. However, discussions in this article are limited solely to the synthesis and stereochemistry of the compounds whose double-bond systems 1 play a decisive role in the generation of their chirality by being explicitly twisted in the molecular environment (e.g. ring) surrounding the unsaturated center.

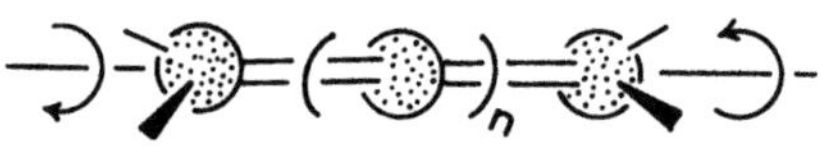

1

Among a variety of compounds whose molecular characteristics fall into this category, our current interests have further limited our discussion to the double-bond systems which may conveniently be classified as follows: 1. chiral (*E*)-cycloalkenes, 2. chiral anti-Bredt rule compounds, 3. trans-doubly bridged ethylenes ("betweenanenes"), 4. bridged allenes, and 5. overcrowded olefins.

II Chiral (*E*)-Cycloalkenes

The chiral nature of (*E*)-cycloalkenes, whose "jump rope" conformational interconversions are restricted because of non-bonding interaction across the ring, was first pointed out in 1952 when Blomquist [1] presented the enantiomeric figures of (*E*)-cyclononene in his paper describing the syntheses of cis- and trans-isomers of cyclononene.

A decade had elapsed, however, before this was first demonstrated by Cope's preparation of optically active (*Z*),(*E*)-1,5-cyclooctadiene [2] followed by his successful optical resolution of (*E*)-cyclooctene [3].

II.1 (*E*)-Cyclooctene

II.1.1 Preparation of Optically Active (*E*)-Cyclooctene

Although the racemic modification of (*E*)-cylooctene had been reported by Ziegler [4] in 1950, the optically active modification was first obtained in 1962 through resolution via chiral Pt(II) complexes containing (+)-1-phenyl-2-aminopropane [3,5]. This pioneering work has been followed by various approaches among which some are excellent as regards their synthetic convenience and stereoselectivity.

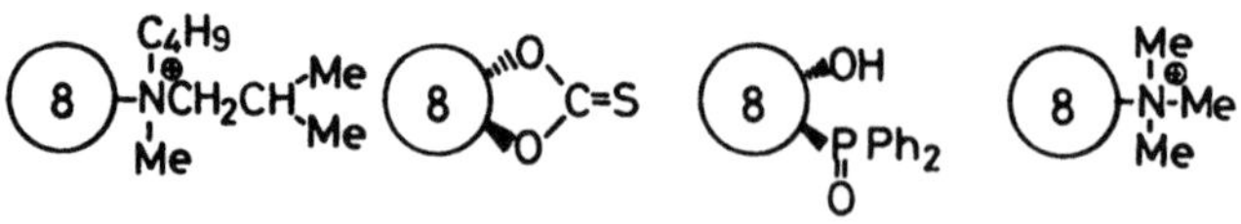

(+)-*2* (+)-*3* *4* *5*

a) The Hofmann elimination of the optically active ammonium salt 2 (1.1–1.4% asymmetric induction) [6].
b) Treatment of the (+)-thioncarbonate 3 with triisooctyl phosphite (high optical purity and isomeric purity) [7].
c) "Destructive asymmetric hydroboration" of (±)-(*E*)-cyclooctene employing the diborane prepared from (+)-pinene (20% optical purity) [8].
d) Reductive elimination of 4 with NaH in DMF [9].
e) The Hofmann elimination of the ammonium salt 5 in 3-aryl-3,5-cholestadiene liquid crystals (poor isomeric purity and 1.3–7.2% e.e.) [10].

II.1.2 Absolute Configuration

Chirality of (*E*)-cycloalkene has been customarily classified as planar in nature, and specified following Cahn, Ingold and Prelog's convention [11,12].

Shortly after the first announcement of optical resolution of (*E*)-cyclooctene, Moscowitz and Mislow [13] published a communication in which, on the basis of their MO calculation, they assigned the (S)-configuration to the (—)-enantiomer. Eventually, this conclusion was proved wrong [14,15] and the opposite configuration was assigned when the absolute configuration of (—)-(*E*)-cyclooctene was shown to be directly correlated with that of (+)-tartaric acid [16a,b].

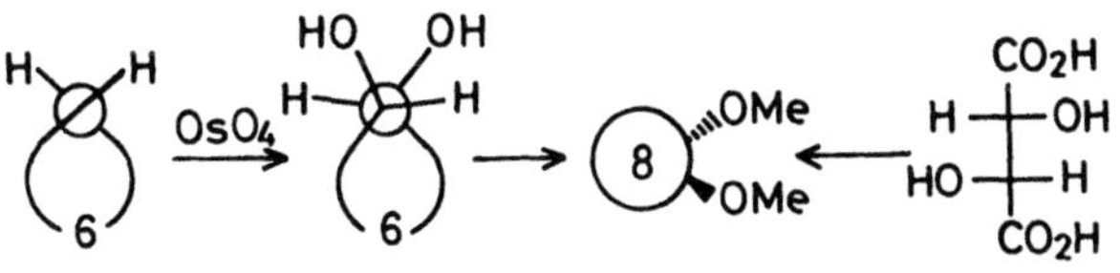

(-)-6 (+)-7 (+)-8 (+)-9

Osmium tetroxide oxidation of (—)-(*E*)-cyclooctene (6) afforded the (+)-diol 7 whose absolute configuration was related to that of (+)-tartaric acid (9) via the (+)-dimethoxy derivative 8. The (R)-configuration assigned by this correlation has been confirmed by a number of direct or indirect approaches.

a) X-ray analysis of (*E*)-dichloro[(—)-(*E*)-cyclooctene][(+)-α-methylbenzylamine] [17].
b) X-ray analysis of the (+)-dibromocarbene adduct 10 prepared from (—)-(*E*)-cyclooctene [18].
c) Correlation of (—)-(*E*)-cyclooctene with (+)-(*E*)-1,2-cyclooctanediol 7 via the (+)-thioncarbonate 3 [7].
d) Through the $[2_\pi + 4_\pi]$addition product 11 obtained from (+)-(*E*)-cyclooctene and butadiene [9,19].

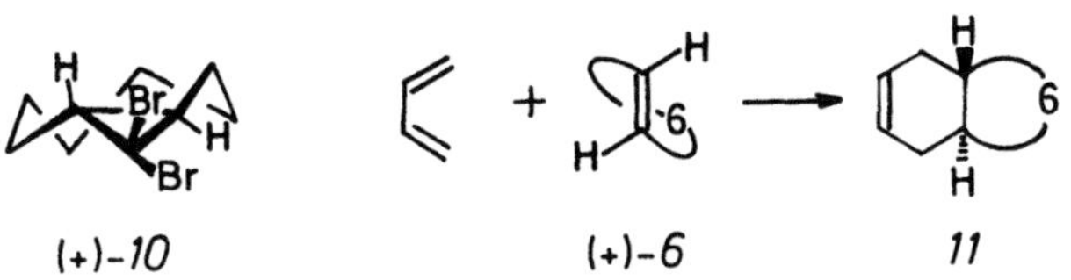

(+)-10 (+)-6 11

II.1.3 Optical Stability of (*E*)-Cycloalkenes

Studies of the rate of racemization [20] of optically active (*E*)-cyclooctene enabled Cope and coworkers to calculate the following $t_{1/2}$: 122 h (132.7 °C), 15 h (156.4 °C), and 1 h (183.9 °C). These values can be compared with the $t_{1/2}$ value of 6 sec (30 °C) exhibited by optically active (*E*)-cyclononene [21], which was prepared by optical resolution via a chiral Pt(II) complex.

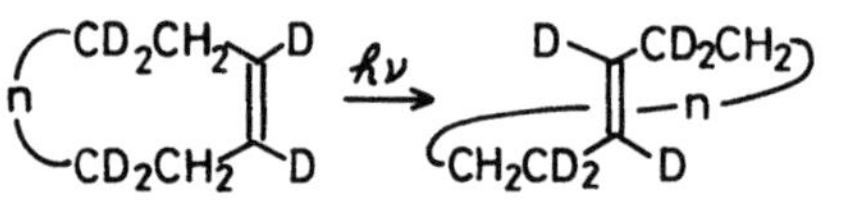

12 13

n	$t_{1/2}$
2	10^5 year
3	10 s
4	10^{-4} s

Although an unsuccessful optical resolution [21] of (*E*)-cyclodecene suggested its optical instability, Robert's dynamic NMR studies [22] of the racemization process in deuterated (*E*)-cycloalkenes 13 succeeded in providing a $t_{1/2}$ value of 10^{-4} sec (room temperature). This optical instability, found in the parent compound, could explain Westen's fruitless attempt to prepare the chiral (*E*)-cyclodecenone 16 from the (+)-methanesulfonate 15 [23].

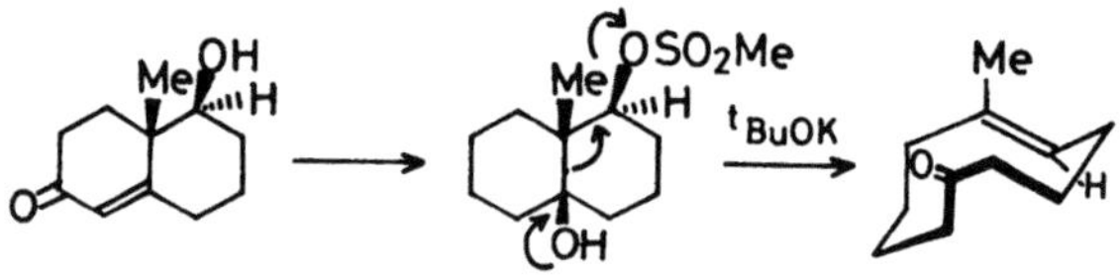

(+)-14 (+)-15 16

The cross-conformation 17 of (—)-(R)-(*E*)-cyclooctene was first disclosed by an X-ray analysis [17] of (*E*)-dichloro [(—)-(*E*)-cyclooctene][(+)-α-methylbenzylamine]-

17

(*R*)-cross 18 (*S*)-chair 19

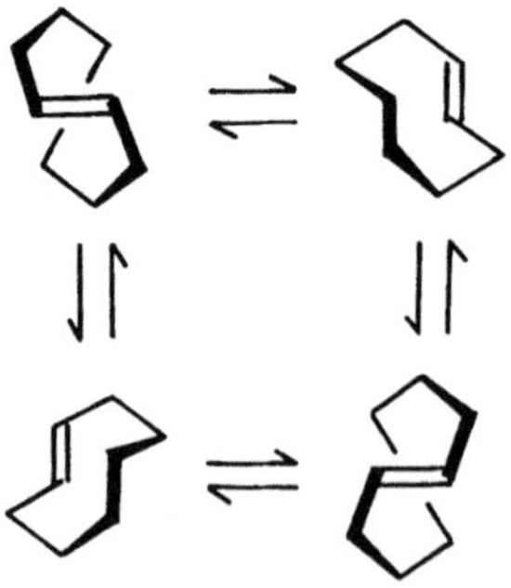

(*R*)-chair 21 (*S*)-cross 20

Pt(II), and this cross structure has been supported by another X-ray analysis [24)] of (*E*)-2-cyclooctenyl-3′,5′-dinitrobenzoate as well as an electron diffraction study [25)] of (*E*)-cyclooctene itself. Force-field calculations [26a, b)] have revealed that the racemization of (*E*)-cyclooctene should follow the pathway: (R)-cross 18→ (S)-chair 19 → (S)-cross 20, and that the former pathway is rate determining.

II.2 Substituted (*E*)-Cycloalkenes

In the attempted optical resolution by complexing with chiral Pt(II) complexes, both the benzo derivatives 22 [27)] and 23 [28a, b)] afforded the corresponding diastereomeric pairs of Pt(II) complexes. But their decomposition to the optically active modifications were found unsuccessful. This optical instability of 22 is especially surprising when this is compared with that of (*E*)-cyclononene [21)] which, though labile, could be resolved under a similar condition (vide supra).

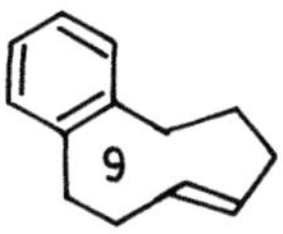

22

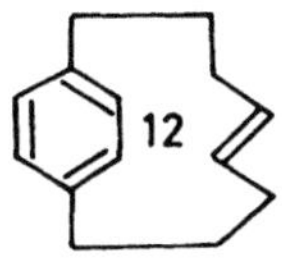

23

Expecting that the introduction of 1,2-dimethyl substituents to (*E*)-cycloalkenes should increase non-bonding interaction across the ring, Marshall and coworkers [29)] prepared (—)-(*E*)-1,2-dimethylcyclodecene (27a) and showed that this compound is optically quite stable. In their synthetic approach to 27a, they started from the β-keto ester 24 which was converted into (+)-25 through a sequence of reactions involving condensation with 3-buten-2-one, $LiAlH_4$ reduction, and resolution via the camphor-

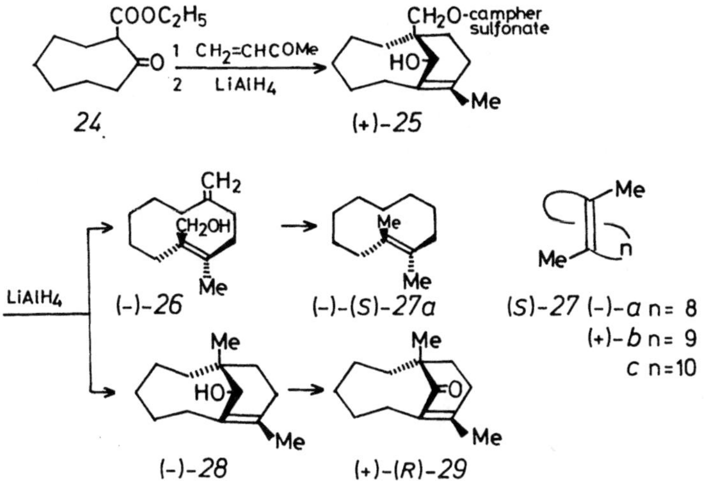

sulfonate. Cleavage of (+)-25 with $LiAlH_4$ yielded the (—)-cyclodecene derivative 26, accompanied by (—)-28. The routine synthetic sequence converted (—)-26 into the (—)-(*E*)-cyclodecene 27a whose (S)-configuration was established by correlating it with that of the by-product (—)-28. This by-product 28 was oxidized to the (+)-α,β-unsaturated ketone 29, which exhibited a (+)-Cotton effect, indicating that the (R)-configuration was present at the quaternary asymmetric carbon atom.

(+)-(S)-(*E*)-1,2-Dimethylcycloundecene (27b) was also prepared by a similar method and was found to be optically stable, but preparation of the cyclododecene derivative 27c in an optically active modification was unsuccessful.

II.3 (*Z*),(*E*)- and (*E*),(*E*)-1,5-Cyclooctadienes

As mentioned earlier, preparation of (*Z*),(*E*)-1,5-cyclooctadiene (31) in an optically active modification [2] first demonstrated the chiral nature of (*E*)-cycloalkenes. In this classical experiment, Cope and coworkers obtained (+)-31 by the Hofmann elimination of the (+)-ammonium salt 30. They were also successful in obtaining (+)-31 by optical resolution of racemic 31 through complexing with a chiral Pt(II) complex [5,30].

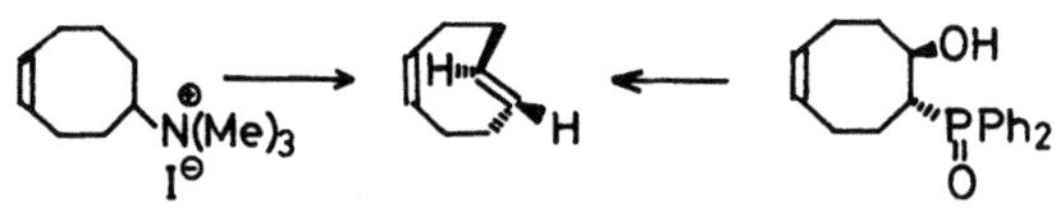

(+)-30 (+)-(S)-31 (-)-32

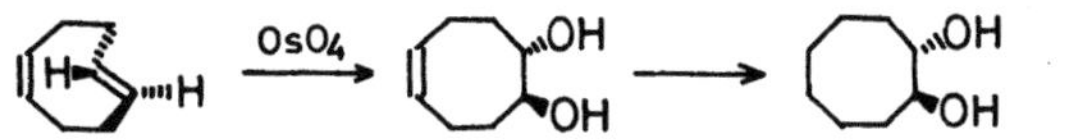

(-)-(R)-31 (+)-33 (+)-(1S,2S)-34

The (+)-enantiomer 31 which has the highest optical rotation ($[\alpha]_D$ + 171°) so far recorded, was reported to be formed by reductive elimination [9,31] of (—)-32 using NaH-DMF. Osmium tetroxide oxidation discriminated the two unsaturated centers in (—)-31, yielding the unsaturated (+)-*cis*-diol 33 whose conversion into (+)-(1S,2S)-34 established the (R)-configuration of (—)-31 [32].

The (*E*),(*E*)-isomer 35 [33] of 1,5-cyclooctadiene was reported to be formed (1.5% yield) when a solid residue, obtained from an UV irradiation product of bis[chloro-((*Z*),(*Z*)-1,5-cyclooctadiene)Copper(I)], was decomposed with NaCN solution. It has been claimed that the same compound 35 was also formed (2.4% yield) by the double Hofmann elimination of (*Z*)- or (*E*)-1,5-bis-(dimethylamino)cyclooctane dimetho-

35 36 37

iodide. The highly strained structure inherent in 35 seems to be reflected in its labile nature toward O_2 and acid. UV irradiation of 35 was reported to give the tricyclic 36. A force-field calculation [26b)] suggested that the cross-conformation 35 of D_2 symmetry appears to be 5.35 kcal/mol more stable than the chair form 37.

II.4 Other Related Systems

An X-ray analysis [34a)] of (*E*),(*E*),(*E*)-1,5,9-cyclododecatriene (38) suggested that this molecule is most comfortable in a twisted chiral conformation of D_3 symmetry, and a dynamic NMR study and a force-field calculation [34b)] estimated $\Delta G^* = 8.6$ kcal/mol for the racemization process between two enantiomeric D_3 conformers. A pseudo-chair conformation of similar D_3 symmetry was also suggested [35)] for 1,5,9-cyclododecatriyne (39), following an *ab initio* STO-3G calculation and photoelectron spectroscopy.

Although the (*E*)-cycloalkene structures have been found in a large variety of natural products especially in macrolides [36a–c)], only a brief comment on the conformational aspects of some sesquiterpenes [37)] will conclude this section.

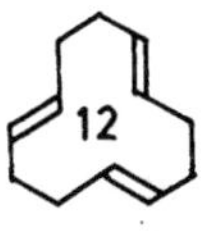

38

39

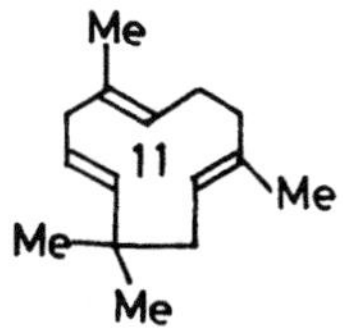

40

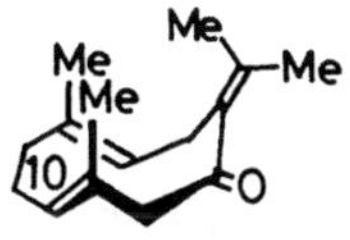

41

While a dynamic NMR study [38)] was carried out to reveal rather complicated patterns of conformational interconversion in some eleven-membered sesquiterpenes including humulene (40) and zermbone, an interesting controversy concerning conformational stability in the germacrone family of sesquiterpenes is worth mentioning here.

Tori and coworkers' dynamic NMR studies [39)] of germacrene concluded that this system is free in conformational inversion, suggesting that the optically active modification should not be isolated because of the low energy barrier between enantiomeric conformers. Hill and coworkers [40)], however, succeeded in preparing (−)-germacrone (41), which has an optical rotation as high as $[\alpha]_D^{33}$ −42.5° (EtOH). Osawa's force-field calculation [41)] of this system gave $\Delta H^* = 23$ kcal/mol as the lowest energy barrier between the enantiomeric conformers, and this seems to mean that optical resolution of germacrone should be feasible at room temperature.

III Chiral anti-Bredt Rule Compounds

Considering the impressive accumulation of review articles [42a–f)] on the Bredt rule and number of papers reporting exotic anti-Bredt rule compounds, it is rather surprising to realize that the chiral nature of anti-Bredt rule compounds had not been explicitly pointed out until Nakazaki's papers [43a, b)] on the synthesis and absolute configuration of the first optically active anti-Bredt rule compound appeared.

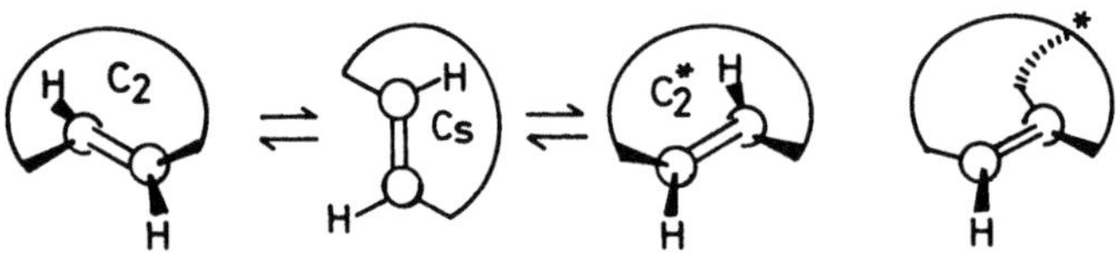

42 *43* *44* *45*

The close relationship [44a, b)] between bridgehead olefins and (*E*)-cycloalkenes had been discussed solely with regard to their strained characteristics, but, a careful examination revealed that this relationship should be extended to their chiral natures.

The chiral C_2-conformer 42 of a (*E*)-cycloalkene can transform into the enantiomeric C_2^*-conformer 44 through a planar C_s-conformer 43, and this "rope jump" racemization can be prevented by anchoring one end of the unsaturated center onto the ring by means of an extra-bridge. This bridging creates a bicyclic anti-Bredt rule compound 45, revealing that all anti-Bredt rule compounds (45) with one double bond are necessarily asymmetric (C_1 symmetry) and have one asymmetric carbon atom.

The synthetic approach towards optically active bicyclo[3.3.1]1(2)-nonene (49) [45a–c)] started from the (—)-(1R,3S)-hydroxycarboxylic acid 47 [46)], whose absolute configuration and optical purity were determinated by converting this into (+)-(R)-3-methylcyclohexanone (46). Conversion of (—)-47 to the Wittig compound 48 followed the routine synthetic sequence, involving protection of the carbonyl group and extension of the side chain. The final intramolecular ring closure of 48 [45c)] was accomplished by heating 48 with NaH in tetraglyme to yield the (—)-(S)-anti-Bredt rule compound 49.

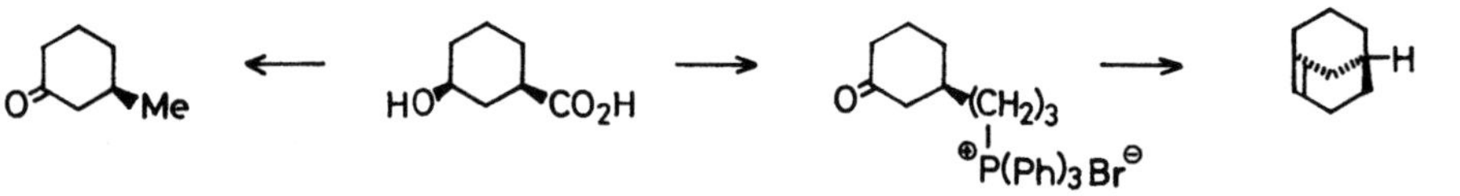

(+)-(*R*)-*46* (-)-(1*R*,3*S*)-*47* *48* (-)-(*S*)-*49*

Comparison of (—)-(R)-(*E*)-cyclooctene (50) and the (—)-(S)-anti-Bredt rule compound 51 would suggest their close structural relationship, and this is reflected in their respective absolute rotation values [5,16b)]: $[\alpha]_{D abs}$ —458° (neat) and —725° ($CHCl_3$) as well as in their respective (—)-Cotton curves: [θ] -1.4×10^5 at 196 nm (cyclohexane) and -13.6×10^5 at 213 nm (isooctane).

These chiroptical properties are compatible with the prediction made by Scott's octant rule [47)] which says that both compounds, which have polymethylene bridges in the (—)-regions (see 52), should exhibit (—)-Cotton effects.

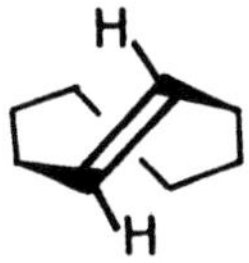

(–)-(*R*)-50

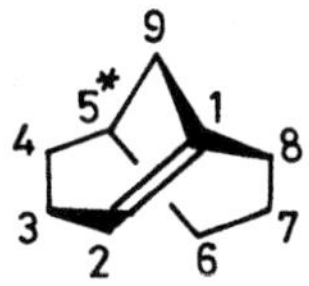

(–)-(*S*)-51

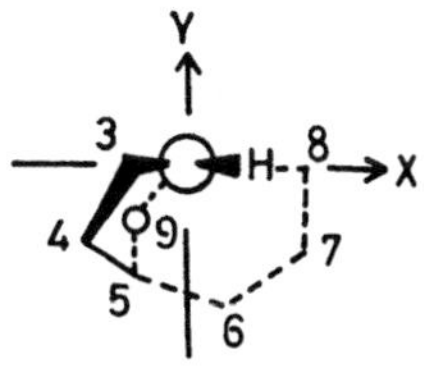

52

The (*E*)-cyclooctene moiety can also be seen in bicyclo [4.2.1]1(8)-nonene (53) and bicyclo[4.2.1]1(2)-nonene (54), "the smallest isolable members" among anti-Bredt rule compounds [48].

The interesting tricyclic compound 55 [49], which has (*E*)-cycloheptene moiety, was found to be very labile and could only be trapped by condensing with diphenylisobenzofuran.

53

54

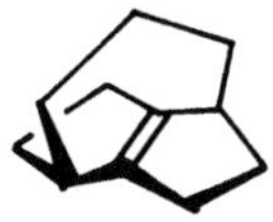

55

Since a "single" anti-Bredt rule compound is neccessarily asymmetric, combination of two of them should afford "double" anti-Bredt rule compounds ("bridgehead dienes") either of C_s (or C_i) symmetry or of C_2 symmetry depending upon the ways in which the enantiomers are combined to make up the molecules.

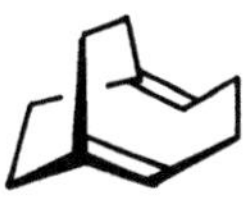

56

57

While the achiral "double" anti-Bredt rule compound 56 [50] of C_s symmetry was isolated in an impure state and was found to be very labile toward O_2 and heat, the chiral 57 of C_2 symmetry was assumed to exist very briefly in the pyrolysis of 3,6-dimethylidene-1,7-octadiene [51].

IV (*E*)-Doubly Bridged Ethylenes ("Betweenanenes")

A hypothetical compound "bis-((*E*)-polymethylene)ethylene" 58 of D_2 (V) symmetry had been formulated by Cahn, Ingold, and Prelog [11] for the sake of illustrating its planar chirality, but what aroused our independent interests in the synthesis and stereochemistry of this type of compound was a close structural relationship between

58 and [8][8]- (59) and [10][8]-paracyclophanes (60) [52], whose synthesis and determination of absolute configuration had been accomplished in our laboratory [53a–e].

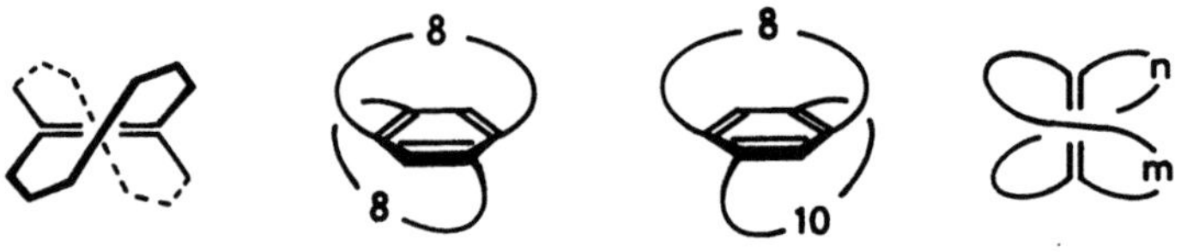

58 (+)-(*S*)-59 (−)-(*R*)-60 61 *a* m=10, n=8
b m=n=10
c m=n= 8

IV.1 Preparation of Betweenanenes

In 1977, almost simultaneously, groups from Osaka University [54a–e] and Northwestern University [55a–g] independently reported their respective syntheses of 61a and 61b, which, according to Marshall's proposal [55a], can be called [10.8] and [10.10] betweenanenes.

Nakazaki's synthetic approach is conspicuous by its remarkable straightforwardness; it has been proved to be so far the simplest synthetic route to the target compounds. In their first synthesis of 61a [54a], Nakazaki and coworkers started from cyclododecyne (62a), whose oligomerization with two molecules of butadiene afforded the bicyclic 63a. The cis[10.8] precursor 64a, obtained by partial catalytic hydrogenation with Raney nickel catalyst, was dissolved in cyclohexane, which contained xylene as photosensitizer, and the solution was irradiated with a medium pressure Hg lamp for 12 h. Examination of the reaction mixture by means of GLC indicated that the product was a 2.4:1 mixture of (*Z*) 64a and (*E*) 61a, and the further study [54b] showed that this ratio could be raised to 1:2 by irradiation of a hexene solution with a low pressure Hg lamp.

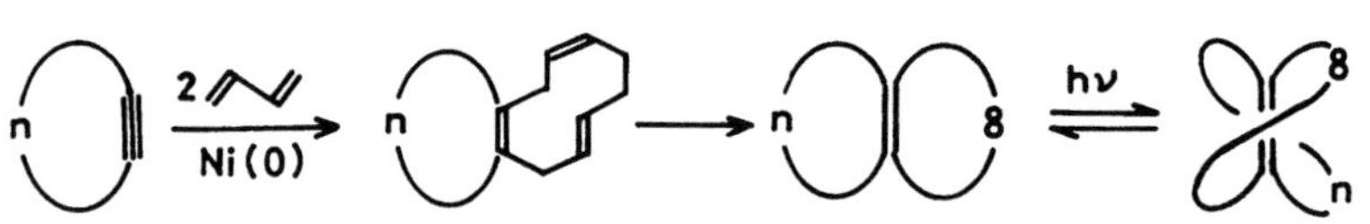

62 *a* n=10 63 *a* n=10 64 *a* n=10 61 *a* n=10
b n= 8 *b* n= 8 *b* n= 8 *c* n= 8

In their original procedure [54a], they took advantage of the inertness of (*E*) 61a toward dichlorocarbene for separating (*Z*) and (*E*) isomers. After the isomeric mixture was stirred with aq NaOH, $CHCl_3$ and cetyltrimethylammonium chloride (phase transfer catalyst), the reaction mixture was chromatographed over SiO_2 gel to yield [10.8]betweenanene (C_2-bicyclo[10.8.0]eicos-1(12)-ene), mp 37–38 °C (61a). This outstanding inertness toward dichlorocarbene was supplemented by another inertness found in catalytic hydrogenation (PtO_2 catalyst, in AcOH and AcOEt, at 60 °C), suggesting the unusually buried nature of the double bond, which is sandwiched between two polymethylene bridges.

In their first paper [54a)], they reported their failure in photoisomerization of the (*Z*)[8.8] precursor 64b which was prepared from cyclodecyne (62b) via the similar synthetic sequence as described for the higher homolog 61a. Irradiation [54b)] with a low pressure Hg lamp, however, saved the situation, yielding a 9:1 mixture of (*Z*)[8.8] and (*E*)[8.8] isomers after 4 h irradiation in cyclohexane.

The isolated [8.8]betweenanene (D_2-bicyclo[8.8.0]octadec-1(10)-ene) (61c), bp 125–127 °C/0.1 mmHg, was found to be converted back to the (*Z*)-(*E*) isomeric mixture by UV irradiation, and was characterized again by its inertness toward dichlorocarbene and catalytic hydrogenation.

The buried nature of the double bonds in both (*E*)[10.8] and (*E*)[8.8] isomers further explains their chromatographic behavior, namely that the (*E*)-isomers were always found in faster moving fractions.

A dramatic change in the UV spectra was observed in going from the cis to the trans series of doubly-bridged ethylenes, and this can most clearly be shown between the pair of (*Z*) and (*E*)[8.8] isomers: (*E*)[8.8], λ_{max} 201.5 nm (log ε 4.02); (*E*)[8.8], λ_{max}222.5 nm (log ε 3.73). This bathochromic shift, as large as 21 nm, accompanied by a marked decrease in the extinction coefficient obviously reflects the unusual nature of double bond in this (*E*)[8.8] isomer 61c, which is caused by out-of-plane bending and rehybridization of the strained π-bond [56)].

The key step in Marshall's approach [55a)] to [10.10]betweenanene (61b) was the stereospecific reductive cleavage of the (*E*) cyclic phosphate 68 with $LiAlH_4$, yielding the (*E*) 69. For the preparation of this strategic intermediate 69, they started from cyclododecane-1,2-dione (65), which was treated with dimethylsulfonium methylide to give a 1.5:1 mixture of the (*E*) 66a and the (*Z*) 66b bis-epoxides. Treatment of the separated (*E*) 66a with allyllithium gave the (*E*)-diol 67 which was then transformed to the cyclic phosphate 68 by reacting with butyllithium followed by phosphodichloridate. Extension of the side chains by the routine sequence of conversions involving hydroboration-oxidation gave the dimethyl dicarboxylate 70, which, after acyloin condensation and removal of the functional groups from the resulting acyloin, yielded the (*E*)[10.10] 61b, mp 64–65 °C.

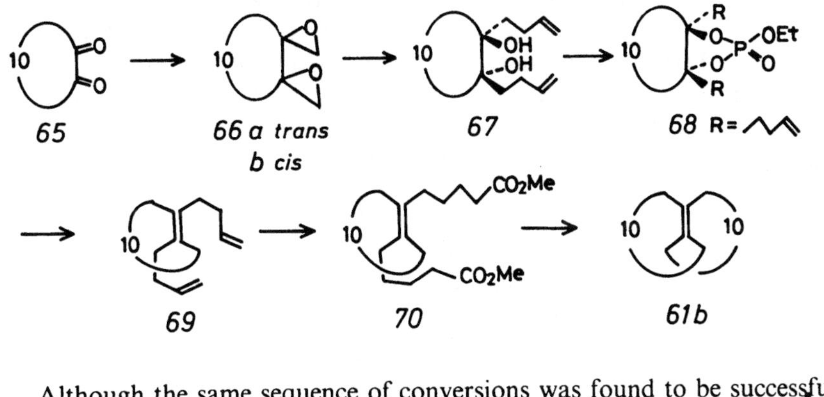

Although the same sequence of conversions was found to be successfully applied to the *cis*-bis-epoxide 66b, yielding the (*Z*)[10.10] isomer, mp 136–138 °C, this method failed in the synthesis of [10.8]betweenanene(61a) because of fruitless acyloin cyclization of the lower homolog of the diester 70 [55g].

In their second short communication (1979) [55b], Marshall reported that the (*E*)[10.10] isomer 61b was immune to epoxidation with *m*-chloroperbenzoic acid (quantitatively recovered after 3 weeks), while the (*Z*)[10.10] isomer was completely converted to epoxide within 3 min reaction time under the same condition.

They applied this newly-found chemical distinction to the problem of isolating both isomers. On treatment of the (*Z*)[10.8] precursor 64a with H_2SO_4-AcOH in benzene for 16 h, they found that the resulting mixture contained a 70:30 mixture of (*E*)[10.8] and unchanged (*Z*)[10.8] compounds, and an identical mixture was found to result from the (*E*)[10.8] isomer 64a upon similar treatment. Pure (*E*)[10.8] was reported to be obtained from these reaction mixtures by selectively converting the (*Z*)[10.8] precursor to the epoxide.

As the first step in extending this novel and facile betweenanene synthesis to the [10.10]series of compounds, they prepared the (*Z*)[10.10] precursor 73 from 1,2-cyclododecanedione (65), using McMurry's ring closure of the dialdehyde 72b as the key step.

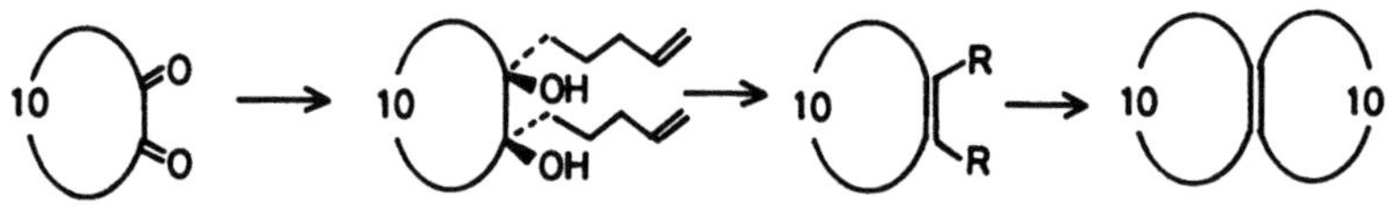

65 *71* *72a* R= *b* R= CHO *73*

Addition of 4-pentenyllithium to the dione 65 gave the *cis*-diol 71 which was converted to the (*Z*)-1,2-disubstituted cyclododecene 72a. Hydroboration-oxidation and chromium trioxide oxidation provided the dialdehyde 72b whose McMurry ring closure, followed by partial catalytic hydrogenation gave the (*Z*)[10.10] precursor 73. Treatment of this (*Z*)-olefin 73 with H_2SO_4-AcOH in benzene was reported to effect conversion into [10.10]betweenanene (61b) of 95% purity and high yield.

They also reported the sensitized photoisomerization of the cis[10.10] precursor 73 to a 50:50 mixture of (*Z*) and (*E*) isomers, and this mixture was said to be converted to virtually pure (*E*)[10.10] isomer on acid treatment. They attributed this facile isomerization to the thermodynamical stability of the (*E*) isomers over the (*Z*) isomers. However, their third short communication [55c] showed that the course of this acidic isomerization had not followed the route reported in their preceding paper.

Expecting that acidic isomerization involving a double Wagner-Meerwein rearrangement would transform the bicyclic olefin 74 into supposedly stable [10.10]betweenanene and its (*Z*)[10.10] isomer, they treated 74 with H_2SO_4-AcOH in benzene only to find that the product was an 85:15 mixture of 75 and 76. Solvolysis of the spiro compound 77 was also found to yield a 60:40 mixture of 75 and 76 which was totally free from the fused (*Z*)[10.10] and (*E*)[10.10] olefins.

These unexpected findings prompted them to re-examine the acidic isomerization of the (*Z*)[10.10] precursor 73, and what they found was that the isomerization product was the fused trisubstituted olefin 78 instead of the previously reported [10.10]

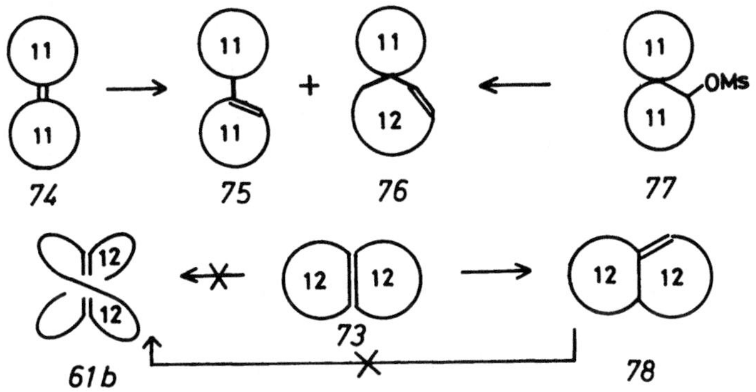

betweenanene (61b). Incidentally, acid treatment effected no change of the olefin 78 even after a prolonged reaction time, and exposure of authentic [10.10]betweenanene to acid failed to give any of the expected rearranged products.

Although these findings had seemed to resolve the conflicting situations, Nickon and coworker [57)] reported an interesting ring-size effect directing the course of this acidic isomerization.

They prepared the higher homolog 85 of Marshall's spiroalcohol 77, and found this gave upon treatment with $HClO_4$ a complex mixture of olefins containing a small amount of the cis-fused (*Z*)82. This unexpected observation encouraged them to explore a novel three-step synthetic route to [11.11]betweenanene (84). McMurry's coupling of cyclododecanone (79) afforded a 90:10 mixture of 80 and 81 which was treated with trifluoroacetic acid(neat) overnight. The heterogeneous reaction mixture was found to contain two fused-ring olefins 82 and 83. Chromatographic separation of (*Z*)[11.11] precursor 82 (22% yield, mp 129–130 °C) and its UV irradiation in heptane gave a 50% yield of [11.11]betweenanene (84), mp 103–104 °C, whose inertness toward catalytic hydrogenation was confirmed.

Marshall's original but rather involved [10.10]betweenanene synthesis has been improved in their second approach which reportedly presented a sixfold improvement in overall yield [55d)].

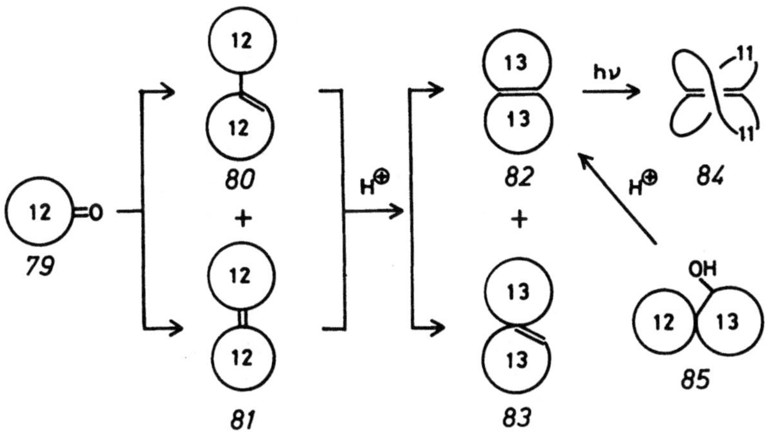

Aminomethylation of cyclododecanone (79) gave 86a which was converted to 1,2-dimethylenecyclododecane (87) via the Wittig condensation, quaternization, and Hofmann elimination. Bromination of 87 to the (*E*)-dibromide 88 (98% yield) was the key step in this approach. The routine synthetic sequence modified the side chains of 88 to provide the (*E*)-dialdehyde 88b whose McMurry ring closure, followed by removal of the newly formed unsaturated center, completed the synthesis.

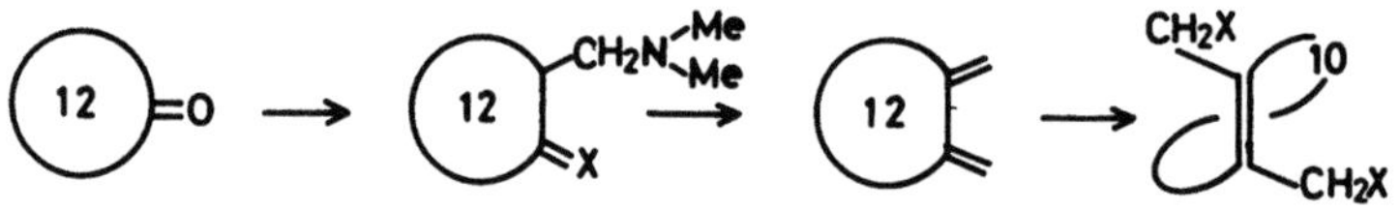

79 — 86 *a* X = O, *b* X = CH_2 — 87 — 88 *a* X = Br, *b* X = $(CH_2)_3CHO$

McMurry's condensation was again fully employed in Marshall's synthesis of "conformationally flexible" [20.10], [22.10], and [26.10]betweenanenes [55e)]. The first McMurry condensation of the diketones 89 afforded the monocyclic olefins 90 whose side chains were modified to the desired dialdehydes 91. The second McMurry condensation, when applied to these dialdehydes 91, brought about ring closure to give the (*Z*)-precursors, which have one extra unsaturated center. Partial catalytic hydrogenation removed this unsaturated center, and the resulting (*Z*)[20.10] 92a, [22.10] 92b, and [26.10]precursors 92c were irradiated, following Nakazaki's procedure [54a, b)], in cyclohexane-xylene solutions with a medium pressure Hg lamp. The (*E*):(*Z*) ratios in the reaction mixtures were 1.1:1; 1.6:1 and 1.3:1 respectively. No yields and no physical properties have been reported on these "conformationally flexible" betweenanenes except that the (*E*) isomers were invariably found in the fast moving zones in chromatography.

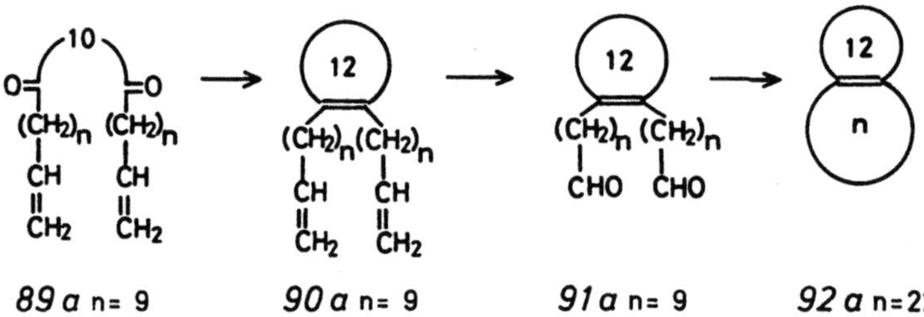

89 *a* n = 9, *b* n = 11 — 90 *a* n = 9, *b* n = 11 — 91 *a* n = 9, *b* n = 10, *c* n = 11 — 92 *a* n = 22, *b* n = 24, *c* n = 28

Finally, a brief comment on the approach [58a, b)] of a group at Bologna University to thia-betweenanenes will be presented here to conclude this section.

1-Bromocyclododecanone (93a) was converted into the spiroketone 94a via 93b, and the Wittig condensation transformed the resulting 94a to the unsaturated 94b.

Methylation with CF_3SO_3Me in dichloromethane gave a 1:1 mixture of the "cis" and "trans" isomeric methosulfonium salts 95 and 96. When this mixture was treated with t-BuOK (0.5 eq) at —40 °C in THF, the product was found to consists solely of the (*Z*)-thia[10.7] salt 97, mp 48–49 °C, which could readily be separated from the unreacted "trans" isomer 96. This "trans" salt, when treated again with the base at —40 °C, gave essentially pure thia [10.7]betweenanene (98), mp 42–43 °C. The buried nature of the double bond in this thia analog was reflected again in its reactivity toward *m*-chloroperbenzoic acid: (*Z*)-thia[10.7] 97 gave an epoxide sulfone, mp 191–192 °C, while the (*E*)-thia[10.7] 98 resisted epoxidation, yielding an unsaturated sulfone, mp 101–102 °C.

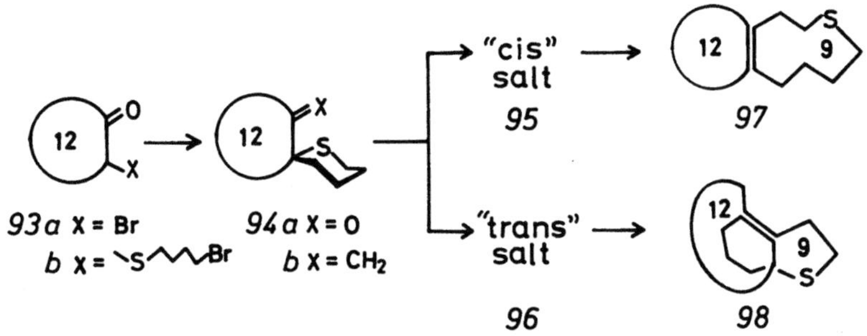

In their full paper [58b)], they reported the preparation of a diastereomeric mixture of ethyl [10.6]betweenanenecarboxylates via a sequence of conversions involving Ramberg-Bäckland's sulfur extrusion reaction.

IV.2 Chemical Characterization

Although various aspects of the chemical differences between (*Z*)[m.n] precursors and [m.n]betweenanenes have been already presented in the preceding sections, this section will summarize these characteristics and add some data, taken mainly from Marshall and Black's short communication [55e)].

a) Reactivity to dichlorocarbene:
   The inertness of [10.8]betweenanene toward dichlorocarbene was the first reaction [54a)] observed which distinguishes between (*Z*)[10.8] precursor and (*E*)[10.8] isomer, and this distinction was further observed in the [8.8]series of compounds [54b)].
b) Catalytic hydrogenation:
   Inertness of betweenanenes toward catalytic hydrogenation was first observed in [10.8] and [8.8]betweenanenes [54a, b)] and confirmed in [11.11] compound [57)].
c) Chromatographic behavior:
   The (*E*) isomers were invariably found in fast moving fractions. This was first reported in the [10.8] and [8.8] series [54b)] and has been confirmed in the [20.10], [22.10] and [26.10] series of compounds [55e)].

d) Photoisomerization:
Photoequilibrium between the (*Z*)-precursors and betweenanenes was found in the [10.8] and [8.8] series [54a,b]. [11.11] [57], [20.10], [22.10], and [26.10]Betweenanenes [55e] have been prepared from the (*Z*)-precursors by means of this photoisomerization procedure.

e) Hydroboration-oxidation:
Hydroboration with excess BH_3 in THF followed by oxidation with alkaline H_2O_2 converts the (*Z*)-99 (n = 20, 22, and 24) into the (*Z*)-fused alcohols 100, and betweenanenes (101) (n = 20, 22, and 24) into the (*E*)-fused alcohols 102 [55e]. The reaction rates were found to be slightly affected by ring size, but almost no effect was observed between cis and trans isomers.

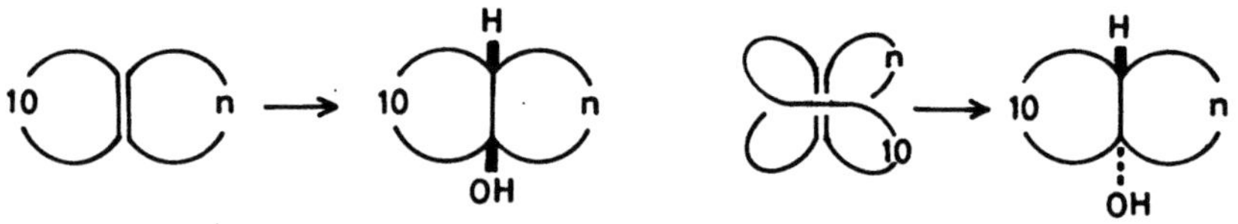

99 100 101 102

f) Epoxidation with *m*-chloroperbenzoic acid:
The (*Z*)-(*E*) discrimination with this reagent was most pronouncedly demonstrated in [10.10]series [55e], where (*Z*)[10.10] reacted within 1 min, while (*E*)[10.10] was recovered unchanged after three weeks. Although this difference in reactivity was further observed in the [20.10] and [22.10] series but to a lesser extent, take out no difference in reactivity between (*Z*)[26.10] precursor and [26.10]betweenanene was observed.

g) Photosensitized oxidation: [55e]
This oxidation converts the (*Z*)- and (*E*)[m.n] compounds into allylic alcohols after treatment with $LiAlH_4$ to reduce the initially formed hydroperoxides. [10.10]Betweenanene was recovered unchanged after 20 h; the reaction of (*Z*)[26.10] precursor was completed within 1 h while that of (*E*)[26.10] required 6 h.

h) Acidic isomerization:
The difference between (*Z*) and (*E*)[10.10] in the reaction pathways during acidic rearrangement was explained earlier [55c] (vide supra): (*Z*)[10.10] transforms into the fused olefin with a double bond in 1(2)-position while (*E*)[10.10] slowly decomposes. Isomerization of (*Z*)[26.10] and (*E*)[26.10] was effected using various acids [55e]. All procedures gave rise to mixtures containing principally same fused 1(2)-unsaturated alkene at relative rate of ca. 2:1 in favor of the (*Z*)[26.10] isomer.

i) Complexing with tetracyanoethylene:
All [m.n]betweenanenes so far prepared in Marshall's laboratory were found to exhibit a wine-red color reaction with tetracyanoethylene [55g], while the corresponding (*Z*) isomers did not.

IV.3 Preparation and Chiroptical Properties of Optically Active Betweenanenes

The Osaka University group [54c,d)] was the first to announce the preparation of betweenanene in an optically active modification and the assignment of the absolute configuration.

The key step in their approach was asymmetric photoisomerization of the α,β-unsaturated (*Z*)-ketone precursor 103b in diethyl (+)-L_g-tartrate. The bromide 103a obtained by N-bromo-succinimide bromination of the (*Z*)[8.8] precursor 64b, was converted into the α,β-unsaturated (*Z*)-ketone 103b by the routine synthetic procedures. Irradiation in a hexane solution with a medium pressure Hg lamp effected the photoisomerization of the (*Z*)-precursor 103b to afford a 1:5.5 mixture of (*Z*)-(*E*)[8.8] ketones. After these preliminary experiments, a neat solution of 103b in diethyl (+)-L_g-tartrate was irradiated for 3 h. Preparative GLC of the resulting 1:7 mixture of (*Z*) and (*E*)[8.8] ketones produced a 38% yield of (*E*)[8.8] ketone 104 enriched in the (—)-enantiomer, $[\alpha]_D^{24}$ —13° (hexane).

64b *103a* X = <(H)(Br) *b* X = O (−)-*104* X = O (−)-(*R*)-*61c* X = H_2

For the preparative purpose of optically active [8.8]betweenanene (61c), the 1:7 mixture of (*Z*) and (*E*)[8.8] ketones slightly enriched in (—)-104 was directly reduced by the Wolff-Kishner reduction. The optically active [8.8]betweenanene which was isolated by taking advantage of the inertness toward dichlorocarbene, was levorotatory, $[\alpha]_D^{24}$ —2.3° (isooctane), $[\theta]_{222.5\ nm}$ -1.8×10^3 (isooctane). The (R)-configuration was assigned to the (—)-modification following Scott's octant rule [47)], and comparison of its rotatory strength with that of (*E*)-cyclooctene suggests ca. 0.5–1.0% optical purity in this specimen. A specimen of much higher optical purity, $[\alpha]_D^{25}$ +10.3° (hexane) was obtained by direct optical resolution of (±)-[8.8] betweenanene through chromatography, employing (—)-poly(triphenylmethyl methacrylate) as the chiral adsorbent [59)].

Since Marshall's strategy [55e)] in preparing optically active betweenanenes was asymmetric epoxidation by (+)-monoperoxycamphoric acid, this approach would be solely effective in [m.n]betweenanenes where the m ring is large enough to sway away, so exposing the buried unsaturated center.

Epoxidation of [22.10] (105b) and [26.10]betweenanenes (105c) with this chiral reagent was quenched at 50% conversion, and the optically active [22.10] and [26.10] betweenanenes, isolated by SiO_2 gel chromatography, exhibited $[\alpha]_D^{22}$ —32.4° (hexane) and —24.7° (hexane) respectively [60)].

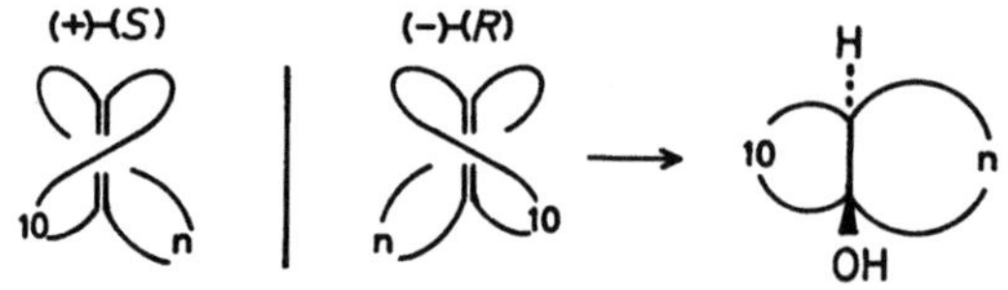

105 a n=20
b n=22
c n=26

106

Both these optically active betweenanenes exhibited a (—)-Cotton effect [47] at 195 nm, indicating their (R)-configuration. Hydroboration-oxidation of these specimens gave the (*E*)-fused alcohols 106 whose respective optical purities (7.6 and 6.0%) were estimated by means of the NMR chiral shift reagent method.

Preparation of the almost optically pure (+)-enantiomer of [10.10]betweenanene (61b) as well as its (R)-configuration were reported in Marshall's recent paper [55f].

In their "stereorational" synthesis of (+)-[10.10] 61b, they reacted the epoxide 107 with a 1:1 3-butenylmagnesium bromide-cuprous iodide complex in dimethylsulfide-THF at low temperature. The predominant S_N2' pathway gave the (±)-(*E*)-allyl alcohol 108 whose Sharpless asymmetric epoxidation in dichloromethane at −23 °C for 10 min provided the corresponding epoxy alcohol and recovered (+)-(R)-allyl alcohol 108 (78% yield and 95% optical purity). The (R)-configuration was assigned following the Sharpless model [61] for allylic alcohol epoxidation.

107 (+)-(*R*)-*108* (+)-(*R*)-*109* (+)-(*R*)-*110*

Extension of the side chain of the (+)-allyl alcohol 108 to the (+)-triene 109 of C_2 symmetry was accomplished by reacting 3-butenylmagnesium bromide-cuprous iodide complex with the phosphate derivative prepared from (+)-108. The optical purity (higher than 90%) of the (+)-triene 109 was estimated by applying the NMR chiral shift reagent method to the saturated diol, which was obtained from 109 through partial catalytic hydrogenation and hydroxylation of the resulting (+)-(*E*)-1,2-dipentylcyclododecene.

The remaining sequence of conversions leading to (+)-(R)-[10.10]betweenanene (61b), mp 85–86 °C, $[\alpha]_D^{29}$ +46.9° ($CHCl_3$), was rather straightforward involving: 1. hydroboration-oxidation, 2. oxidation with pyridinium chlorochromate, 3. McMurry's ring closure, and 4. removal of the newly-formed double bond by partial catalytic hydrogenation. The (R)-configuration of this (+)-modification was further confirmed from its negative Cotton effect, and the calculated $[M]_D$ 140° was found comparable with those of the various 1,2-disubstituted (*E*)-cycloalkenes so far prepared in his laboratory.

V Bridged Allenes

The number of the sp carbon atoms interposed between the two terminal sp^2 carbon atoms conveniently classifies the cumulated double bond systems into two categories: 1. even-numbered cumulenes of D_{2h} symmetry (e.g. ethylene), and 2. odd-numbered cumulenes of D_{2d} symmetry (e.g. allene).

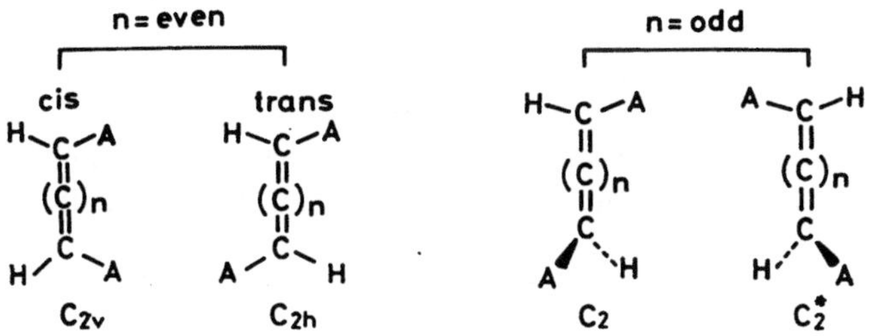

111 *112* *113* *114*

The symmetries characteristic of the parent hydrocarbons, determine the symmetry of the 1,n+2-disubstituted derivatives, 111, 112, 113, and 114, where A represents an achiral substituent.

While the even-numbered cumulene gives (*Z*)-111 of C_{2v} symmetry and (*E*)-112 of C_{2h} symmetry, the odd-numbered cumulene gives two enantiomers 113 and 114 both of C_2 symmetry. Bridging these two substituents A and A in 113 or 114 by a chain gives rise to a single-bridged allene of C_2 symmetry whose chirality can be specified following the axial chirality rule [11].

V.1 Single-Bridged Allenes

As was the case with chiral (*E*)-cycloalkenes, Cope and coworkers [62] were the first to prepare single-bridged allene in an optically active modification [63].

Optical resolution employing a chiral Pt(II) complex enabled them to obtain (—)-1,2-cyclononadiene (the enantiomer of 116), $[\alpha]_D^{24}$ —71° (neat) which was estimated to have an optical purity of 44%. In their second approach, (—)-(R)-(*E*)-cyclooctene (6) was treated with dibromocarbene to yield the (+)-adduct 115, which was further treated with methyllithium to give (+)-116, $[\alpha]_D^{25}$ +138° (neat) (ca. 85% optical purity) [64].

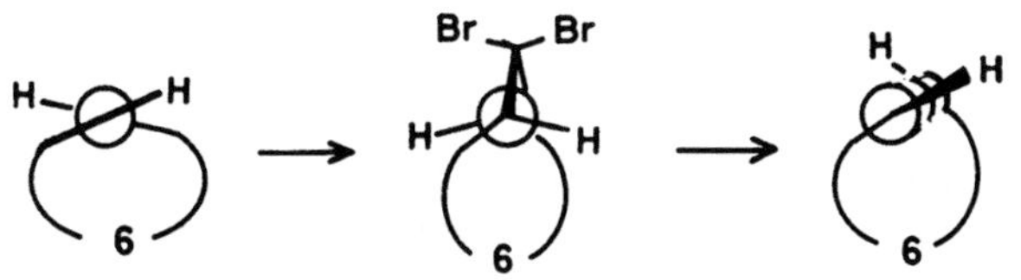

(—)-(*R*)-*6* (+)-*115* (+)-(*R*)-*116*

Lowe's chiroptical rule [65] concerning chiral allenes tells us that the enantiomers which exhibit dextro-rotation in the D line region invariably have the (S)-configuration.

By applying this rule to (+)-1,2-cyclononadiene (116) the (S)-configuration can be assigned. However its (+)-Cotton effect was found to be incompatible with the (—)-Cotton effects commonly observed in (+)-non-cyclic allenes. This discrepancy prompted Moore and coworkers [66] to re-examine the absolute configuration of this chiral monocyclic allene (116).

(+)-1,2-cyclononadiene (116) was converted into the dibromocarbene adduct 117 whose thermodynamically controlled Na-liq NH_3 reduction gave the (+)-trans-cyclopropane derivative 118. Comparison of its (+)-Cotton effect with the (+)-Cotton effects exhibited both by (—)-(1R,2R)-1,2-dimethylcyclopropane and (+)-(1R,2R)-bicyclo[6.1.0]nonane of known configurations enabled the (1R,9R)-configuration to be assigned to the cyclopropane intermediate 118 and this led eventually to the (R)-configuration of (+)-1,2-cyclononadiene (116). This absolute configuration has been supported by means of CD spectroscopy [67,68], oxymercuration [69], asymmetric hydroboration [70], and oxythallation [71].

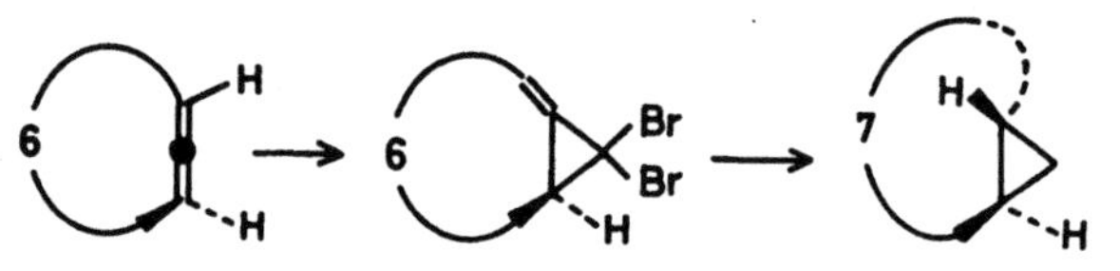

(+)-(*R*)-*116* *117* (+)-(1*R*,9*R*)-*118*

V.2 Monocyclic Diallenes

Symmetrical incorporation of two units of 113 or 114 into one ring should yield either the monocyclic diallenes of D_2 symmetry or of C_{2h} symmetry, depending upon the ways of combining the two enantiomers. Preparation of monocyclic diallenes in these two modifications was accomplished in Sondheimer's laboratory [72a,b].

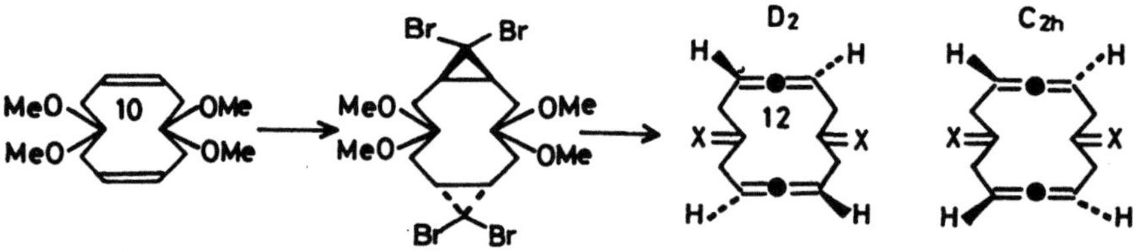

119 *120* *121a* X=(OMe)₂ / *b* X=O *122a* X=(OMe)₂ / *b* X=O

The double dibromocarbene adduct 120 obtained from 119 was shown to be predominantly the anti-form [73], and its reaction with (—)-sparteine-methyllithium reagent at —10 °C was reported to yield a mixture of 121a of D_2 symmetry, $[\alpha]_D$ +24.4° (hexane), and a small amount of the meso-form 122a, $[\alpha]_D$ 0°. Both these isomers were hydrolyzed to the corresponding diketones 121b and 122b. Although optical activity of 121a suggests its D_2 symmetry, its absolute configuration remains to be determined.

1,2,6,7-Cyclodecatetraene, mp 36 °C, a 10-membered monocyclic diallene, was prepared [75,76] and was shown by means of X-ray analysis [77] to posses a meso-form. Force-field calculations were carried out in order to study possible conformations of this mesoform and the dl-form of this compound [78a, b].

Sondheimer [74a, b] further reported the preparation of 3,4,5,10,11,12-cyclotetradecahexaene-1,8-dione which has two units of 111 or 112 (n = 2) in a 14-membered ring, but again its stereochemistry remained unknown.

V.3 Doubly-Bridged Allenes

A hypothetical doubly-bridged allene 123 of D_2 symmetry was first formulated in Cahn, Ingold, and Prelog's classical paper [11] in which they summarized their novel proposal for specifying various molecular chiralities. Despite the close structural resemblance between 123 and [m.m]betweenanene, Cahn, Ingold and Prelog's rules shows that the chirality of these two classes of compounds are to be specified as axial and planar respectively.

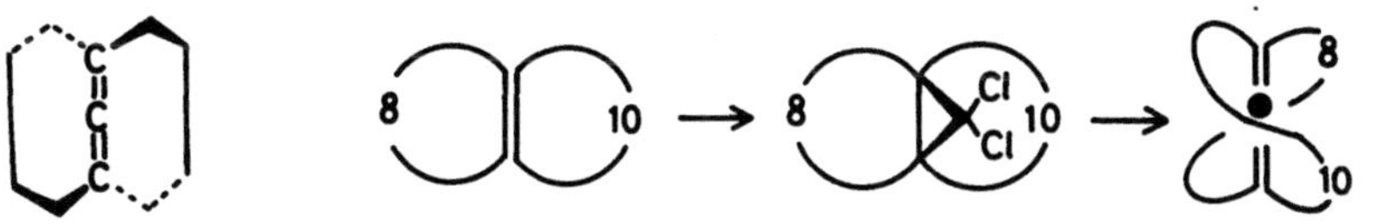

123 *64a* *124* (+)-(*R*)-*125* .

The racemic mixture and an optically active form of a doubly-bridged allene of this type was prepared in Nakazaki's laboratory [79a, b].

They treated the dichlorocarbene adduct 124, obtained from the (*Z*)[10.8] precursor 64a, with (—)-sparteine-methyllithium reagent at —80 °C. SiO_2-gel chromatography of the reaction mixture indicated a 9% yield of (+)-125, mp 59–64 °C, $[\alpha]_D^{18}$ +4.3° (hexane), whose (R)-configuration and ca. 4% optical purity were estimated from its Cotton curve [68]: $[\theta]$ $+1.4 \times 10^2$ (220 nm); -2.0×10^2 (231 nm).

VI Overcrowded Olefins

Since discussions in the preceding sections are limited solely to the twisted double bond systems whose unsaturated centers are constrained within the ring, this last section will give some example of systems in which the double bonds are twisted by crowdedness around these unsaturated centers.

McMurry's condensation of one of the enantiomers of a chiral ketone 126 would give two diastereomeric pairs (127 and 128), which both have C_2 symmetry. Guided by this analysis, Wynberg and coworkers [80] condensed (+)-camphor by this method

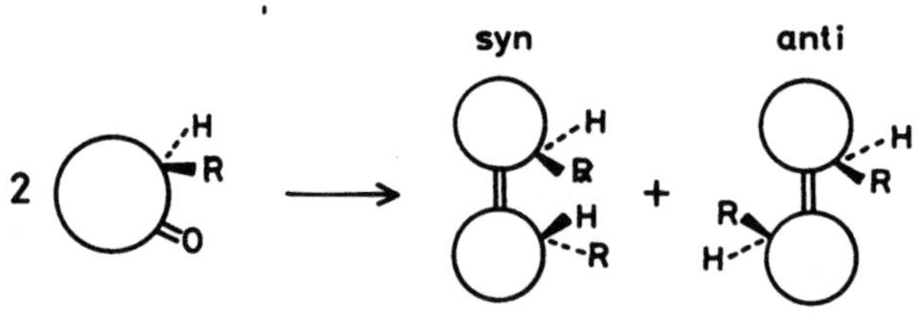

126 127 128

and isolated a syn-isomer (an oil) and an anti-isomer, mp 100 °C, whose structure was determined by converting this anti-isomer into an epoxide of C_2 symmetry.

The structure of the analogous syn-2,2′-bifenchylidene E [81] was studied by X-ray analysis.

McMurry's condensation of 129 was reported [82] to give the syn-isomer 130 and anti-isomer 131, and both could be resolved into the corresponding optically active modifications. This proved their chiral structure, indicating that the antiisomer 131 should be a C_2-conformer instead of a C_i-conformer.

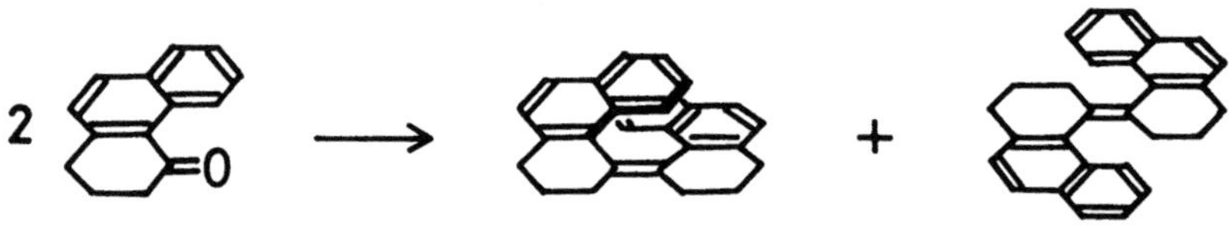

129 syn-130 anti-131

VII References

1. a) Blomquist, A. T., Liu, L. H., Bohrer, J. C.: J. Am. Chem. Soc. *74*, 3643 (1952)
   b) Prelog, V.: Bedeutung der vielgliedrigen Ringverbindungen für die theoretische Organische Chemie (Sir Todd, A. ed.), Perspectives in Organic Chemistry, Interscience Publ. Inc., New York 1956, p. 129
2. Cope, A. C., Howell, C. F., Knowles, A.: J. Am. Chem. Soc. *84*, 3190 (1962)
3. Cope, A. C., Ganellin, C. R., Johnson, H. W. Jr.: ibid. *84*, 3191 (1962)
4. Ziegler, K., Wilms, H.: Liebigs Ann. Chem. *567*, 1 (1950)
5. Cope, A. C., Ganellin, C. R., Johnson, H. W. Jr., Van Auken, T. V., Winkler, H. J. S.: J. Am. Chem. Soc. *85*, 3276 (1963)
6. Cope, A. C., Funke, W. R., Jones, F. N.: ibid. *88*, 4693 (1966)
7. Corey, E. J., Shulman, J. I.: Tetrahedron Lett. *1968*, 3655
8. Waters, W. L.: J. Org. Chem. *36*, 1569 (1971)
9. Newton, P. F., Whitham, G. H.: J. C. S. Perkin I *1979*, 3072
10. Seuron, P., Solladie, G.: J. Org. Chem. *45*, 715 (1980). The enantiomeric equilibrium of *trans*-cyclooctene in cholesteric liquid crystals is also reported.
11. Cahn, R. S., Ingold, C. K., Prelog, V.: Angew. Chem. Int. Ed. Engl. *5*, 385 (1966)
12. IUPAC Tentative Rules for the Nomenclature of Organic Chemistry. Section E. Fundamental Stereochemistry. J. Org. Chem. *35*, 2849 (1970)
13. Moscowitz, A., Mislow, K.: J. Am. Chem. Soc. *84*, 4605 (1962). Although they follwed the axial instead of planar chirality convention in their R,S-specification for chiral *trans*-cyclooctene, both procedures give the same R,S-notation in this case.

14. Yaris, M., Moscowitz, A., Berry, R. S.: J. Chem. Phys. *49*, 3150 (1968)
15. Levin, C. C., Hoffmann, R.: J. Am. Chem. Soc. *94*, 3446 (1972)
16. a) Cope, A. C., Mehta, A. S.: ibid. *86*, 1268 (1964)
    b) Cope, A. C., Mehta, A. S.: ibid. *86*, 5626 (1964)
17. Manor, P. C., Shoemaker, D. P., Parkes, A. S.: ibid. *92*, 5260 (1970)
18. Bach, R. D., Mazur, U., Brummel, R. N., Lin, L.-H.: ibid. *93*, 7120 (1971)
19. Newton, P. F., Whitham, G. H.: J. C. S. Perkin I *1979*, 3067
20. Cope, A. C., Pawson, B. A.: J. Am. Chem. Soc. *87*, 3649 (1965)
21. Cope, A. C., Banholzer, K., Keller, H., Pawson, B. A., Whang, J. J., Winkler, H. J. S.: ibid. *87*, 3644 (1965)
22. Binsch, G., Roberts, J. D.: ibid. *87*, 5157 (1965)
23. Westen, H. H.: Helv. Chim. Acta *47*, 575 (1964)
24. Gavin, R. M. Jr., Wang, Z. F.: J. Am. Chem. Soc. *95*, 1425 (1973)
25. Ermer, O.: Angew. Chem. Int. Ed. Engl. *13*, 604 (1974)
26. a) Ermer, O., Lifson, S.: J. Am. Chem. Soc. *95*, 4121 (1973)
    b) Allinger, N. L., Sprague, J. T.: ibid. *94*, 5734 (1972)
    c) Ermer, O.: Tetrahedron *31*, 1849 (1975)
27. Cope, A. C., Fordice, M. W.: J. Am. Chem. Soc. *89*, 6187 (1967)
28. a) Cope, A. C., Pawson, B. A.: ibid. *90*, 636 (1968)
    b) Whiteside, G. M., Pawson, B. A., Cope, A. C.: ibid. *90*, 639 (1968)
29. Marshall, J. A., Konicek, T. R., Flynn, K. E.: ibid. *102*, 3287 (1980)
30. Cope, A. C., Hecht, J. K., Johnson, H. W. Jr., Keller, H., Winkler, H. J. S.: ibid. *88*, 761 (1966)
31. Newton, P. F., Whitham, G. H.: J. C. S. Perkin I *1977*, 2629
32. Leitich, J.: Tetrahedron Lett. *1978*, 3589
33. Whiteside, G. M., Goe, G. L., Cope, A. C.: J. Am. Chem. Soc. *89*, 7136 (1967)
34. a) Bogdanovic, B., Heimbach, P., Kröner, M., Wilke, G.: Liebigs Ann. Chem. *727*, 143 (1969)
    b) Anet, F. A. L., Rawdah, T. N.: J. Am. Chem. Soc. *100*, 5003 (1978)
35. Houk, K. N., Strozier, R. W., Santiago, C., Gandour, R. W., Vollhardt, K. P. C.: ibid. *101*, 5183 (1979)
36. a) Masamune, S., Bates, G. S., Corcoran, J. W.: Angew. Chem. Int. Ed. Engl. *16*, 585 (1977)
    b) Weinheimer, A. J., Chang, C. W. J., Matson, J. A.: Fortsch. Chem. Org. Naturst. *36*, 285 (1979)
    c) Mallam, A. K., Puar, M. S., Rossman, R. R.: J. Am. Chem. Soc. *103*, 3938 (1981)
37. Yoshioka, H., Mabry, T. J., Timmerman, B. N.: Sesquiterpene Lactones, University Tokyo Press, Tokyo 1973, p. 7
38. Dev, S., Anderson, J. E., Cormier, V., Damodaran, N. P., Roberts, J. D.: J. Am. Chem. Soc. *90*, 1246 (1968)
39. Horibe, I., Shigemoto, H., Tori, K.: Tetrahedron Lett. *1975*, 2849
40. Hill, R. K., Fracheboud, M. G., Sawada, S., Carlson, R. M., Yan, S.-J.: ibid. *1978*, 945
41. Shirahama, H., Osawa, E., Matsumoto, T.: ibid. *1979*, 2245
42. a) Fawcett, F. S.: Chem. Rev. *47*, 219 (1950)
    b) Köbrich, G.: Angew. Chem. Int. Ed. Engl. *12*, 464 (1973)
    c) Buchanan, G. L.: Chem. Soc. Rev. *3*, 41 (1974)
    d) Keese, R.: Angew. Chem. Int. Ed. Engl. *14*, 528 (1975)
    e) Liebman, J. F., Greerberg, A.: Chem. Rev. *76*, 311 (1976)
    f) Shea, K. J.: Tetrahedron *36*, 1683 (1980)
43. a) Nakazaki, M., Naemura, K., Nakahara, S.: J. C. S. Chem. Commun. *1979*, 82
    b) Nakazaki, M., Naemura, K., Nakahara, S.: J. Org. Chem. *44*, 2438 (1979)
44. a) Wiseman, J. R.: J. Am. Chem. Soc. *89*, 5966 (1967)
    b) Wiseman, J. R., Pletcher, W. A.: ibid. *92*, 956 (1970)
45. a) Marshall, J. A., Faubl, H.: ibid. *89*, 5965 (1967)
    b) Marshall, J. A., Faubl, H.: ibid. *92*, 948 (1970)
    c) Becker, K. B.: Helv. Chim. Acta *60*, 81 (1977)
46. Klyne, W., Buckingham, J.: Atlas of Stereochemistry, Chapman and Hall, London 1974, p. 42; and the references cited therein.
47. Scott, A. I., Wrixon, A. D.: Tetrahedron *26*, 3695 (1970)
48. Wiseman, J. R., Chan, H.-F., Ahola, C. J.: J. Am. Chem. Soc. *91*, 2812 (1969)

49. Greenhouse, R., Ravindranathan, T., Borden, W. T.: ibid. *98*, 6738 (1976)
50. Wiseman, J. R., Vanderbilt, J. J.: ibid. *100*, 7730 (1978)
51. Shea, K. J., Wise, S.: Tetrahedron Lett. *1978*, 2283
52. Nomenclature of [m][n]paracyclophane: Smith, B. H.: Bridged Aromatic Compounds, Academic Press, New York 1964 p. 13
53. a) Nakazaki, M., Yamamoto, K., Tanaka, S.: Tetrahedron Lett. *1971*, 341
 b) Nakazaki, M., Yamamoto, K., Tanaka, S.: J. C. S. Chem. Commun. *1972*, 433
 c) Nakazaki, M., Yamamoto, K.: Chem. Lett. *1974*, 1051
 d) Nakazaki, M., Yamamoto, K., Tanaka, S.: J. Org. Chem. *41*, 4081 (1976)
 e) Nakazaki, M., Yamamoto, K., Ito, M., Tanaka, S.: ibid. *42*, 3468 (1977)
54. a) Nakazaki, M., Yamamoto, K., Yanagi, J.: J. C. S. Chem. Commun. *1977*, 346
 b) Nakazaki, M., Yamamoto, K., Yanagi, J.: J. Am. Chem. Soc. *101*, 147 (1979)
 c) Nakazaki, M., Yamamoto, K., Maeda, M.: J. C. S. Chem. Commun. *1980*, 294
 d) Nakazaki, M., Yamamoto, K., Maeda, M.: J. Org. Chem. *45*, 3229 (1980)
 e) Nakazaki, M., Yamamoto, K.: J. Synth. Org. Chem. Japan (Yuki Gosei Kagaku Kyokai Si) *39*, 624 (1981)
55. a) Marshall, J. A., Lewellyn, M.: J. Am. Chem. Soc. *99*, 3508 (1977)
 b) Marshall, J. A., Bierenbaum, R. E., Chung, K.-H.: Tetrahedron Lett. *1979*, 2081
 c) Marshall, J. A., Black, T. H., Shone, R. L.: ibid. *1979*, 4737
 d) Marshall, J. A., Chung, K.-H.: J. Org. Chem. *44*, 1566 (1979)
 e) Marshall, J. A., Black, T. H.: J. Am. Chem. Soc. *102*, 7581 (1980)
 f) Marshall, J. A., Flynn, K. E.: ibid. *105*, 3360 (1983)
 g) Marshall, J. A.: Acc. Chem. Res. *13*, 213 (1980)
56. Allinger, N. L.: J. Am. Chem. Soc. *80*, 1953 (1958)
57. Nickon, A., Zurer, P. S. J.: Tetrahedron Lett. *1980*, 3527
58. a) Ceré, V., Paolucci, C., Pollicino, S., Sandri, E., Fava, A.: J. C. S. Chem. Commun. *1980*, 755
 b) Ceré, V., Paolucci, C., Pollicino, S., Sandri, E., Fava, A.: J. Org. Chem. *46*, 486 (1981)
59. Yuki, H., Okamoto, Y., Okamoto, I.: J. Am. Chem. Soc. *102*, 6356 (1980)
60. These rotatory strengths, when combined with the reported optical purities, give $[\alpha]_{abs}$ 426° and 412° for [22.10] and [26.10] betweenanenes, respectively. Since these values automatically indicate their ridiculously high $[M]_{abs}$ ca. 2000, the $[\alpha]_D$ apparently are to be reduced to 1/10 th of the reported values.
61. Katsuki, T., Sharpless, K. B.: J. Am. Chem. Soc. *102*, 5974 (1980)
62. Cope, A. C., Moore, W. R., Bach, R. D., Winkler, H. J. S.: ibid. *92*, 1243 (1970)
63. For a review surveying the synthesis and chirality of optically active allenes, see: Rossi, R., Diversi, P.: Synthesis *1973*, 25
64. An erroneous interpretation of the steric pathway of this reaction led Krow to assign the opposite (S)-configuration: Krow, G.: Topic in Stereochemistry *5*, 39 (1970)
65. Lowe, G.: J. C. S. Chem. Commun. *1965*, 411
66. Moore, W. R., Anderson, H. W., Clark, S. D., Ozretich, T. M.: J. Am. Chem. Soc. *93*, 4932 (1971)
67. Rauk, A., Drake, A. F., Mason, S. F.: ibid. *101*, 2284 (1979)
68. Crabbé, P., Velarde, E., Anderson, H. W., Clark, S. D., Moore, W. R., Drake, A. F., Mason, S. F.: J. C. S. Chem. Commun. *1971*, 1261
69. Bach, R. D., Mazur, U., Brummel, R. N., Lin, L.-H.: J. Am. Chem. Soc. *93*, 7120 (1971)
70. Moore, W. R., Anderson, H. W., Clark, S. D.: ibid. *95*, 835 (1973)
71. Bach, R. D., Holubka, J. W.: ibid. *96*, 7814 (1974)
72. a) Garratt, P. J., Nicolaou, K. C., Sondheimer, F.: J. C. S. Chem. Commun. *1970*, 1219
 b) Garratt, P. J., Nicolaou, K. C., Sondheimer, F.: J. Am. Chem. Soc. *95*, 4582 (1973)
73. For the X-ray analysis of the tetrachloro analog of this anti-form 126, see: Baker, R. W., Pauling, P. J.: J. C. S. Perkin II *1972*, 1451
74. a) Garratt, P. J., Nicolaou, K. C., Sondheimer, F.: J. C. S. Chem. Commun. *1971*, 1018
 b) Garratt, P. J., Nicolaou, K. C., Sondheimer, F.: J. Org. Chem. *38*, 2715 (1973)
75. Skattebøl, L.: Acta Chem. Scand. *17*, 1683 (1963)
76. Dehmlow, E. V., Ezimora, G. C.: Tetrahedron Lett. *1970*, 4047
77. Irgartinger, H., Jäger, H.-U.: ibid. *1976*, 3595

78. a) Anet, F. A. L., Yavari, I.: J. C. S. Chem. Commun. *1975*, 927
    b) Anet, F. A. L., Yavari, I.: J. Am. Chem. Soc. *99*, 7640 (1977)
79. a) Nakazaki, M., Yamamoto, K., Maeda, M.: Chem. Lett. *1981*, 1035
    b) Nakazaki, M., Yamamoto, K., Maeda, M., Sato, O., Tsutsui, T.: J. Org. Chem. *47*, 1435 (1982)
80. Wynberg, H., Lammertsma, K., Hulshof, L. A.: Tetrahedron Lett. *1975*, 3749
81. Pilati, T., Simonetta, M.: J. C. S. Perkin II *1977*, 1435
82. Feringa, B., Wynberg, H.: J. Am. Chem. Soc. *99*, 602 (1977)

Planar Chiral Molecular Structures

Karl Schlögl

Institut für Organische Chemie der Universität Wien, A-1090 Wien

Table of Contents

1 Introduction

Chirality — the topological property of handedness of a given structure — is one of the most fascinating features of stereochemistry. This term, first coined by Lord Kelvin in 1886, was introduced into chemistry in 1964 [1,2]. It was soon widely accepted (e.g. 1965 in the field of metallocene chemistry) [3] to substitute the old and mostly incorrectly used terms asymmetry and dissymmetry and has been since the subject of many discussions, interpretations, considerations and factorizations [4,5,6]. For the specification of molecular chirality according to the principle of topology three elements of chirality (stereogenic units) [7] have been defined: In addition to the vast and long known field of centrochiral compounds, the axis and the plane of chirality were introduced in 1956 [8] (see also Ref. [4]) in order to specify an increasing number of chiral compounds devoid of a chiral center. These three stereogenic units are the basis of the generally accepted CIP (*Cahn-Ingold-Prelog*) system [9].

1.1 Scope

This survey deals with organic structures possessing a plane of chirality (see 1.2.) and reviews the more recent results covering approximately the last ten years. Whereas obviously it would be an impossible task to deal with centro or even axial chiral compounds in one review, the field of planar chiral structures can still be surveyed, especially as there is some excellent background material covering several aspects [10-18] (but to date no comprehensive survey has been published). These reviews will serve as the basis for some discussions presented in this article.

1.2 Planar Chirality

Whereas a center or an axis of chirality can by clearly defined [2,7] there is still some ambiguity in respect to the specification of a plane of chirality. It therefore seems somewhat difficult to define the scope of planar chiral compounds which at a first glance include mostly rather rigid aromatic compounds of special interest from synthetic, structural, spectroscopic and especially chiroptical points of view.

The first definition in 1956 "a chiral plane is caused if a plane of symmetry is destroyed in such a way that chirality arises only by the difference of both sides of the plane" [8] (i.e. a plane of chirality differentiates two dimensions from the third one) was and is sufficient for the types of chiral (and optically active) compounds known at this time which are not covered by the other two stereogenic units (centre and axis of chirality): the cyclophanes such as the so-called ansacompounds (see 2.4.) or [2.2]paracyclophane derivatives (see 2.5.). It is, however, not elaborate enough for the classification of other organic structures such as the large group of chiral metallocenes (especially ferrocenes) [19] and perhaps also for bridged [10] and [14]-anulenes and (*E*)-cyclooctene (see 3. and 4.) which are not planar in the sense of the above definition. Consequently, (*E*)-cyclooctene and related alkenes could and were also regarded as axial chiral [10,20,21], anelated [2.2]metacyclophanes also als helical (cf. section 2.7.1) [81b].

Unsymmetrically disubstituted (and hence chiral) metallocenes (with point group C_1) such as the ferrocene *1* [22] provide a rather special problem: they were first defined as planar chiral [4,19] and the specification of molecular chirality (descriptors R_p and S_p) applied accordingly [19]. Several authors still classify these structures as planar chiral [21,23,24].

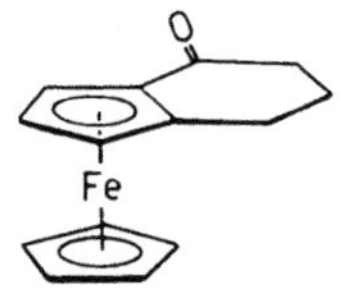

(+)(1S)-*1*

Cahn, *Ingold* and *Prelog*, however, in accordance with the present author, prefer to specify this chirality according to the rules for centrochiral compounds [2,19,25] with the metal atoms linked to ring atoms by a formal single bond. This formalism has been the subject of some controversy [6,21] and has led to new definitions for chiral planes including also metallocenes. In order to avoid these problems we prefer the term "metallocene chirality" [26] using, however, the descriptors (*R*) and (*S*) as specified for centrochirality [19,25]. Anyway, chiral metallocenes will be omitted from this survey since the field has become far too large to be covered within one limited article.

Recently new definitions for the chiral plane were put forward [6,21,27] and appropriate procedures for the application of the descriptors $(R)_p$ and $(S)_p$ proposed [21,27]. They will not be discussed in this article since we shall follow the revised and refined proposal as presented in Ref. [7]:[1,2]

A plane of chirality can be depicted by stereomorphous figures of five points, where Z lies outside the plane ABY (respresenting the plane of chirality; see Fig. 1). If the two-dimensional asymmetric trigonal atom X (usually a sp^2-hybridized carbon atom) is omitted together with its three bonds and then ABY and Z are connected by lines, one obtains a tetrahedron of symmetry C_s. Only if its two topologically equivalent corners A and B are differentiated by substitution with appropriate ligands a tetrahedral stereogenic unit results which can be specified in the usual way [2].

The descriptors (*S*) or (*R*) are determined by the left or righthanded sense of the sequence A—Y—Z (where A has priority over B) or alternatively by the sign of the torsional angle of the helical sequence of A—X—Y—Z: (−) = (*S*) and (+) = (*R*).

This leads to the same assignments as the procedure proposed in 1966 [2] and generally accepted since [9,10]: the pilot atom (corresponding to Z) marks the side of the

1 Some of the treatments on chirality (e.g. [6,7]) are not easy to follow — especially for the organic chemist; they are rather formal and do not or hardly provide actual structures as examples.

2 K. Mislow and J. Siegel have recently completed a very elaborate and fascinating paper on "The Theoretical Foundation of Factorization in Stereochemistry" which advances a scheme for a novel classification of chiral structures and abandons the concept of axial and planar chirality. The contents of this work can, however, not be discussed in this survey.

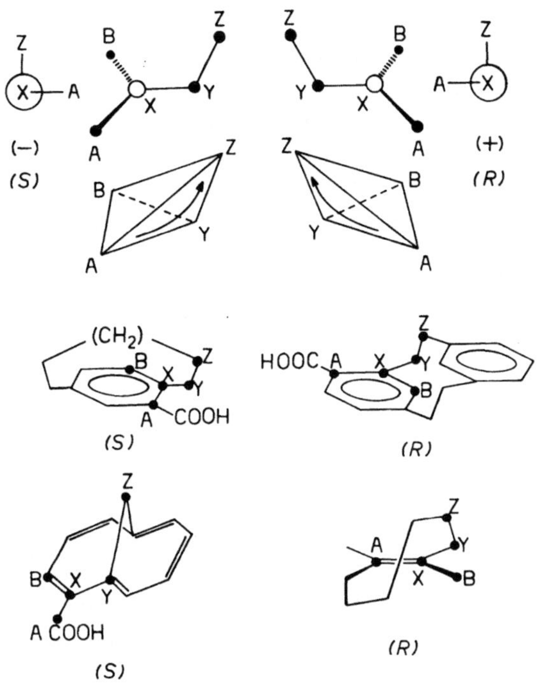

Fig. 1. Plane of Chirality: Enantiomorphous figures and corresponding torsional angles A—X—Y—Z [7]. Molecular examples ([n]paracyclophane, [2.2]metacyclophane, bridged[10]anulene and (*E*)-cyclooctene) with descriptors (*R*, *S*)

plane in which a sequence Y—X—A is observed (again with A having priority over B) which specifies the planar chirality (according to its left or righthanded turn).[3]

On the basis of this definition (Fig. 1) the following classes of chiral compounds will be treated in this article: cyclophanes, bridged anulenes and [8]anulenes, and (*E*)-cyclooctene and related structures. As mentioned above, metallocenes will be excluded.

2 Cyclophanes

2.1 Introduction and Definitions

Phanes are compounds with at least one aromatic nucleus and at least one bridge [28]. Such structures can incorporate a great number of arenes — carbophanes contain carbocycles such as benzene or naphthalene, heterophanes heteroaromatic rings including metallocenes — and a wide variety of bridges spanning the ring (s) in different positions: from one to several simple alkylidene residues (e.g. hexamethylene or

3 If however the C_1-tetrahedron (ABYZ) were treated in the usual way for specifying the chirality (by viewing towards the corner with lowest priority and determining the (*R*) or (*S*) sequence of the three corners left according to the sequence rule [2]) this might reverse several assignments (as have been deduced so far) and would result in much confusion.

ethylene bridging two benzene rings up to the so-called "superphane" with six bridges)[29] to more complex bridges containing different building blocks or even additional phane residues to give multilayered structures (see 2.6.2.). Thus, phanes with their unique molecular geometry provide a large playground for research and are favoured model compounds for structural and stereochemical studies. Background information on the following topics can be drawn from excellent reviews: cyclophane chemistry [11], synthesis of [2.2]phanes [30], stereochemistry of [2.2]meta [12] and [2.2]-metaparacyclophanes [13], rigid cyclophanes for illustrating stereochemical principles [14], steric interactions in organic chemistry [16], chemistry of multilayered cyclophanes [17] (some aspects of these compounds are also treated in ref. [15]) chemical behaviour of multibridged cyclophanes [31].

2.2 Nomenclature

The nomenclature of phanes is simple [32]: It mainly defines the length(s) of the bridge(s) and their positions in the ring as illustrated by the following examples: [6]paracyclophane (*2*) 1,10-dioxa [10]paracyclophane (*6*), [2.2]paracyclophane (*7*), [2.2]meta- (*8*) and [2.2]metaparacyclophane (*9*) and [8](2,5)pyridinophane (*10*). The formulae show also the kind of projection used in this survey to illustrate stereochemical relations.

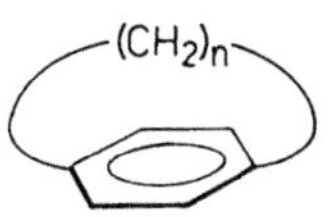

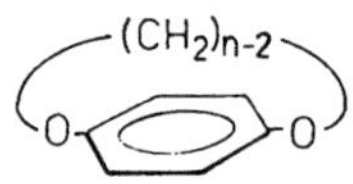

6

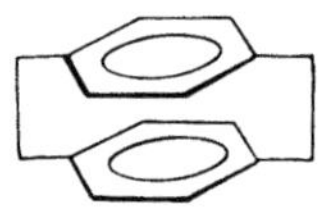

7

	n
2	6
3	7
4	8
5	10

8

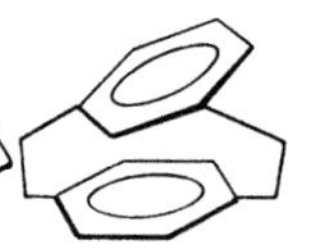

9

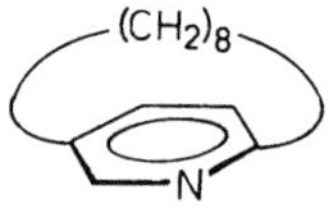

10

2.3 Molecular Geometry

If the bridges are short enough as for instance in *2*, *3* or *7* and *8* or if large substituents are attached to the aromatic ring (as e.g. in *19*), the torsional barrier is high enough to prevent rotation of the bridge around the arene moiety. If in addition the sides of the symmetry plane bisecting the ring (e.g. in *2–5*, whereby conformational changes of the bridge are disregarded and therefore idealized molecular models are assumed as in many other cases) are differentiated by appropriate substitution, planar chirality is created (see 1.2. and Fig. 1). Consequently, compounds of this type may be resolved into enantiomers as was first demonstrated for the carboxylic acid of dioxa[12]paracyclophane (*19*)[4] in 1941 [33], of [2.2]paracyclophane (*23*) in 1955 [34] and of [2.2]metacyclophane (*55*) in 1972 [35].

Moreover, with short bridges the benzene rings are distorted to some extent to adopt boat-like "conformations" such as in [2.2] or in [6]paracyclophane (*7*, *2*) or even chair-like structures as in layered cyclophanes (see 2.6.2. and 2.7.2.). It seems quite obvious that the torsional barriers are strongly dependent on the length of the bridge(s) thereby influencing the optical stability of enantiomers to a large extent as already shown in 1947 [36] and 1958 [37] for paracyclophanes of type *16–19* and *26*. This demonstrates once again — as, for example, in the case of torsional isomeric biphenyls — the temperature dependent border-line between conformation and configuration.

2.4 [n]Paracyclophanes

The parent compounds and their (chiral) derivatives have been prepared mainly by classical methods: The ansacompounds of type *6* are accessible from hydroquinone derivatives by reaction with appropriate α,ω-dibromoalkanes [33, 36, 38]. [10]Paracyclophane (*5*) was prepared by forming the bridge either by acyloin condensation of the corresponding benzenebisalkanoic acid and subsequent Clemmensen-reduction of the acyloin [39] or by Friedel-Crafts cyclization of 10-phenyldecanoic acid and Huang-Minlon-reduction of the resulting ketone [40]. For a general approach to [7]- and [8]paracyclophanes (*3*, *4*) see Ref. [41]. The smallest [n]paracyclophane (*2*) described so far has a six-membered carbon bridge (n = 6) and is known since 1974. It was obtained by flash pyrolysis of the corresponding lithium salt of the tosylhydrazone

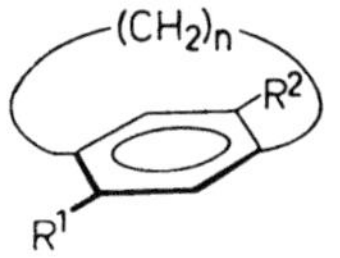

	n	R^1	R^2
11	6	COOH	H
12	7	COOH	H
(−)-*13*	8	COOH	H
(+)-*14*	10	COOH	H
(−)-*15*	10	COOH	CH_3

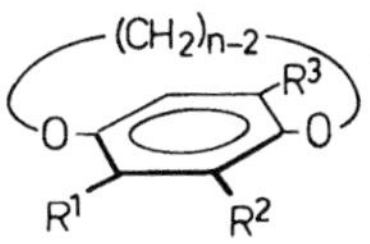

	n	R^1	R^2	R^3
16	10	COOH	H	H
17	11	COOH	H	H
18	12	COOH	H	H
19	12	COOH	H	Br
20	12	H	COOH	COOH

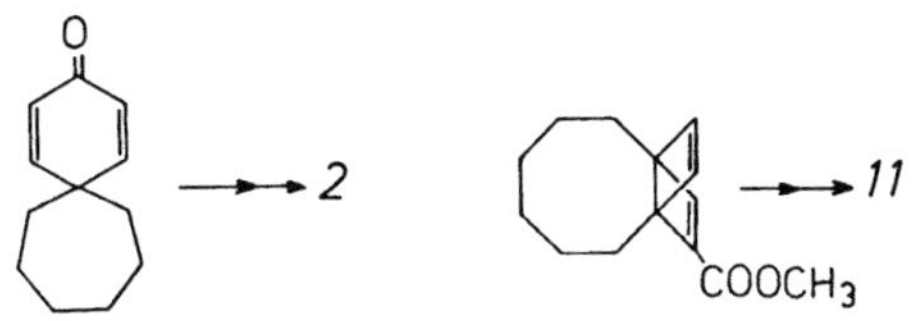

21 *22*

4 If not stated otherwise, for chiral compounds only *one* enantiomer is shown.

of the bicyclic ketone *21* [42]. An efficient synthesis of its carboxylic acid analogue *11* utilizes the facile dewarbenzene — benzene valence isomerization of the [6.2.2]-propelladiene *22* to its phane-isomer *11* [43].

Obviously carboxy derivatives such as *11–19* are simple chiral structures suitable for optical resolutions through diastereomeric salts. For this purpose carboxylic groups have been introduced into [10]- and [8]paracyclophane either by chloromethylation and oxidation of the carboxaldehydes obtained thereof [39,44] or by lithiation and subsequent carboxylation [40]. Electrophilic substitution of strained paracyclophanes is not advisable since it may initiate rearrangement to the more stable metacyclophanes. Carboxy[7]paracyclophane (*12*) was first prepared in 1972 by ring contraction of a diazoketone derived from 4-carboxy[8]paracyclophane (*13*) [45].

The distortion of the benzene rings from planarity is strongest in the [6]paracyclophanes (*2, 11*) where according to an X-ray structure the para carbon atoms of the ring are lifted out of the plane by appr. 20° [43]. With increasing length of the bridge the inversion barrier is decreased which is reflected in the increasing optical lability. In contrast to a prior note [40] (see also Ref. [44]) [10]paracyclophanecarboxylic acid (*14*) and its derivatives are optically stable at room temperature [39,40]; however no kinetic studies have been performed. The dioxa-derivative *17* (with n = 11) racemizes above 70 °C, dioxa[10]paracyclophanecarboxylic acid (*16*) is optically stable (in accordance with the corresponding carbophane *14*) [40], whereas no optical resolution could be accomplished in case of n = 12 (*18*) [16,36] (except if the large bromine was introduced, *19*) [33].

Optical resolution of these and related carboxylic acids were achieved using salt formation with alkaloids (strychnine, brucine, cinchonidine) [33,39,44] or with optically active amines [1-phenyl- or 1-(β-naphthyl)ethylamine] [40,44]. The following rotations $[\alpha]_D$ have been reported: [8]paracyclophanecarboxylic acid (*13*) +18° (chloroform) [44]; [10]homologue (14) +80° (chloroform) [39] and +67° (chloroform) [40]; its methyl-derivative (*15*) —28° (methanol) [44]. Dioxa[10]paracyclophanecarboxylic acid (*16*) +104° (ethanol) [36] and bromo-dioxa[12]paracyclophanecarboxylic acid (*19*) —37° (acetone) [33].

At least for *14* the usual methods for determining the enantiomeric purity (especially NMR-methods) failed. From *14* and *15* several optically active derivatives were prepared [40,44] and their chiroptical properties [especially the circular dichroism (CD) spectra of derivatives of *14*] [40] recorded.

Diastereomeric amides of 1,12-dioxa[12]paracyclophanedicarboxylic acid (*20*, as the first example of diastereomers with a plane of pseudo-asymmetry) could be separated by chromatography on silicagel; on the basis of their ^{1}H-NMR spectra tentative configurations were assigned to the amides [38].

The absolute configurations (chiralities) of carbophanes will be discussed in chapter 2.9.

2.5 [m.n.]Paracyclophanes

Again a carboxylic acid (*23*) was the first representative of optically active (and hence chiral) [2.2]paracyclophanes and related [m.n]homologues thereby proving the highly hindered rotation and opening up a new and large field of stereochemical research on torsional isomerism. Syntheses of [2.2]paracyclophanes have been reviewed [30]. Recent advances are the photoextrusion of SO_2 from appropriate bis-sulfones [47] or the photochemical ring contraction of diselena [3.3]cyclophanes with hexamethylphosphorous triamide [48].

1,6-Elimination of p-[(trimethylsilyl)methyl]benzyltrimethylammonium iodide with fluoride leads *via* the intermediate p-quinodimethene to [2.2]paracyclophane (*7*). This very versatile method can also be applied for the syntheses of [2.2](2,5)furano- and thiophenophane [48a].

As in the case of [n]paracyclophanes the torsional barrier(s) as well as the distortion of the benzene rings (as reflected for example in optical properties) depend on the size of the bridges, i.e. on m and n. Both reach their maximum if m = n = 2. Up to [3.4]paracyclophanecarboxylic acid (*26*) a separation into enantiomers could be accomplished [34, 37], whereas the [4.4] homologue is optically too labile for a resolution [37].

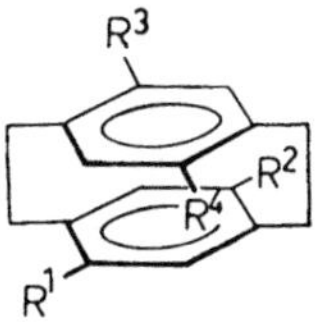

	R^1	R^2–R^4
(−)-*23*	COOH	H
(−)-*24*	$COOCH_3$	H
25	CH_3	CH_3

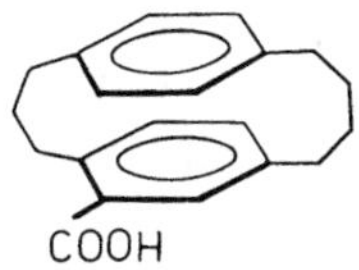

26

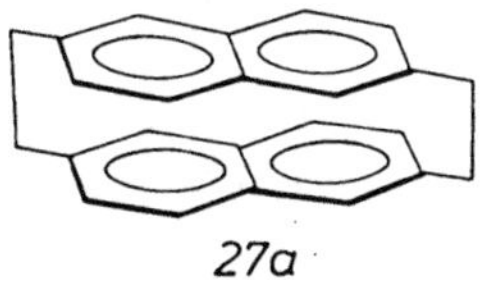

27a

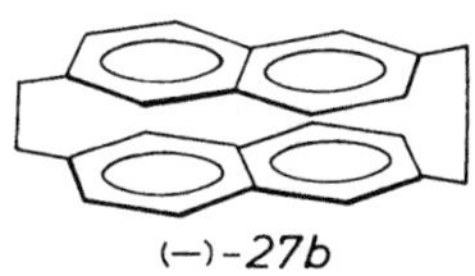

(−)-*27b*

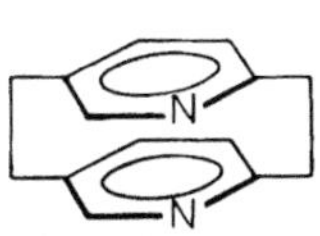

28a

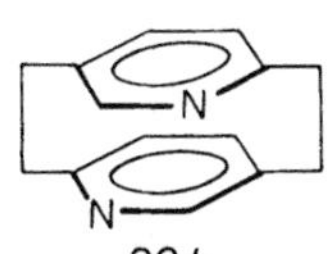

28b

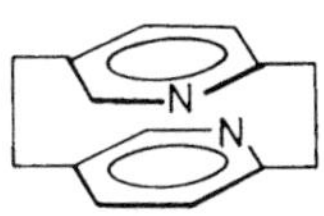

28c

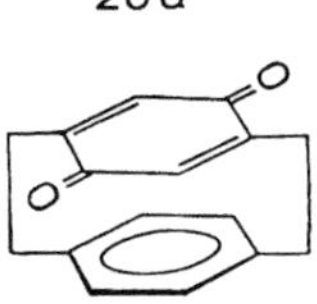

29

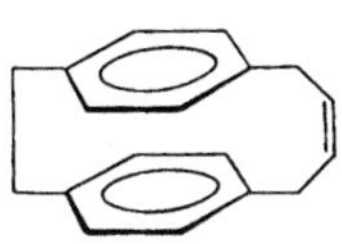

30

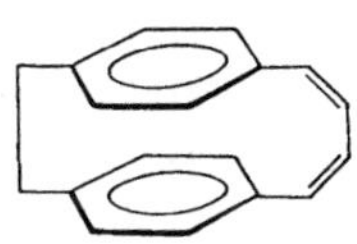

31

A variety of chiral [m.n]cyclophanes has been described, including [2.2](2,6)-naphthalenophane (*27*) [49, 50] and the corresponding diene [49], or [2.2](2,5)pyridinophanes (*28*) [51]. In both cases (*27*, *28*) achiral and chiral isomers (*a*, *b*) were isolated and their structures assigned mainly by ^{1}H-nmr spectroscopy. The chiral structure of *27b* was confirmed by optical resolution [49]. Some of these cyclophanes, such as the chiral *29* were prepared to study intraanular donor-acceptor properties [52]. Only recently syntheses of many chiral electron donor-acceptor [2.2] and [3.3]paracyclophanes have been reported together with the X-ray structures and electron absorption spectra of the racemates [52a]. *30* and *31* are according to their ^{1}H-nmr spectra, a rapidly equilibrating mixture of two C_2-conformers [14]. Another chiral [2.2]paracyclophane devoid of functional groups is the tetramethyl-derivative *25*, the only isomer formed on dimerization of appropriate 2,5-dimethyl-p-xylene derivatives [53].

The key compound [2.2]paracyclophanecarboxylic acid (*23*) was resolved *via* its brucine [34] or 1-phenylethylamine salts [54]. Both methods gave acids of comparable optical rotation ($[\alpha]_D$ 160–164° in chloroform) which are enantiomerically pure according to high pressure liquid chromatography (HPLC) of the diastereomeric amides with (—)-1-(α-naphthyl)ethylamine [55]. Many optically active derivatives have been prepared by simple chemical transformations of the carboxylic group [54, 56, 57]; their chiroptical properties (especially their CD-spectra) were reported and discussed with respect to their configuration [54] (see 2.9.). The methylester *24* undergoes thermal racemization at about 200 °C *via* a diradical intermediate [58], whereas three other derivatives (methyl, dimethyl, hydroxymethyl) as well as *24* undergo facile photolytic racemization at two wavelengths in various solvents [56, 59]; this process can be explained by a mechanism involving bond breaking to give (achiral) benzyl radicals as intermediates [56].

The hydrocarbon *25* has been partially resolved by asymmetric complexation with Newman's reagent [TAPA; (—)-α(2,4,5,7-tetranitro-9-fluorenylideneaminooxy)propionic acid] thereby establishing its chiral D_2-structure [53]. Similarily, the naphthalenophane *27b* could be resolved by chromatography on silicagel coated with (—)-TAPA [49] and recently also by HPLC on optically active poly(triphenylmethyl methacrylate) [49a] which also proved to be very useful for the optical resolution of many other axial and planarchiral aromatic compounds [49b].

Syntheses of cyclophanes with more than two bridges have led to the ultimate achievement of a fully bridged [2.2.2.2.2.2](1,2,3,4,5,6)cyclophane, the so-called *superphane* [29]. It is of course achiral as well as the three isomeric [2.2.2]cyclophanes [31].

2.6 Layered Paracyclophanes

Another approach to chiral phanes with interesting structures (mainly with respect to the distortion of benzene rings and resulting chiroptical properties) is either the double bridging of benzene or [2.2]paracyclophane on both sides of the planes or the stacking of more than two aromatic rings to give multi-layered cyclophanes. This amounted to an optically active [2.2]paracyclophane containing as much as three paracyclophane units, i.e. six layered benzene rings. Several aspects of such layered phanes have been reviewed [15, 17].

2.6.1 [m][n]Paracyclophanes

After the syntheses of racemic doubly bridged [8][8]- and [8][10]paracyclophanes (*32* and *33*) with D_2- and C_2-symmetry respectively [60], the optically active compounds were prepared [44,61] starting from the previously mentioned (+)-[8]paracyclophanecarboxylic acid (*13*, see 2.4.), the absolute configuration of which had been established as (*S*) by correlation with (+)(*S*)[2.2]paracyclophanecarboxylic acid (*23*, see 2.9.). After transforming the carboxyl group into a methyl group the second eight-membered bridge was constructed *via* the [2.2] benzenofuranocyclophane *34* by opening up the furane ring with sulfuric acid and reduction of the intermediate diketone to give (+)-*32* with known chirality (*S*)[5]. Pyrolysis of an intermediate Hofmann base on the other hand afforded the doubly [8][8]bridged levorotatory [2.2]paracyclophane *35*, albeit only in 2.5% yield [44].

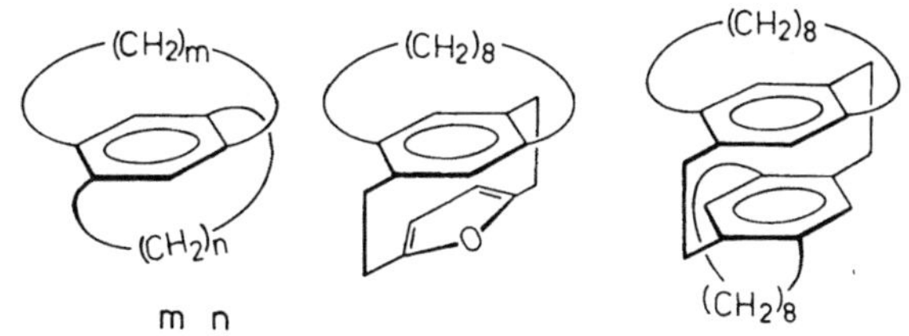

	m	n		
(−)-*32*	8	8	*34*	(+)-*35*
(−)-*33*	8	10		

By a reaction sequence similar to the one outlined above for the [8][8]cyclophane *32* also the levorotating [8][10]paracyclophane *33* was obtained starting from (—)-methyl[10]paracyclophanecarboxylic acid *15*[6]. Opposite Cotton effects of (+)-*32* and (—)-*33* indicated that they had opposite chiralities, and hence it followed that both (+)-*32* and (+)-*33* had the same chirality, namely (*S*). It should be noted that for (—)-*14* the chirality (*S*) had been established [40] (cf. also Ref. [62] and sections 2.9.2. and 2.9.3.); it would be somewhat surprising that its levorotatory methyl derivative *15* had (*R*) chirality as deduced from the above sequence (—)-*15* → (—)(*R*)-*33* [44]. Moreover, for levorotatory [m][n]paracyclophanes (*S*)chirality had been proposed by application of a sector rule [63] (see also 2.9.4.).

2.6.2 Layered [2.2]Paracyclophanes (Chochins)

Ingenious approaches have been developed for the syntheses of multi-layered [2.2]-paracyclophanes and especially for optically active representatives with known chiralities. For such "gyrochiral" structures with C_2- or D_2-symmetry the genetic name [n]chochins was proposed (meaning an oldfashioned Japanese lantern) [64] where n represents the number of layers. Parts of their chemistry have been reviewed [17].

For the syntheses of such chochins both the 1,6-Hofmann elimination reactions of suitable [2.2]paracyclophane derivatives with p-xylyl ammoniumbases (e.g. *36* and *37* — as used for the first synthesis of a [4]chochin *43* [65] — and the photodesulfurization method [17,66] have proved applicable.

5 Chirality — especially together with the description (*R*) or (*S*) — will be used in this survey in the sense of "absolute chirality" and "absolute configuration", respectively (see also Ref. [2,7] and Sect. 2.9.).

6 The authors [44] remarked that the optical instability of [10]paracyclophanecarboxylic acid (*14*) had caused them to use the methyl derivative *15* instead, although it had been established years before that *14* was optically stable [39,40].

Optically active chochins were prepared by the Hofmann route [64] starting from (*R*)(—)-4-methyl[2.2]paracyclophane (*38*) with known chirality [54,67] (see 2.9.). Introduction of the trimethyl-ammoniomethyl group (*via* acetylation and subsequent transformations) afforded (—)-*39* which was then cross-coupled with the ammonium base *36* to give a mixture of [2.2]paracyclophane and the levorotatory[3] and [4]chochins (*40, 43*) with (*R*)-chirality; in these cases the descriptors (*R*) and (*S*) specify the planar chirality of the inner rings(s) as shown in Fig. 2, in accordance with the rules presented in Section 1.2.

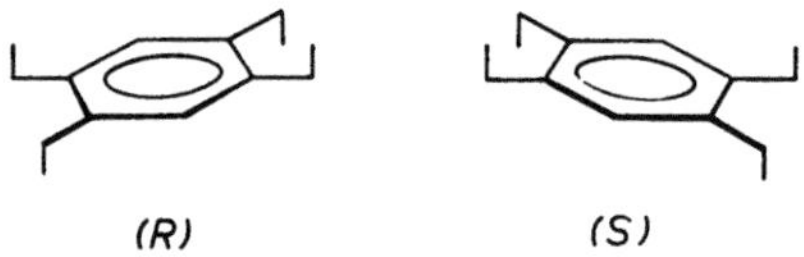

(*R*) (*S*)

Fig. 2. Planar chirality of the inner benzene rings in [n]chochins [64]

Cross coupling between (—)-*39* and the base *37* and subsequent transformation of the methyl group in the triple layered phane *41* yielded the [3]chochin derivative *42* which after a second cross coupling with (—)-*39* gave a separable mixture of levorotatory [4]-, [5]- and [6]chochins (*43, 44* and *45*) [64].

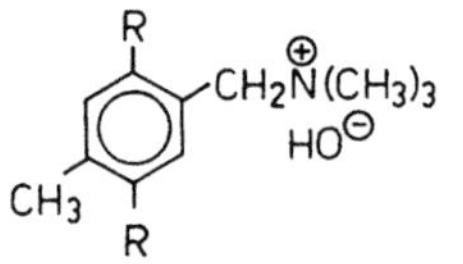

	R
36	H
37	CH_3

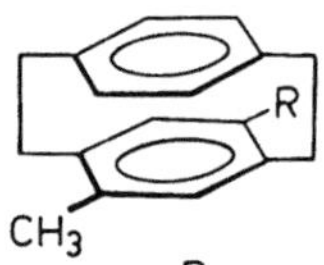

	R
(—)-*38*	H
(—)-*39*	$CH_2\overset{\oplus}{N}(CH_3)_3\overset{\ominus}{O}H$

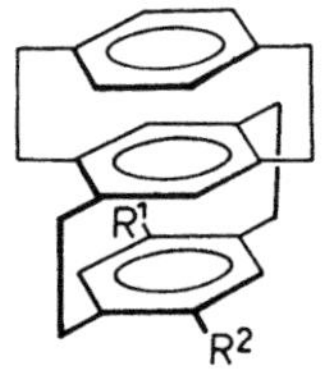

	R^1	R^2
(—)-*40*	H	H
(—)-*41*	CH_3	CH_3
(—)-*42*	CH_3	$CH_2\overset{\oplus}{N}(CH_3)_3\overset{\ominus}{O}H$

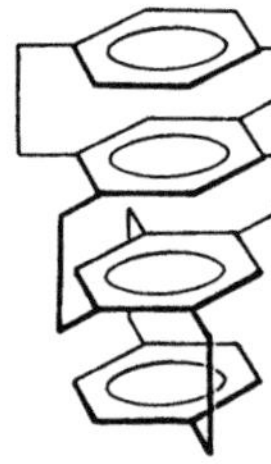

(—)-*43*

Another approach to the construction of optically active chochins of known chirality — the so-called "chiral recognition principle" — involves the coupling of the optically active [2.2]paracyclophane derivative (*R*)-*46* [16] with the racemic mixture of the [3]chochin derivative *47*. From the two diastereomeric [5]chochins which could be expected thereof, the one with (*R*)(*S*)(*S*)-chirality (see Fig. 2) because of eclipsed steric interactions is thermodynamically less stable than the (*R*)(*R*)(*R*)-isomer with D_2-symmetry. The latter (*44*) was indeed the only product isolated from the mixture [which also contained (—)-[4]chochin obtained from *46*] [64]. The sixfold-layered phane ([6]chochin *45*) also formed during this reaction was however optically inactive.

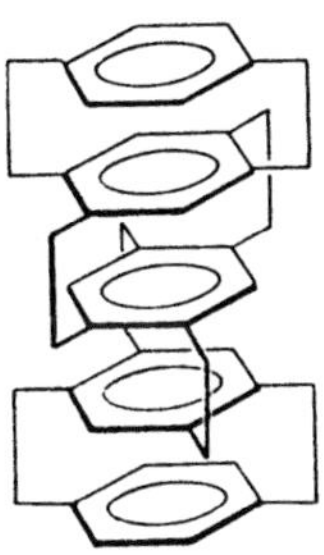

(−)-*44*

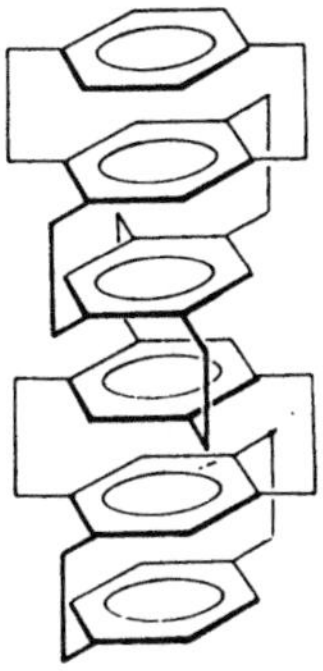

(−)-*45*

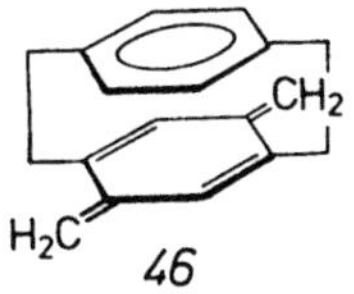

46

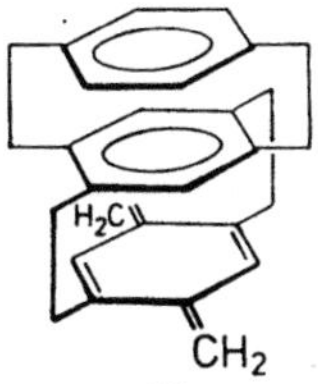

47

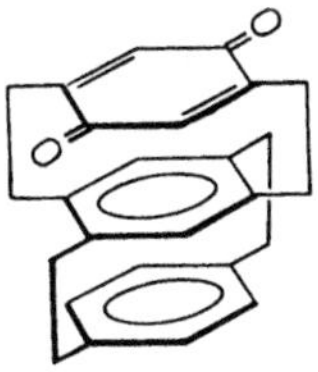

48

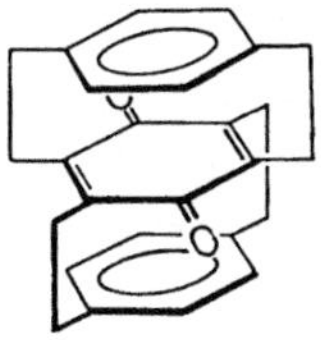

49

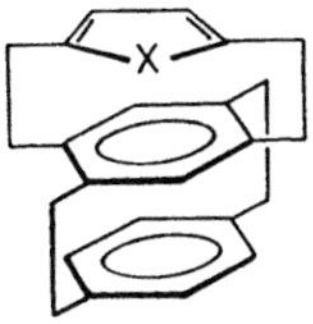

	X
50	O
51	S

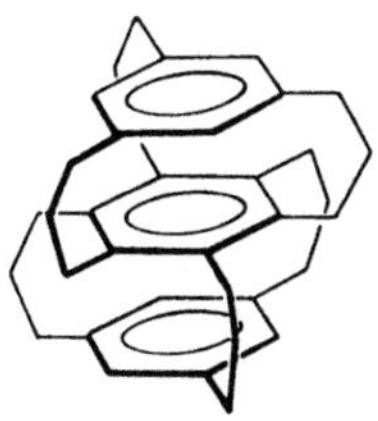

52

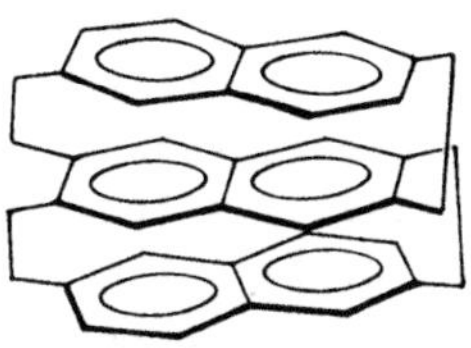

52a

Table 1. Possible stereoisomeric [n]chochins [64]

n	Total number	Isomers with symmetry		
		D_2 (chiral)	C_{2h} (achiral)	C_2 (chiral)
3	2	2	—	—
4	3	2	1	—
5	6	4	—	2
6	10	4	2	4
7	20	8	—	12
8	36	8	4	24

In Table 1 the numbers of possibly chiral and achiral stereoisomers of various [n]chochins are compiled [64]; their chiroptical properties are treated in Section 2.9.

Triple layered paracyclophanes such as *48* and *49* [68], the furano and thiophenophanes *50* and *51* [69], the triplelayered [3.3]paracyclophane *52* [70] or [2.2]naphthalenophane *52a* [70a] are further examples, but so far no optical resolutions of the chiral structures seem to have been described.

2.7 Metacyclophanes

From this important subgroup of cyclophanes [2.2]metacyclophane (*8*) and its derivatives (including [3.3]precursors for synthetic purposes) as well as some related carbo- and heterophanes have attracted much attention. Syntheses [30], the stereochemistry as well as several reactions and spectra [12] have been reviewed earlier.

Several peptide alkaloids are either meta-(Zizyphine A, Mucronine, Maitansine) or paracyclophanes (Frangulanine) where n ranges from 10 to 19 [70b]. They will not be included in this review.

2.7.1 [2.2]- and [3.2]Metacyclophanes

Owing to the interactions of the intra-anular hydrogens H-8 and H-16 in the central ten-membered ring [2.2]metacyclophane (*8*) adopts both in the crystal and in solution a rigid stair-like (anti) conformation with the boat-shaped benzene rings lying in parallel planes, as has been established by X-ray structure and ^{1}H-nmr analyses (cf. Ref. [12] for background information). The above mentioned protons are also excellent probes for conformational studies through ^{1}H-nmr spectroscopy [71].

The high barrier for the inversion process of the ten-membered ring in [2.2]metacyclophanes (with $\Delta G^{\ddagger}$ 132 kJ $\cdot$ mol^{-1} as deduced from racemization studies) [72] is lowered with increasing bridge lengths: It amounts to 73, 50, 60 and less than 38 kJ in [3.2]-, [3.3]-, [4.2]- and [4.3]metacyclophanes according to dynamic nmr studies [73]. From these phanes however, only in the [3.3]-homologue the benzene rings adopt the *syn*-arrangement whereas all others have the *anti*-configuration as in [2.2]metacyclophane itself [73]. Most of these metacyclophanes can conveniently be prepared by carbene insertion into appropriately semiprotected [2.2]- and [3.2]metacyclo-

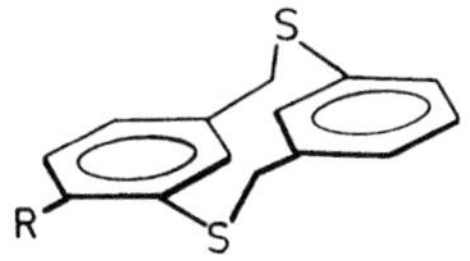

53

54

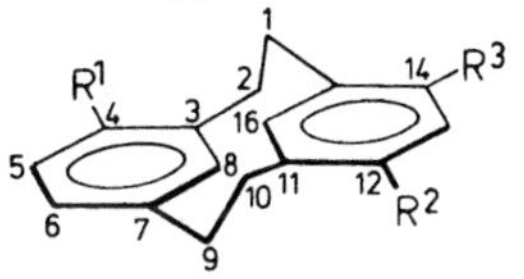

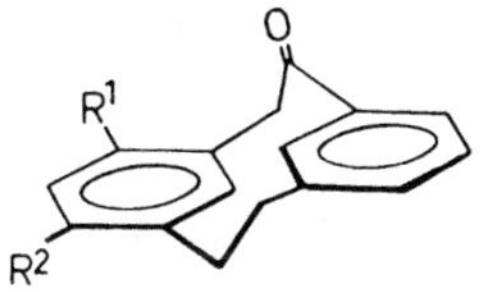

	R^1	R^2	R^3
8	H	H	H
(−)-55	COOH	H	H
(−)-56	COOH	H	COOH
(−)-57	$COOCH_3$	H	$COOCH_3$
(−)-58	COOH	H	Br
59	Br	H	Br
(−)-60	C≡N	Br	H

	R^1	R^2
(−)-61	H	H
(−)-62	CH_3	H
(−)-63	H	CH_3

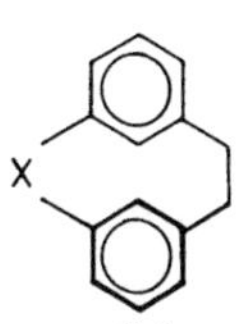

64

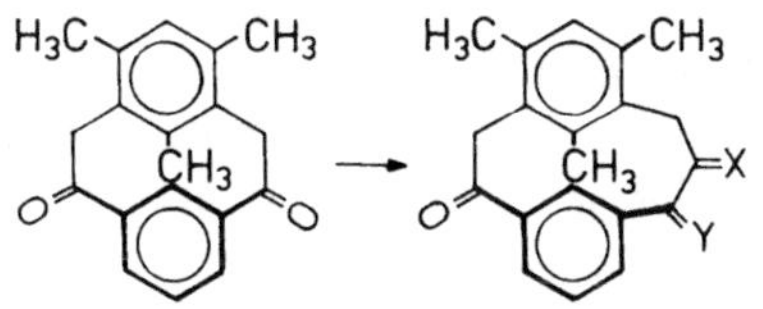

65

	X	Y
66	H_2	O
67	O	H_2

X

a 

b 

X

c 

d 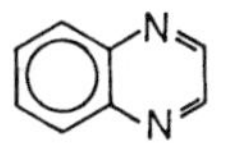

phanediones [74]. The barriers are also reduced in [2.2]heterophanes such as in the dithiaderivative *53* (to 107 kJ · mol^{-1}) [75] and in [2.2](2,6)pyridinophane (*54*) (to 62 kJ — obviously because of the absence of intra-anular hydrogens at the aromatic rings) [76].

All structures mentioned become chiral by appropriate mono-or disubstitution (C_1 or C_2 in metacyclophanes with m = n = 2). However, according to the variable racemization (= inversion) barriers only for [2.2]metacyclophane derivatives conformationally stable enantiomers result which permit optical resolutions.

This was first experimentally verified for the [2.2]metacyclophane-4-carboxylic acid (*55*) which had to be prepared by an elaborate 7-step synthesis [77] in order to avoid an electrophilic substitution which might have led to a transanular ring closure (as had been observed in so many cases of [2.2]metacyclophanes) [12]. The resolution of *55* was accomplished via salt formation with (+)-1-phenylethylamine and gave the levorotatory acid ($[\alpha]_D$ −9° in $CHCl_3$) which then was transformed into several optically active derivatives. The enantiomeric purity of *55* (and therefore of all compounds correlated with it) was confirmed by nmr spectroscopy of the diastereomeric esters with (—)-1-phenylethanol [77] as well as by HPLC of its diasteromeric naphthylamides [55].

Similarily, the 4,14-dicarboxylic acid *56* with C_2-symmetry could also be resolved *via* its 1-phenylethylamine salts and its configuration unambiguously correlated with the monocarboxylic acid *55* through the monobromo derivative *58* [78]. Accordingly *55* and *56* with the same sign of optical rotation have the same chirality. Many racemic and optically active homo- and heterodisubstituted 4,12- and 4,14-disubstituted [2.2]metacyclophanes have been prepared and chemically correlated [78,79] mainly to study their chiroptical properties [78]. Whereas 4,12-homodisubstituted compounds have a center of inversion (S_2-symmetry) and are therefore achiral "meso-forms", the corresponding 4,14-isomers are chiral with C_2-symmetry. All heterodisubstituted products are chiral (C_1-symmetry; see also Section 2.9.4 for the discussion of their chiroptical properties and their use as models for the application of the theory of chirality functions).

Chirality can also be introduced into the [2.2]metacyclophane skeleton by differentiating the two bridges: this has been shown by the preparation of the optically active 1-oxoderivative *61* ($[\alpha]_D$ —440° in $CHCl_3$) *via* the corresponding equatorial-positioned alcohol and its resolution through diastereomeric urethanes with (+)-1-phenylethylisocyanate [80]. From racemization studies for the ketone *61*, an inversion barrier ΔG^{*} of 104 kJ · mol^{-1} was deduced, whereas the barrier of the corresponding alcohol (with an sp^3-hybridized carbon in the bridge) is considerably higher [80]: 138 kJ — similar to that of 132 kJ for [2.2]metacyclophane derivatives [72].

[2.2]Metacyclophanes with aromatic structures anulated to the tenmembered central ring such as *64a* [81a] and *64b–d* [81b] represent further chiral structures with different bridges. They have been named "arenicenes" and classified as helical rather than planarchiral structures [81b].

The helical skeleton is revealed by X-ray structures [81b]. This once again reveals the ambiguity of the assignments of the elements of chirality (see Section 1.2.). *64a* [81a] to *64c* and *64d* [81b] could be resolved by chromatography on triacetylcellulose; whereas for *64a* and *64d* complete separations of the enantiomers were accomplished (*64a*: $[\alpha]_{365}$ ±320° [81a], *64d*: $[\alpha]_{436}$ ±2096° [81b]), *64c* was only partly resolved ($[\alpha]_{546}$ —65°) [81b]. The racemization barriers ΔG^{*} decrease from *64a* (125 kJ · mol^{-1}) [81a] to *64d* (122 kJ) and *64c* (115 kJ) [81b]. Application of the Bijvoet method established the helicity (—)(*P*) for *64c*; CD-correlations make the same helicity also for *64a*, *b* and *d* highly probable [81b].

From other approaches to optically active [2.2]metacyclophanes the following are noteworthy: as just mentioned for *64* (medium pressure) liquid chromatography on microcrystalline triacetylcellulose (cf. Ref. [82]) in ethanol or ether (practicable also at lower temperatures) is a very efficient and successful method for the optical resolution of many axial and planar chiral (aromatic) compounds [83]. In many cases baseline-separations can be achieved and thereby both enantiomers obtained with known enantiomeric purity and in amounts sufficient for further investigations, especially for studying their chiroptical properties (see also 3.2 and 3.3). The disubstituted [2.2]metacyclophanes *57* and *59* (which had been previously correlated to many other derivatives) [78,79] were first resolved by this method [83].

The racemic 4- and 6-methyl[2.2]metacyclophane-1-ones (*62* and *63*, prepared by unambiguous multi-step syntheses) [84] were incubated with *Rhodoturula rubra* to afford mixtures of the levorotatory ketone *62* and the corresponding epimeric (axial- and equatorial-positioned) (—)-carbinols from racemic *62*, whereas the isomeric *63* gave only the isomeric (—)-alcohols — all with high enantiomeric purities [84]. Reduction of the (+)-ketone *62* through its dithioacetal derivative gave (+)-4′-methyl[2.2]-metacyclophane to which, on the basis of its chiroptical properties, the chirality (*R*) was assigned [84] — in contrast to a previous (tentative) proposal [63] (see also 2.9.4).

An ingenious experimental example for the reverse of the socalled "la coupe de Roi" (i.e. the dividing of finite geometric objects into isometric segments) — namely the assembly of two homochiral compounds into an achiral one — was provided by making use of appropriate [2.2]metacyclophanes. Whereas self-coupling of (+)-4-(bromomethyl)-6-(mercaptomethyl)[2.2]metacyclophane (cf. formulas *55* to *60* for the numbering) led stereoselectively to the achiral C_{2v} "cis"-dimer 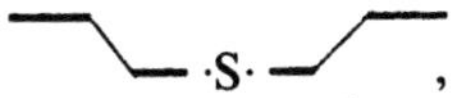

, the racemate afforded an 1:1-mixture of the "cis"- and the C_{2h} "trans"-dimer

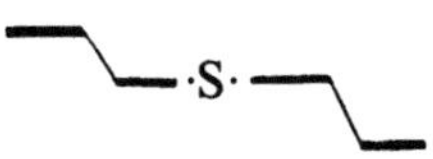

[84a].

Also asymmetric syntheses have recently been successfully employed for the preparation of optically active [3.2]metacyclophanes (see also 2.7.2):

When the above-mentioned ring expansion with diazomethane [74] of trimethyl-dioxo[2.2]metacyclophane *65* (methylation was necessary to increase the inversion barrier to > 130 kJ) was performed in the presence of optically active alcohols at —60 °C, asymmetric induction occurred to an extent of ca. 40% ee (enantiomeric excess; as determined by nmr-spectroscopy in the presence of chiral shift reagents) [85]. (+)-Dibutyl tartrate favoured the dextrorotatory diketone *66* ($[\alpha]_D$ 160° for the optically pure product) — the isomeric *67* was formed only with 3% ee; (—)-ethyl lactate on the other hand led to an excess of (+)-*67* ($[\alpha]_D$ +240°) but gave (+)-*66* with only 10% ee [85].

Several other chiral [2.2]metacyclophanes, such as *68* and *69* [86], *70* [87] have been described, but not resolved. There should be a good chance for these and other related compounds, e.g. *71* or *72* [88], metaheterocyclophanes (such as the flavinocyclophane *73* [89] and some others [12]) to be resolvable by chromatography (e.g. on triacetyl-cellulose).

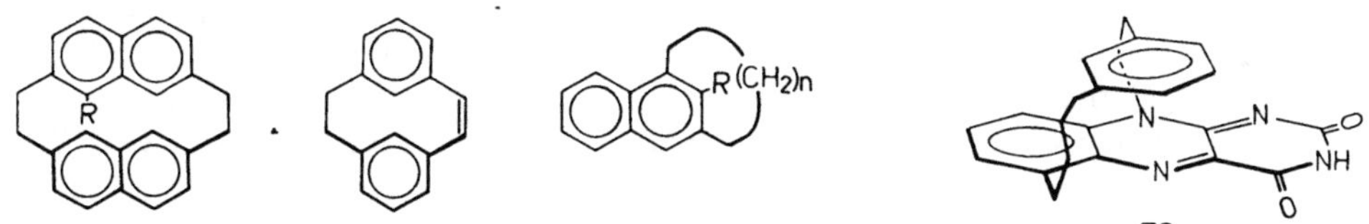

73

R: *68* NO_2; *69* NH_2

70

R: *71* Cl, *72* Br; n = 3,4

2.7.2 Layered Metacyclophanes

Triple-layered [2.2]metacyclophanes (tribenzaspiro[5.5]undecaphanes) [90] have been described by several groups [17,90,91]. Two interconvertible isomers *74* and *75* were isolated (cf. Ref. [17] and [90] for their nomenclature) with one (*75* with C_{2h}-symmetry) being thermodynamically more stable by approximately 16 kJ · mol^{-1} as established by isomerization experiments [90] and having a chair-shaped benzene ring in the center (for an X-ray structure see Ref. [92]). Such isomers are also possible for the corresponding quadruple-layered [2.2]metacyclophanes, of which one isomer (*u*, *d*, *u* [17] and *o*, *u*, *o*, respectively [90]) is the most stable [93]. So far, however, no chiral ring-substituted derivatives seem to have been described. From the diketone of the triple-layered[2.2]metacyclophane (tribenzaspiro[5.5]undecaphane-2,4-dione) [90,92] by an asymmetric carbene insertion [74] (catalyzed by (—)-ethyl lactate) in analogy to *66* and *67* from *65* (*vide supra*) the optically active layered [3.2]metacyclophanediones *76* and *77* were obtained [85].

74

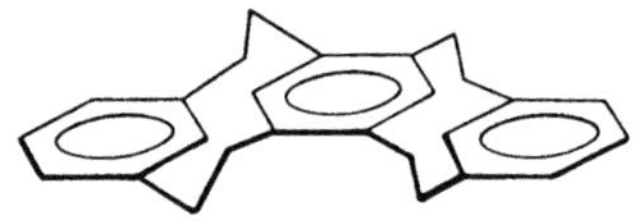

75

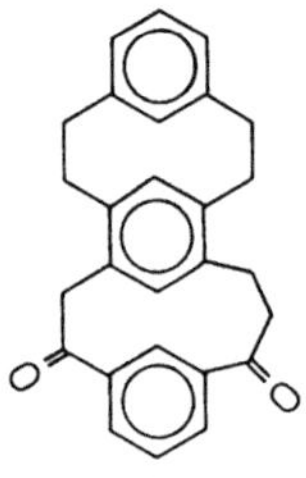

76

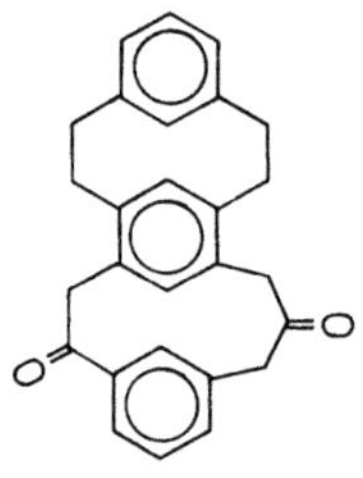

77

Absolute chiralities and chiroptical properties of metacyclophanes will be discussed in Section 2.9.

2.8 Metaparacyclophanes

[2.2]Paracyclophane (*7*) easily rearranges to the parent skeleton, [2.2]metaparacyclophane (*9*), in the presence of $HAlCl_4$ [94], whereas irradiation of the latter gives 42% of [2.2]metacyclophane [95], thus releasing some of the strain and π-electronic repulsion. Stereochemical features of [2.2]metaparacyclophanes have been reviewed [13]. In *9* the parasubstituted benzene ring is bridged by seven carbon atoms and adopts a boat-shaped conformation with a stronger deviation from planarity than in [2.2]-paracyclophane, whereas the meta-bridged ring has a chair-like form. The ring inversion barrier amounts to approximately 92 kJ · mol^{-1} [96a] (or 84 kJ) [96b], according to dynamic ^{1}H-nmr studies, as compared with 132 kJ in [2.2]metacyclophanes [72]. This inversion has been discussed in some detail and is relevant for racemization processes. Two modes are feasible: 1) rocking, with the metaphenylene ring flipping on top of the para-bridged one (whereby both rings move simultaneously) and 2) rotation of the parasubstituted benzene ring around its central axis C-11–C-14 [13].

[2.2]Metaparacyclophane becomes chiral if substituents are introduced in ring positions 4,6,12 or 13 (or in the bridge positions 1, 2, 9 and 10). As deduced from nmr studies, substituents in the positions 12 or 13 hardly influence the inversion barrier [13,96]. Studies with the optically active ester *79* derived from the 12-carboxylic acid *78* (resolved via its brucine salt; $[\alpha]_{546}$ +34° in $CHCl_3$) [96] revealed that *79* is optically stable for 25 h at 200 °C thereby proving that at least one of the above mentioned processes must be hindered. Thus it is more stable than [2.2]paracyclophanecarboxylic acid methylester (*24*) which racemizes at about 200 °C [56]. On the basis of nmr and model studies it was proposed that process (2) may be excluded [13,96].

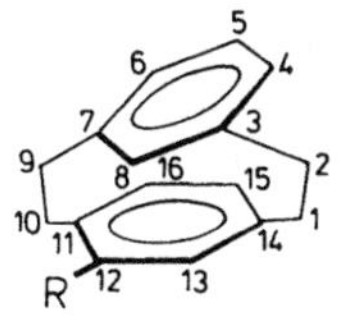

	R
9	H
(+)-*78*	COOH
(+)-*79*	$COOCH_3$
(−)-*80*	CH_3

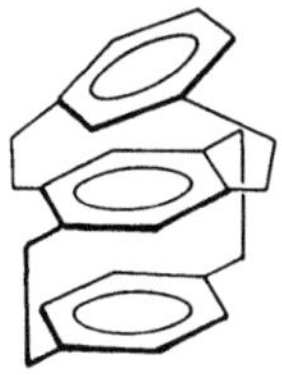

81

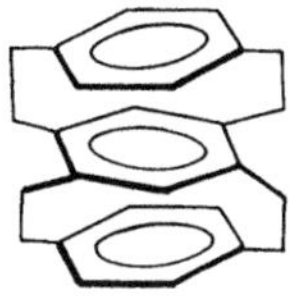

82

83

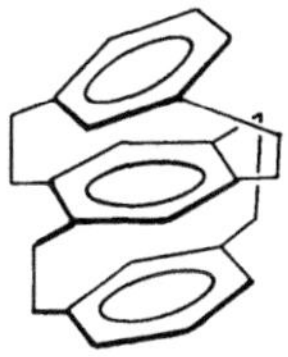

84

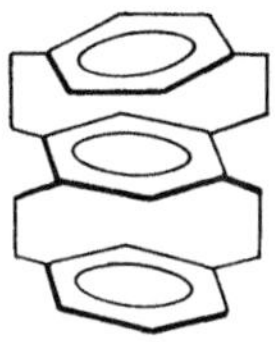

85

(+)(*S*)-4-Methyl[2.2]paracyclophane [(+)-*38*] could be rearranged with $HAlCl_4$ to the optically pure 12-methyl[2.2]metaparacyclophane (*80*, $[\alpha]_{546}$ +27°). The (—)-enantiomer was also accesible from the (+)-carboxylic acid *78* through hydroxymethyl and bromomethyl derivatives [95]. Its rotation (—26°) proved that no racemization had occured during the rearrangement of (+)-*38*. According to the proposed mechanism the chirality (*S*) was assigned to (+)-*80* [95]. Thus the chiralities of the carboxylic acid *78* and its derivatives also seemed to have been established as (+)(*R*). Irradiation of (—)-*80* on the other hand led to 86% racemization and a mixture of isomeric methyl[2.2]metacyclophanes [95].

Besides these compounds, no other optically active metaparacyclophanes seem to have been described so far. In analogy to [2.2]paracyclophane itself and its methyl-derivative, the triple-layered [3]chochin *40* also undergoes rearrangement with $HAlCl_4$ to afford the triple-layered metaparacyclophane *81* [97], whereas with $SnCl_4$, $TiCl_4$ or BF_3 in the presence of hydrochloric acid an approximately equimolar mixture of *82* and *83* was obtained in good yields [17]; *81* and its stereoisomers *82*, *84* and *85* were also prepared [78] by conventional methods, as described in Section 2.6.2 for layered [2.2]paracyclophanes. Compounds *81* and *84* — again in analogy to [2.2]metaparacyclophane itself — exhibit a dynamic mobility of the meta-bridged ring as established by temperature-dependent nmr spectroscopy [99]. From these structures, *81* and *84* are chiral and might be resolvable by asymmetric chromatography (e.g. on triacetylcellulose).

2.9 Chiralities[7] and Chiroptical Properties of Cyclophanes

Absolute configurations (chiralities) of phanes — together with those of several other planar chiral structures — have been compiled in the "Atlas of Stereochemistry" [25] (for previous surveys see Refs. [10] and [100]).

As usual in stereochemical research, four main approaches have been applied to the problem of assigning chiralities to optically active cyclophanes. They are listed in order of their reliabilities: i) anomalous X-ray diffraction (*Bijvoet* method), ii) chemical correlations with compounds of known chiralities (preferably established by the *Bijvoet* method), iii) kinetic resolutions and/or asymmetric syntheses, iv) interpretation of chiroptical properties (mainly circular dichroism) on the basis of (sector) rules including theoretical methods.

2.9.1 Bijvoet Method

So far the chiralities of only three cyclophanes have been established by this unambiguous method.

For [2.2]paracyclophane-4-carboxylic acid (*23*) as (—)(*R*): This result has been mentioned in a footnote in Ref. [101] but seems never to have been published (see also Ref. [61]). The chirality of this acid was correlated *via* its (—)-aldehyde with a levorotatory hexahelicene derivative which, according to the paracyclophane moiety at the terminal, had to adopt (*M*)-helicity. Its chiroptical properties are comparable to those of hexahelicene itself [101]. For the (—)-bromoderivative of the latter the (*M*)-helicity was established by the *Bijvoet*-method [102]. In a later study, (—)paracyclophane-hexahelicene prepared from (—)-1,4-dimethylhexahelicene with known chirality (which in turn was obtained with approximately 12% enantiomeric purity by asymmetric chromatography) confirmed these results. It should be mentioned that [2.2]paracyclophane-4-carboxylic acid (*23*) was the first planar chiral cyclophane whose chirality was determined [104] (see also Ref. [54]). The results justmentioned confirmed the assignment (+)(*S*).

The chiralities of two [2.2]metacyclophanes were determined in 1973 and 1981: X-ray structure of a (—)-meta-bromobenzoate derived from levorotatory 1-oxo[2.2]-metacyclophane (*61*) (*via* the equatorial-positioned alcohol) confirmed the chirality (*S*) for (—)-*61* as had been deduced also from chiroptical properties [105].

(+)-12-Bromo[2.2]metacyclophane-4-carbonitril (*60*, which had previously been correlated with many optically active 4-monosubstituted, 4,14-homo-, 4,12- and 4,14-heterodisubstituted [2.2]metacyclophanes) [77,78] was chosen as a nicely crystalline reference substance [106]: its chirality (*R*) as determined by the *Bijvoet*-method [106] contrasts with the previously assigned chirality based on results of a kinetic resolution [77] (*vide infra*), but agrees with the assignment by the coupled oscillator method (to the bisester *57*) [107] and by optical comparison and correlation [84].

2.9.2 Chemical Correlations

An unambiguous correlation of the ansacompound dioxa[11]paracyclophane-carboxylic acid (gentisic nonamethylene ether, *17*) with cis-3-hydroxy-cyclohexane-

---

7 See also footnote[5] on p. 37.

carboxylic acid *89* of known absolute configuration (+)(1*S*, 3*R*) is an excellent example of a direct correlation of planar to centrochirality [62]: cis-hydrogenation of (—)-*17* [36] and ring cleavage of the bridge in the cis-cyclohexane derivative *86* with t-BuOK in dimethyl sulfoxide furnished (after hydrogenation and subsequent esterification of the carboxylic acid *87*) the key intermediate (—)-*88*. Its dextrorotatory enantiomer on the other hand was accessible from (+)-cis-(1*S*, 3*R*)-*89* by reaction with the bromide *90* *via* a series of straight forward steps. This established the chirality (—)(*S*) for *17* as had been proposed previously on the basis of a sector rule [63] (see also 2.9.4). By comparison of the chiroptical properties this result confirmed also the chiralities of related dioxa[n]paracyclophanes and [n]paracyclophanes (e.g. of *14*, the homologue with a tencarbon bridge) [63,108].

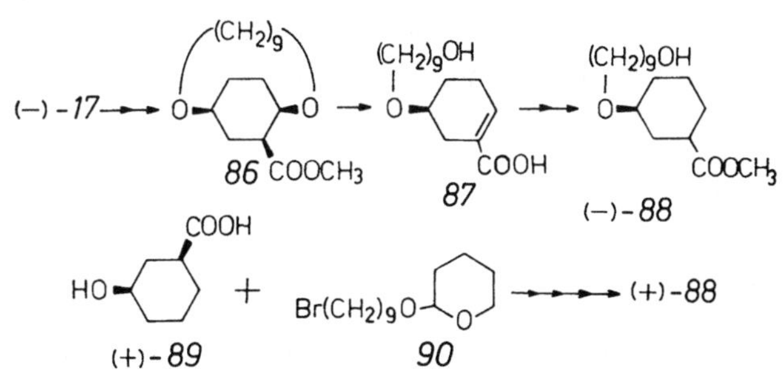

For chemical correlations of planar chiral cyclophanes with centrochiral derivatives for the purpose of applying *Horeau*'s method [100], see the following section.

2.9.3 Kinetic Resolutions

The first assignment of the chirality (absolute configuration) to a planar chiral phane: ([2.2]paracyclophanecarboxylic acid *23*, in 1968) [104] was deduced from the results of a kinetic resolution of its (racemic) anhydride with (—)-1-phenylethylamine and is based on the related topology of *23* and 2-methyl-metallocene-1-carboxylic acids [19]. For these chiral compounds, this method had given (correct) results, as confirmed afterwards by the *Bijvoet* method [109]. Since this method has been reviewed in some detail [19,100] it will not be discussed in this survey.

The chirality (—)(*R*) for *23* (and 20 derivatives thereof), as deduced by kinetic resolution [54,104], was also confirmed by the application of *Horeau*'s method (kinetic resolution of 2-phenylbutanoic anhydride with optically active carbinols) [100] to the two epimeric carbinols (—)-*92* and (+)-*93* derived from (—)-*23* via (—)-*91* and

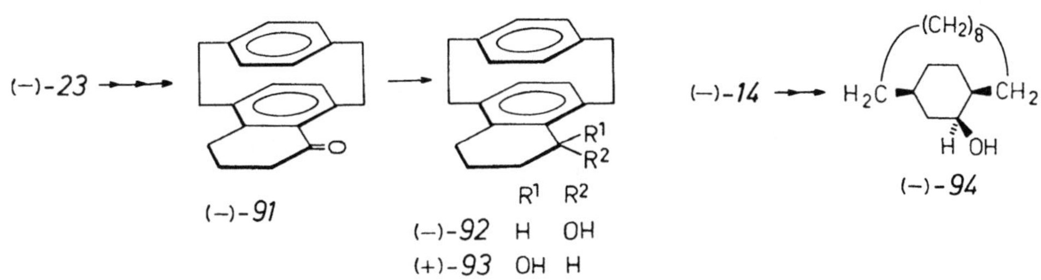

subsequently also both by theoretical considerations based on the exciton theory [67] and by the indirect application of the *Bijvoet* method, as outlined in the previous Section 2.9.1 (cf. Ref. [101]).

The known chirality of *23* and of the corresponding methyl derivative *38* has served as the basis for several configurational assignments especially for [m][n] [44] and layered [2.2]paracyclophanes [64] as well as for the interpretation of rearrangement reactions to optically active [2.2]metaparacyclophanes [95] and for other stereochemical investigations [57,103] (see 2.6 and 2.8).

In analogy to *23*, the chiralities of [2.2]meta- and [10]paracyclophanecarboxylic acids were also deduced from the results of kinetic resolutions [40,77]. For the application of *Horeau*'s method, (—)-[10]paracyclophanecarboxylic acid (*14*) was transformed by stereoselective hydrogenation and subsequent sodium borohydride reduction of an intermediate cyclohexanone into the (—)-cis-cyclohexanol *94* which on reaction with racemic 2-phenylbutanoic anhydride afforded a 15% excess of the levorotatory acid thereby proving (in agreement with the kinetic resolution of the anhydride of *14*, *vide supra*) the chirality (*S*) for (—)-*14* and all its derivatives [40]. Optical comparison with dioxa[10]paracyclophanecarboxylic acid (*16*) confirmed this result [63,108].

The assignment of [(—)(*S*)] to [2.2]metacyclophanecarboxylic acid (*55*) on the basis of the kinetic resolution of its anhydride had to be revised to (—)(*R*) after application of the *Bijvoet*-method to a derivative (*60*) [106] (see 2.9.1.). (See also the results presented in Ref. [84] and discussed in Section 2.7.1). It therefore seems as if the general topology of [2.2]metacyclophanes is sufficiently different from that of paracyclophanes and 1,2-disubstituted metallocenes and therefore does not permit the application of the above-mentioned kinetic resolution method.

A tentative configurational assignment (*R*) to (+)-[2.2](2,6)naphthalenophane (*27b*) was proposed on an assumed stereochemic model of the relative stabilities of its complexes with (+)(*S*)- and (—)(*R*)-TAPA at the silicagel surface [49].

2.9.4 Chiroptical Methods

The electron absorption spectra of cyclophanes, especially paracyclophanes, have been studied in great detail, mainly with respect to the distortion of the benzene rings and the electronic transanular effects which cause the spectra to be much more complex than those of corresponding substituted benzenes. Optically active para- and metacyclophanes offer excellent models for testing several of the theoretical assumptions by studying the circular dichroism (CD) spectra and correlating the signs and positions of the Cotton effects with the chiralities of these strained compounds. For a detailed treatment of these phenomena see: [10]paracyclophanes [40], [2.2]paracyclophanes [54,67], [2.2]metacyclophanes [78,110], [m.n]paracyclophanes [44] and layered [2.2]paracyclophanes [64].

The chirality of [2.2]paracyclophane derivatives has been deduced as being (—)(*R*) on the basis of the exciton theory of coupled oscillators [67] and confirmed by experimental results (see 2.9.1 and 2.9.3). In these compounds a negative Cotton effect at 270 nm (corresponding to the p-band) seems to be specific for the (*R*)-chirality [54].

From the optical and chiroptical properties of [2.2]metacyclophanes, compared with those of [10]paracyclophanes [40], it was concluded that in the former an inherently achiral aromatic chromophore perturbed by vibrations is presumably responsible for their optical activity, as displayed in the CD-spectra [110]. Optically active [2.2]-metacyclophanes (with skeleton symmetry C_{2h}) served also as models for studies [111] in connection with the algebraic theory of chirality functions, which previously had been tested with chiral compounds of D_{2d} and T_d skeleton symmetry [112]. Discrepancies between experimental and theoretical values (studied on more than 30 optically active derivatives) [111b] and the consequences with respect to observations of the chirality phenomenon [111a] have been discussed.

Based on the CD-spectra of various optically active carbophanes, a sector rule for the correlation of the sign of the ^1L_b-Cotton effect of the benzene chromophore with the absolute chiralities of these phanes (as well as for chiral dihydro-9,10-ethanoanthracenes) has been proposed [63].

While for several phanes such as [n]paracyclophanes, dioxa[n]para or [2.2]metaparacyclophanes, the assignments by this rule were in agreement with the experimental results, they had to be revised for [2.2]meta- and [m][n]paracyclophanes after the application of more reliable methods. These discrepancies were discussed in previous sections (2.6.1, 2.9.1). A quadrant rule proposed for cyclic ketones includes also the configurational assignment to oxo[2.2]metacyclophanes (*61–63*) and stereoisomeric alcohols derived therefrom [113].

The chirality of the D_2-twist benzene rings (see Ref. [114] for an X-ray structure analysis) occurring in doubly bridged and layered paracyclophanes (2.6.) could be related to the sign of their Cotton effects in the 240–360 nm region as follows: a $(PMP)_2$-twist (as shown in Fig. 3, *A*) gives rise to a positive Cotton effect, while, on the hand, the mirror image $(MPM)_2$ structure (*B*) gives rise to a negative Cotton effect (for the specification of the chiralities *A* and *B* see Ref. [64]).

A (PMP) (+)

B (MPM) (–)

Fig. 3. Chirality of D_2-twist benzene and signs of Cotton effects [64]

3 Bridged Anulenes

3.1 Introduction and Definitions

Bridged [10]- and [14]anulenes[8] are of considerable interest not only because of their aromaticity (complying with *Hückel*'s rule) but also because of their (rigid) molecular geometry. The most simple representative 1,6-methano[10]anulene (*95*) first obtained in 1964 [115], as well as others with oxido or imino bridges (cf. *103*), have C_{2v}-symmetry, but become chiral (C_1) by substitution in any position. Such chiral anulenes can be classified as planar chiral under the assumption that the ten-

8 Anulene is derived from the latin word *anulus* (small ring) and should therefore be spelled, in contrast to the general custom, with *one* n.

(or fourteen-) membered ring is planar. According to X-ray studies, however, it is puckered [116], with the bridgehead carbons (C-1 and C-6 in *95*) lying above this plane. Because of the obvious analogy to the cyclophanes, however, (with an aromatic ring bridged in "para"-positions) it seems justified to treat anulene derivatives as planar chiral as shown in Fig. 1 (This specification had been confirmed with the late Dr. *R. S. Cahn* in 1971; see footnote in Ref. [117]). Consequently the bridge atom has to be regarded as pilot atom for specifying the planar chirality (Z in Fig. 1). Many aspects of bridged anulenes, especially their syntheses, reactions, molecular geometries and spectral properties have been reviewed elsewhere [118a–d].

Spiniferin-1 is a naturally occurring furanosequiterpene with the skeleton of 1,6-methano[10]anulene, isolated from a sponge found in the Bay of Naples. The total synthesis of the racemate was recently reported [118e].

The discussion in this article shall be restricted to optically active derivatives.

3.2 Bridged [10]Anulenes and [10]Azaanulenes

Apart from a footnote in a review article [118b] the first optically active bridged [10]-anulene was described in 1971 thus experimentally proving their chirality [119]: The carboxylic acid *96* was resolved through i s salts with (+)- and (—)-1-phenylethyl-amine ([α]$_D$ +250° in ethanol *via* the (—)-amine [117,119]). The enantiomeric purity

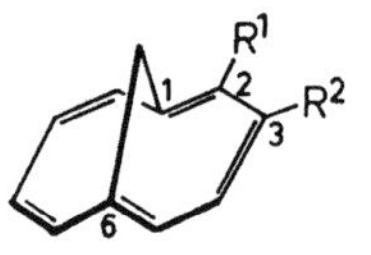

	R¹	R²
95	H	H
(+)-*96*	COOH	H
(+)-*97*	$COOCH_3$	H
(+)-*98*	CHO	H
(+)-*99*	$COCH_3$	H
(−)-*100*	CH_3	H
101	H	$COOCH_3$

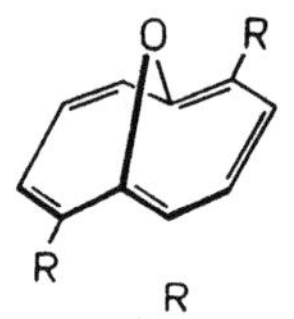

	R
103	H
104	Br

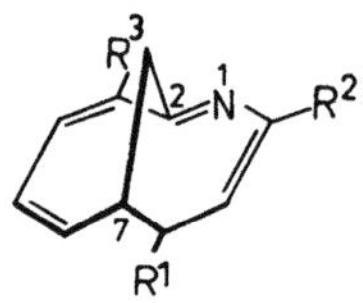

	R¹	R²	R³
107	H	H	H
108	CH_3	H	H
109	H	CH_3	H
110	$COOCH_3$	H	H
111	H	$COOCH_3$	H
112	H	CHO	H
113	H	H	Br

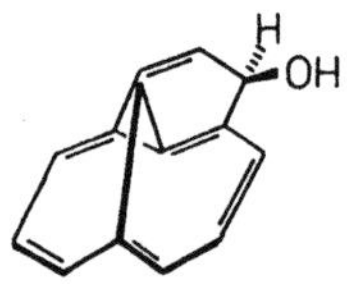

(+)-*102*

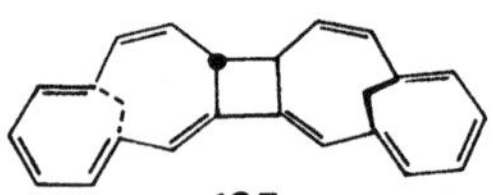

105

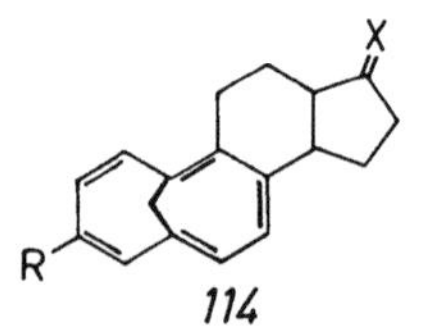

114

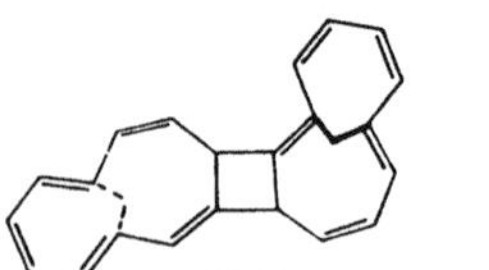

106

was confirmed by nmr-spectroscopy of the diastereomeric phenethylamides [117]. Several derivatives were prepared from (+)-*96*, some of them in connection with the determination of the absolute chirality. (For the chromatographic resolution of the methylester *97* see Fig. 4). From the aldehyde (+)-*98* *via* its tosylhydrazone a carbene could be generated which dimerizes to afford *105* as the only product having *anti*-configuration of the bridges, whereas the racemic aldehyde had given a mixture of *syn* and *anti-105* together with its isomer *106* [120].

Introduction of nitrogen into the anulene ring (e.g. of *95*) leads to a methano-azaanulene *107* [121] with C_1-symmetry which is therefore chiral (like its mono- or disubstituted derivatives) [118]. The low basicity of *107* ($pK_a \approx 3.20$) prevented its optical resolution by conventional methods (e.g. through salts with optically active acids). Excellent results were obtained, however, (as also in the case of the two isomeric carbocyclic methylesters *97* and *101* and of several derivatives of azaanulene) by chromatography on microcrystalline triacetyl cellulose in ethanol at 7 bar [122] (see also Section 2.7.1). In many cases base line separations were accomplished to give both (optically pure) enantiomers. Enantiomeric relations were confirmed in all cases by recording the CD-spectra of both fractions. Some results of these separations are shown in Fig. 4 together with the optical rotations ($[\alpha]_D$ in ethanol) of the enantiomers.

The close similarity of the CD-curves of methano[10]azaanulene (*107*) itself, its isomeric methylderivatives (e.g. *109*) and the 3-bromo-product *113* on the one hand and of the methylesters *110* or *111* on the other (typical CD-spectra are shown in Fig. 5) permitted the correlation of relative configurations within each of these groups: compounds with the same sign of rotation seem to have the same chiralities [122].

A chemical correlation of the (—)-methyl derivative *109* with the (—)-methylester *111* was possible by selenium dioxide oxidation of the former and subsequent silver

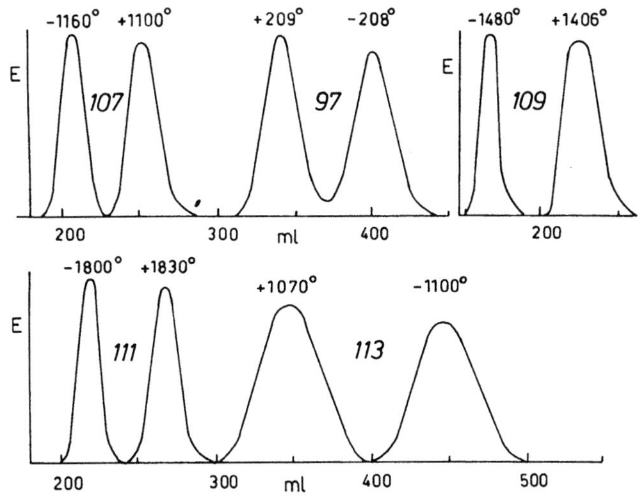

Fig. 4. Separation of enantiomeric methano[10]anulenes (*97*) and [10]azaanulenes (*107*, *109*, *111* and *113*) on triacetylcellulose in ethanol at 7 bar and 40 °C. Optical rotations $[\alpha]_D^{20}$ in ethanol [122]

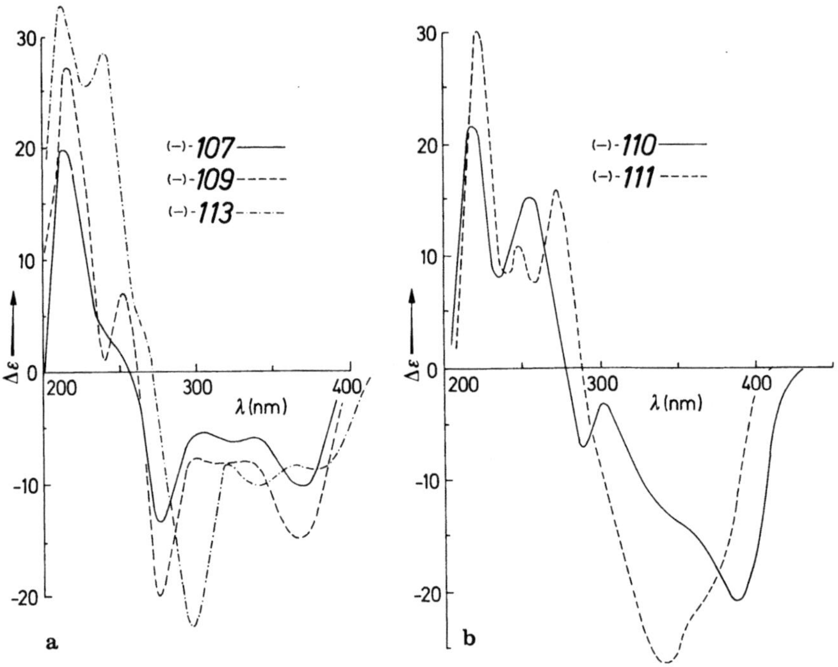

Fig. 5. Circulardichroism spectra of methano[10]azaanulenes in ethanol [122]

oxide oxidation of the intermediate aldehyde (*112*) to give (after esterification) the latter [122].

The chirality of methano[10]anulene-2-carboxylic acid (*96*) was derived from results of a kinetic resolution as outlined for cyclophanes in Section 2.9.3. The reaction of the anhydride of *96* with (—)-1-phenylethylamine as well as the kinetic resolution of *102* (unambiguously correlated with *96 via* the acetyl derivative *99*) with (+)-2-phenylbutanoic anhydride (*Horeau*'s method) — affording an excess of (+)-carbinol *102* — led to the assignment of the descriptor (*S*) to (+)-*96* and all its derivatives [117].

Comparison of the CD-spectra of (—)(R)-*2*-methyl-methano[10]anulene (*100*) [117] and of (—)-8-methyl-methano[10]azaanulene (*108*) [122] allowed a rather tentative assignment of the descriptor (*R*) to (—)-*108* and — on the basis of the above-mentioned correlation — also to (—)-*107* and to other levorotatory [10]azaanulenes, as shown in the following:

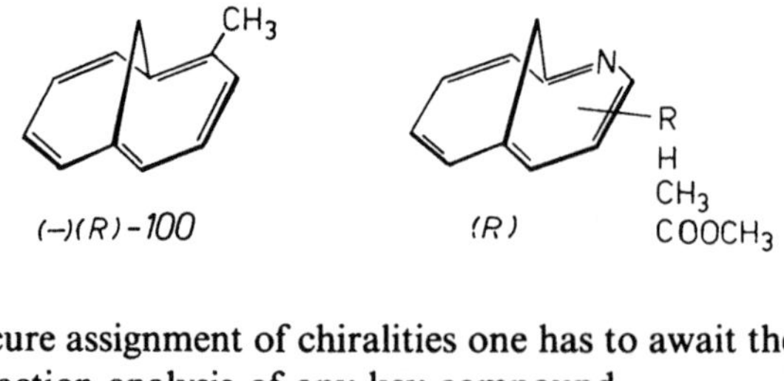

For a secure assignment of chiralities one has to await the results of an anomalous X-ray-diffraction analysis of any key compound.

From the several other chiral [10]anulenes described so far (mainly by *E. Vogel* and his group) [118] (for a tetracyclic derivative *114* with a steroid skeleton, cf. Ref. [123]), only dibromo-oxido[10]anulene *104* has been (partially) resolved by chromatography on triacetylcellulose [124]. Although no apparent peak separation could be observed, the first and last peak fractions exhibited enantiomeric CD-curves, thus indicating that enantiomers (albeit with unknown enantiomeric purity) had been isolated.

3.3 Bridged [14]Anulenes

With regard to the stability of the skeletons of bridged anulenes, the possibility of a bridge inversion is of considerable interest. The availability of optically active bridged [10]- and [14]anulenes offered the chance to study these phenomena by racemization experiments:

Both the bisoxido- and bisimino-(cis)bridged dibromo[14]anulenes *115* and *116* with C_2-symmetry could be quantitatively separated into their enantiomers by chromatography on triacetylcellulose in ethanol at 1.7 bar [124]. The dextrorotatory enantiomers with $[\alpha]_{546}$ values of 1700° (*115*) and 1500° (*116*) in ethanol were eluted first. They exhibit very similar CD-curves with strong Cotton effects around 260

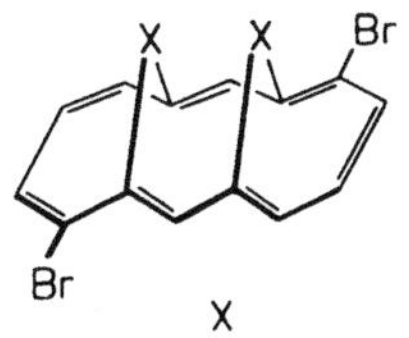

X
115 O
116 NH

and 330 nm (Δε −16 to −28 and +30), thus establishing that (+)-*115* and (+)-*116* have the same chirality[9]. Racemization studies in dodecane revealed for the bisimino-[14]anulene *116* an optical stability up to 250 °C (for 2 h) whereas the bisoxido-compound *115* racemizes at 150 °C (2 h) to the extent of 91–92%. Assuming that the inversion process proceeds *via* a trans-intermediate (with C_i-symmetry) by first order kinetics — and not by the highly improbable simultaneous inversion of both bridges — the free activation enthalpy $\Delta G^{\neq}$ of this process amounts to 134 kJ · mol^{-1}. This is in good agreement with the value of 143 kJ calculated for the unsubstituted bisoxido[14]anulene by an SCF-MO-method (based on Dewar's MNDO-method) [125]. For *116* the barrier is higher than 174 kJ.

The dibromo-oxido[10]anulene *104* (for its partial resolution see the previous Section 3.2) on the other hand is optically stable up to 250 °C (for 2 h) [124] with a calculated $\Delta G^{\neq}$-value of ca. 255 kJ · mol^{-1} [125]. Obviously the ten-membered ring is more rigid than the fourteen-membered one — as one might expect.

3.4 [8]Anulenes

Nonbridged and nonaromatic anulenes such as derivatives of cyclooctatetraene ([8]anulene) (cf. *117* to *121*) are chiral owing to the nonplanarity of the eight-membered

9 Added in proof: for (+)-*116* the chirality (S) — as shown in the formula — has been established only recently by the *Bijvoet*-method (J. Lex, unpublished).

ring which in cyclooctatetraene itself (with D_{2d}-symmetry) has an inversion barrier of ca. 60 kJ · mol^{-1} [126]. They might be considered as axial chiral (like *E*-cyclooctene, Section 4.2, because of the obvious twist) or as planar chiral: The latter specification seems to be much more appropriate in the light of the definitions outlined in Section 1.2. Thus, if in *120* or *121* the highly substituted double bond is regarded as the plane A—Y—B (see Fig. 1), one of the sp^2-hybridized carbons represents Z as depicted in Fig. 6.

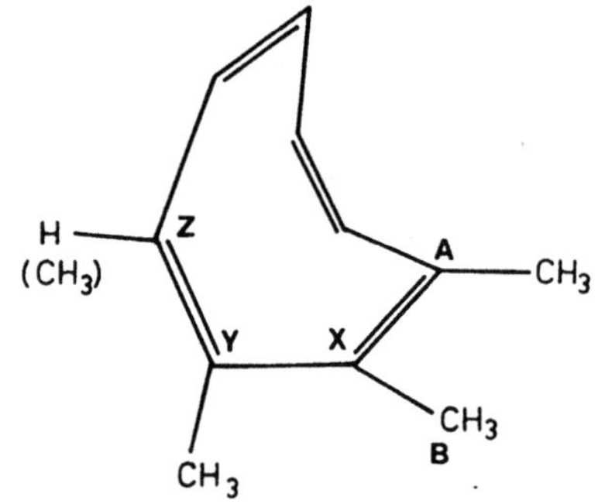

(−)(S)

Fig. 6. Planar chirality of cyclooctatetraene derivatives *120* and *121* (cf. also Fig. 1)

The first optically active example was *117* ($[\alpha]_D$ +71° in ethanol; through its brucine salt) for which a racemization barrier of 113 kJ · mol^{-1} was found [127]. In order to establish the mode of racemization the optically active derivatives, *118* and *119* were prepared through their cinchonidine and brucine salts, respectively ($[\alpha]_D$ —21 and —78°, respectively) [128]. From racemization experiments with the methylesters of *118* and *119* (with barriers of 130 and 105 kJ · mol^{-1}, respectively) it could be concluded that a biradical rearrangement predominates in *118*, whereas for *117* and *119* a simple ring inversion process is in accordance with the results [128].

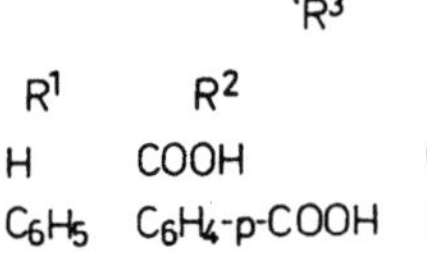

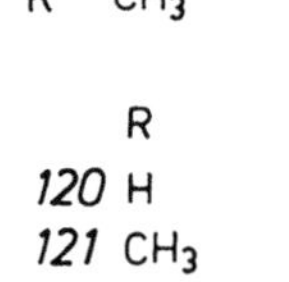

	R^1	R^2	R^3	R^4
117	H	COOH	COOH	Br
118	C_6H_5	C_6H_4-p-COOH	H	H
119	C_6H_5	H	C_6H_4-p-COOH	H

	R
120	H
121	CH_3

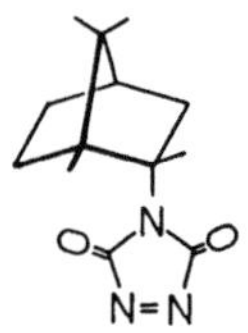

(−)-*122*

(+)-*123*

Related studies were undertaken with the methylcyclooctatetraenes *120* and *121* (point groups C_1 and C_2, resp.). After previous preparations from optically active cyclohexene derivatives which provided the chiralities (—)(*S*) (according to the specification given in Fig. 6) by application of the *Bijvoet*-technique to precursors [129], *120* and *121* could be obtained in high enantiomeric purities as follows [130]: Cycloaddition of an enantiomerically pure triazolinedione *122* to the racemates and subsequent alkaline hydrolysis followed by manganese dioxide oxidation of the separated diastereomeric adducts afforded levorotatory methyl cyclooctatetraenes: $[\alpha]_D$ —161° (*120*) and —310° (*121*). The values found for racemization and inversion ($\Delta G^{\neq}$ 103 and 130 kJ · mol^{-1} for *120* and *121*) are in accordance with a racemization mechanism through a planar D_{4h}-intermediate [131].

Dibenzocyclooctine (*123*), isomeric to cyclooctatetraene with respect to its basic skeleton, represents a borderline case between chiral anulenes and cyclooctenes (see the following Section 4.). Whereas the chiral C_2-conformers of tetramethyl cyclooctine easily interconvert with a barrier of $\Delta G^{\neq} \approx 53$ kJ · mol^{-1}, this barrier is increased to ≈84 kJ by anelation of two benzene rings (in *123*) as shown by ^{1}H-nmr coalescence studies [133]. Its partial resolution was accomplished by chromatography on a silica column coated with (—)-TAPA (*Newman*'s reagent) at —28 °C in pentane to give enantiomers with a maximal optical rotation of 66° at 365 nm [133]. This seems to be the first case of an optically labile compound (with a low inversion barrier of ca. 84 kJ = 20 kcal · mol^{-1}) which has been resolved by a direct method (see for instance Ref. [134]). Racemization studies at +2 °C established a barrier of ≈52 kJ in excellent agreement with the value found from nmr spectroscopy. For (+)-*123* the chirality, as shown in the formula [i.e. (*S*)], was proposed in analogy to the arguments presented in Ref. [49] for [2.2](2,6)naphthalenophane (*27b*, see 2.5).

4 (*E*)-Cycloalkenes and Related Structures

4.1 Introduction and Definitions

(*E*)-cyclooctene (*124*) and related rigid cycloalkenes can occur in two enantiomeric conformations which may be separable by appropriate means if the barrier of equilibration is high enough. They are usually classified as planar chiral (cf. the arguments presented in Section 1.2). The (*E*)-bridging introduces a considerable twist of the double bond with a torsion from planarity of about 20° in (*E*)-cyclooctene (*124*) [135]. A detailed discussion of the role of this torsion in twisted olefins, which mainly contributes to the sign of the Cotton effects, has been published [136].

4.2 (*E*)-Cyclooctene

The first experimental verification of these concepts in 1963 by the resolution of *124* through chiral platinum complexes [137] was soon followed by the assignment of the chirality (—)(*R*) by an ingenious chemical correlation to a centrochiral compound, namely (+)-tartaric acid [138]. This was confirmed in 1970 by application of the *Bijvoet*-method to the platinum complex used for optical resolution [139], but is in contrast to an earlier prediction based on an optical model [140].

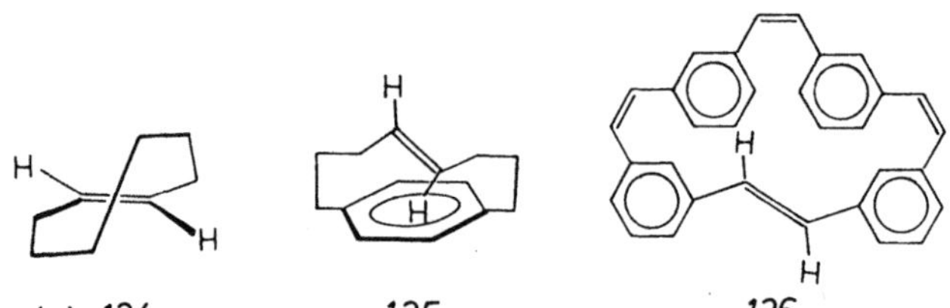

(−)-*124* *125* *126*

An increase in the size of the ring reduces the conformational stability to such an extent that trans-cyclononene and cyclodecene are optically instable with the rotational barrier decreasing from 150 (trans-cyclooctene) to 84 (trans-cyclononene) and 42 kJ · mol^{-1} (trans-cyclodecene) [140a)]. The same is true also for *125* (which can of course also be regarded as an [8]paracyclophane). It racemizes immediately after the cleavage of its diastereomeric platinum complexes [141)]. A fast conformational equilibrium was also established for the cis,cis,cis,trans-metacyclophane tetraene *126* from nmr-studies [142)]. Alkyl substituents on the double bond of trans-cycloalkenes, however, increase the barrier: Thus, (E)-1,2-dimethyl-cyclodecene and -undecene are optically stable ($[\alpha]_D$ −118° and +29°, resp., chiralities (S)) whereas (E)-1,2-dimethyl-cyclododecene racemizes at or below roomtemperature [142a)].

4.3 1,2-Cyclononadiene, Twisted Alkenes, Betweenanenes

Introduction of the allene structure into cycloalkanes such as in 1,2-cyclononadiene (*127*) provides another approach to chiral cycloalkenes of sufficient enantiomeric stability. Although *127* has to be classified as an axial chiral compound like other C_2-allenes it is included in this survey because of its obvious relation to (*E*)-cyclooctene as also can be seen from chemical correlations (*vide infra*). Racemic *127* was resolved either through diastereomeric platinum complexes [143)] or by ring enlargement *via* the dibromocarbene adduct *128* of optically active (*E*)-cyclooctene (see 4.2) with methyllithium [143)] — a method already used for the preparation of racemic *127*. The first method afforded a product of 44% enantiomeric purity whereas *127* obtained from (*E*)-cyclooctene had a rotation $[\alpha]_D$ of 170–175°. The chirality of *127* was established by correlation with (+)(*S*)-(*E*)-cyclooctene which in a stereoselective reaction with dibromocarbene afforded (−)-dibromo-trans-bicyclo[6.1.0]nonane (*128*) [144)]. Its absolute stereochemistry was determined by the *Bijvoet*-method as (1*R*, 8*R*) and served as a key intermediate for the correlation with *127*: ring expansion induced

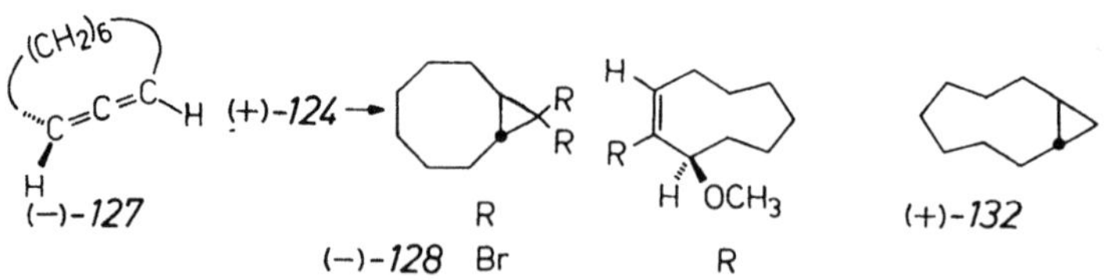

(−)-*127*

	R
(−)-*128*	Br
(+)-*129*	H

	R
(+)-*130*	Br
(−)-*131*	H

(+)-*132*

by silver perchlorate gave (+)-(*Z*)-bromomethoxy cyclononene *130*, which after reduction yielded (—)-*131*, which, on the other hand, is also accessible from (—)-*127*, thus proving the (axial)chirality of the latter as (—)(*S*) [144].

In another approach [145], dibromocarbene was added to (+)-cyclononadiene *127* to give (after reduction of the "outside" adduct) (+)-trans-bicyclo[7.1.0]decane (*132*). From the dextrorotatory enantiomer of the above-mentioned dibromo-derivative *128*, (+)-trans-bicyclo[6.1.0]nonane (*129*) was accessible. Comparison of the chiroptical properties of (+)-*129* and (+)-*132* led to the conclusion that *127* had the (+)(*R*)(axial) chirality [145] in agreement with the results reported above [144].

Apart from (*E*)-bridging, alkenes can be twisted and therefore made chiral by introducing bulky substituents either into (*Z*) or (*E*)-positions[10]. The racemic stereoisomers *134* and *135* were prepared by reductive coupling of the ketone *133* [147]. Whereas the (*Z*)-isomer resembles the helicenes in its molecular geometry, the enantiometric (*E*)-isomer(s) *135* may be regarded as planarchiral. It was resolved by HPLC on alumina coated with (+)-TAPA, furnishing the levorotatory enantiomer, $[\alpha]_{578}$ —508° (hexane), thus also establishing the C_2-structure *135* of the (*E*)-isomer (the other possible stereoisomer with C_i-symmetry apparently was not formed in the coupling reaction [147].

The so-called [m][n]betweenanenes *136* and *137* (for a review see Ref. [18]) are (*E*)-cycloalkenes with the double bond situated between the two (*E*)-bridges. With two bridges of equal lengths (m = n) they have D_2-symmetry similar to the topologically related [m][n]paracyclophanes (see Section 2.6.1).

The original synthetic route involving cyclization of appropriately bisfunctionalized 1,2-disubstituted (*E*)-cycloalkenes [148] is rather inefficient because of the difficult accessibility of the starting materials. A much more attractive approach involves a highly diastereoselective preparation of the latter and has enabled the synthesis of optically active [10][10]betweenane (*137*) whose enantiomeric purity [149] is high, as compared with earlier reports, e.g. for the synthesis of the (—)-[8][8]-homologue *136* [150]. The latter ($[\alpha]_D$ —2.3°) had been obtained by irradiation of the corresponding (*Z*)-enone in diethyl(+)tartrate, affording a 1:7 (*E*)/(*Z*)-mixture and by reduction of the (*E*)-ketone (isolated by preparative GC). CD-comparison with (—)(*R*)-(*E*)-cyclooctene (see 4.2) indicated a (*R*)-chirality for the levorotatory enantiomer. This seems to be in contrast to (+)(*R*) established for [10][10]betweenanene (*137*) as follows [149]: its stereodifferentiating synthesis started from the epoxide of 2-methylene-cyclodecanone *138* which on treatment with 3-butenylmagnesium bromide/cuprous iodide yielded the (*E*)-product *139* with over 98% diastereoselectivity. Its kinetic resolution by the asymmetric *Sharpless*-epoxidation yielded (+)(*R*)-*139* (with 95% ee according to nmr-shift-experiments) which then was transformed through the (+)-triene *140* (with an enantiomeric purity higher than 90%) and the diol (+)-*141* into the dextrorotatory bisaldehyde *142*. By means of a titanium-induced intramolecular reductive coupling, the latter yielded a 4:1 mixture of the (*E*)- and (*Z*)-dienes *143*, which was finally hydrogenated (whereby only the "outer" double bond was reduced) to give (+)(*R*)-[10][10] betweenanene (*137*, $[\alpha]_D$ +47° in $CHCl_3$). Its chirality follows

10 Atropisomeric butadienes (with bulky substituents) (see for example Ref. [146]) clearly belong to the group of axial chiral structures and are not discussed in this review.

from the enantioselective *Sharpless*-epoxidation and from the comparison of its CD-spectrum with that of the precursor (+)-*140*, both of which exhibit negative Cotton effects. The enantiomeric purity of (+)-*137* is higher than 90%, as deduced from (+)-*140*.

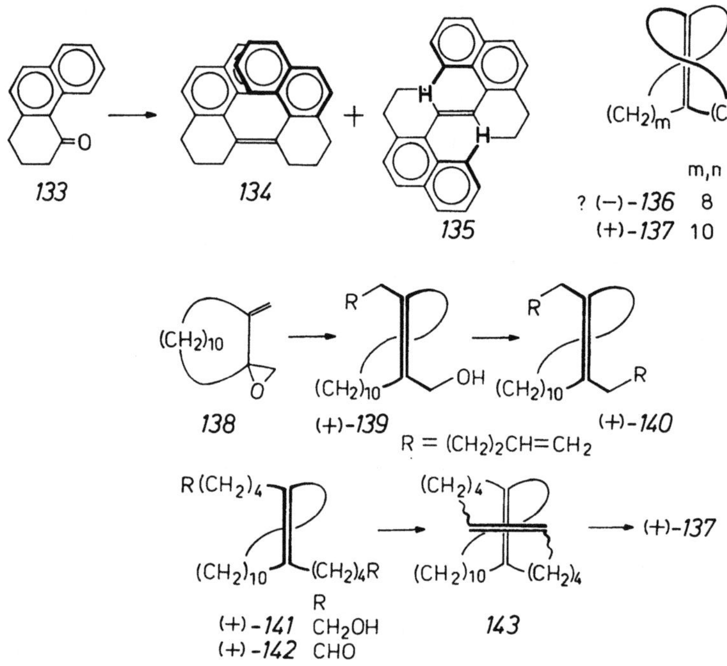

5 Acknowledgement

Stimulating discussion with several colleagues, especially Profs. F. A. L. Anet (UCLA), K. Mislow (Princeton), M. Nakazaki (Osaka) and V. Prelog (Zürich), to whom I am very grateful, have clearly revealed that the subject of planar chirality is still or again a topic of some controversy. This has been demonstrated e.g. in the paper by K. Mislow and J. Siegel (see footnote [2] on p. 30) a preprint of which has been kindly submitted to me by Prof. K. Mislow.

My thanks are also expressed to my collaborators for their excellent work; their names are given in the references. Finally, I am most grateful to the Japan Society for the Promotion of Science for a generous travel grant which gave me the opportinity to meet several colleagues and discuss with them problems pertinent to this review.

6 References

(References considered up to December 1983)

1. Cahn, R. S.: J. Chem. Educ. *41*, 116 (1964)
2. Cahn, R. S., Ingold, C. K., Prelog, V.: Angew. Chem. *78*, 413 (1966); Int. Ed. Engl. *5*, 385 (1966)
3. Falk, H., Schlögl, K.: Monatsh. Chem. *96*, 266, 1065 (1965)
4. Prelog, V.: J. Mol. Catalysis *1*, 159 (1975/76)
5. Mislow, K., Bickart, P.: Isr. J. Chem. *15*, 1 (1976/77)

6. Hirschmann, H., Hanson, K. R.: Top. Stereochem. (Ed. Eliel, E. L., Allinger, N. L.), Vol. *14*, 183 (Wiley-Interscience, New York 1983)
7. Prelog, V., Helmchen, G.: Angew. Chem. *94*, 614 (1982); Int. Ed. Engl. *21*, 567 (1982)
8. Cahn, R. S., Ingold, C. K., Prelog, V.: Experientia *12*, 81 (1956)
9. See for instance the IUPAC rules, section E (Fundamental Stereochemistry) in: J. Org. Chem. *35*, 2849 (1970)
10. Krow, G.: Top. Stereochem. (Ed. E. L. Eliel, N. L. Allinger), Vol. *5*, 31 (Wiley-Interscience, New York 1970)
11. Cram, D. J., Cram, J. M.: Acc. Chem. Res. *4*, 204 (1971)
12. Vögtle, F., Neumann, P.: Angew. Chem. *84*, 75 (1972); Int. Ed. Engl. *11*, 73 (1972)
13. Vögtle, F., Neumann, P.: Chimia *26*, 64 (1972)
14. Cram, D. J., Hornby, R. B., Truesdale, E. A., Reich, H. J., Delton, M. H., Cram, J. M.: Tetrahedron *30*, 1757 (1974)
15. Farina, M., Morandi, C.: ibid. *30*, 1819 (1974)
16. Förster, H., Vögtle, F.: Angew. Chem. *89*, 443 (1977); Int. Ed. Engl. *16*, 429 (1977)
17. Misumi, S., Otsubo, T.: Acc. Chem. Res. *11*, 251 (1978)
18. Marshall, J. A.: ibid. *13*, 213 (1980)
19. Schlögl, K.: Top. Stereochem. (Ed. E. L. Eliel, N. L. Allinger), Vol. *1*, 39 (Wiley-Interscience, New York 1967)
20. Moscowitz, A., Mislow, K.: J. Am. Chem. Soc. *84*, 4605 (1962)
21. Tresckow, E. v.: Thesis, Munich 1982
22. Schlögl, K., Falk, H.: Angew. Chem. *76*, 570 (1964); Int. Ed. Engl. *3*, 512 (1964)
23. Marquarding, D., Klusacek, H., Gokel, G., Hoffmann, P., Ugi, I.: J. Am. Chem. Soc. *92*, 53 (1970)
24. Loim, N. M., Laryukova, M. V., Andianov, V. G., Tsiryapkin, V. A., Parnes, Z. N., Struchkov, Yu. T.: J. Organometal. Chem. *248*, 73 (1983)
24. a. Hayashi, T., Kumada, M.: Acc. Chem. Res. *15*, 395 (1982)
25. Klyne, W., Buckingham, J.: Atlas of Stereochemistry, Second Ed., Vol. *1*, 222, 216; Vol. *2*, 110–115 (Chapman and Hall, London 1978)
26. See for instance: Kalchhauser, H., Schlögl, K., Weissensteiner, W., Werner, A.: J. Chem. Soc., Perkin Trans. 1 *1983*, 1723; Schlögl, K., Werner, A., Widhalm, M.: ibid. *1983*, 1731
27. Lemiere, G. L., Alderweireldt, F. C.: J. Org. Chem. *45*, 4175 (1980)
28. Vögtle, F.: Tetrahedron Lett. *1969*, 3193
29. Sekine, Y., Brown, M., Boekelheide, V.: J. Am. Chem. Soc. *101*, 3126 (1979); Sekine, Y., Boekelheide, V.: ibid. *103*, 1777 (1981)
30. Vögtle, F., Neumann, P.: Synthesis *1973*, 85
31. Vögtle, F., Hohner, G.: Top. Curr. Chem. *74*, 1 (1978); Kleinschroth, J., Hopf, H.: Angew. Chem. *94*, 485 (1982); Int. Ed. Engl. *21*, 469 (1982). Boekelheide, V.: Acc. Chem. Res. *13*, 65 (1980)
32. Vögtle, F., Neumann, P.: Tetrahedron *26*, 5847 (1970)
33. Lüttringhaus, A., Gralheer, H.: Liebigs Ann. Chem. *550*, 67 (1941)
34. Cram, D. J.: J. Am. Chem. Soc. *77*, 1179 (1955); Cram, D. J., Allinger, N. L.: ibid. *77*, 6289 (1955)
35. Kainradl, B., Langer, E., Lehner, H., Schlögl, K.: Liebigs Ann. Chem. *766*, 16 (1972)
36. Lüttringhaus, A., Gralheer, H.: ibid. *557*, 108, 112 (1947); Lüttringhaus, A., Eyring, G.: ibid. *604*, 111 (1957)
37. Cram, D. J., Reeves, R. H.: J. Am. Chem. Soc. *80*, 3094 (1958); Cram, D. J., Wechter, W. J., Kierstead, R. W.: ibid. *80*, 3126 (1958)
38. Helmchen, G., Prelog, V.: Helv. Chim. Acta *55*, 2612 (1972)
39. Blomquist, A. T., Stahl, R. E., Meinwald, Y. C., Smith, B. H.: J. Org. Chem. *26*, 1687 (1961)
40. Eberhardt, H., Schlögl, K.: Liebigs Ann. Chem. *760*, 157 (1972)
41. Gassman, P. G., Bailey, T. F., Hoye, R. C.: J. Org. Chem. *45*, 2923 (1980)
42. Kane, V. V., Wolf, A. D., Jones, jr., M.: J. Am. Chem. Soc. *96*, 2643 (1974)
43. Tobe, Y., Kakiuchi, K., Odaira, Y., Hosaki, T., Kai, Y., Kasai, N.: ibid. *105*, 1376 (1983)
44. Nakazaki, M., Yamamoto, K., Ito, M., Tanaka, S.: J. Org. Chem. *42*, 3468 (1977)
45. Allinger, N. L., Walter, T. J.: J. Am. Chem. Soc. *94*, 9267 (1972)
46. Blomquist, A. T., Smith, B. H.: ibid. *82*, 2073 (1960)
47. Givens, R. S., Olsen, R. J., Wylie, P. L.: J. Org. Chem. *44*, 1608 (1979)

48. Higuchi, H., Kugimiya, M., Otsubo, T., Sakata, Y., Misumi, S.: Tetrahedron Lett. *1983*, 2593
48. a. Ito, Y., Miyata, S., Nakatsuka, M., Saegusa, T.: J. Org. Chem. *46*, 1043 (1981)
49. Haenel, M., Staab, H. A.: Chem. Ber. *106*, 2203 (1973)
49. a. Cf.: Okamoto, Y., Honda, S., Okamoto, I., Yuki, H., Murata, S., Noyori, R., Tanaka, H.: J. Am. Chem. Soc. *103*, 6971 (1981), and references cited therein
49. b. Prof. Okamoto, Y. (Osaka): private communication
50. Blank, N. E., Haenel, M. W.: Chem. Ber. *116*, 827 (1983)
51. Bruhin, J., Jenny, W.: Chimia *25*, 238, 308 (1971)
52. Cram, D. J., Day, A. C.: J. Org. Chem. *31*, 1277 (1966)
52. a. Staab, H. A., Gabel, G., Henke, H. E., Hinz, R., Knaus, G. H., Krieger, C., Reimann-Haas, R., Ulrich, P.: Chem. Ber. *116*, 2785, 2808, 2827, 2835 (1983)
53. Longone, D. T., Reetz, M. T.: J. Chem. Soc., Chem. Commun. *1967*, 46
54. Falk, H., Reich-Rohrwig, P., Schlögl, K.: Tetrahedron *26*, 511 (1970)
55. Eberhardt, R., Glotzmann, C., Lehner, H., Schlögl, K.: Tetrahedron Lett. *1974*, 4365
56. Delton, M. H., Cram, D. J.: J. Am. Chem. Soc. *94*, 2471 (1972)
57. Guest, A., Hoffman, P. H., Nugent, M. J.: ibid. *94*, 4241 (1972)
58. Reich, H. J., Cram, D. J.: ibid. *89*, 3078 (1967); ibid. *91*, 3517 (1969)
59. Delton, M. H., Cram, D. J.: ibid. *92*, 7623 (1970)
60. Nakazaki, M., Yamamoto, K., Tanaka, S., Ito, M.: Tetrahedron Lett. *1971*, 341; J. Chem. Soc., Chem. Commun. *1972*, 434
61. Yamamoto, K., Nakazaki, M.: Chemistry Lett. *1974*, 1051
62. Schwartz, L. H., Bathija, B. L.: J. Am. Chem. Soc. *98*, 5344 (1976)
63. Langer, E., Lehner, H., Schlögl, K.: Tetrahedron *29*, 2473 (1973)
64. Nakazaki, M., Yamamoto, K., Tanaka, S., Kametani, H.: J. Org. Chem. *42*, 287 (1977)
65. Longone, D. T., Chow, H. S.: J. Am. Chem. Soc. *92*, 994 (1970)
66. Otsubo, T., Kitasawe, M., Misumi, S.: Chemistry Lett. *1977*, 977
67. Nugent, M. J., Weigang, jr., O. E.: J. Am. Chem. Soc. *91*, 4556 (1969)
68. Tatemitsu, H., Otsubo, T., Sakata, S., Misumi, S.: Tetrahedron Lett. *1975*, 3059
69. Mizogami, S., Otsubo, T., Sakata, S., Misumi, S.: ibid. *1971*, 2791
70. Hubert, A. J.: J. Chem. Soc. (C) *1967*, 13
70. a. Otsubo, T., Ogura, F., Misumi, S.: Tetrahedron Lett. *1983*, 4851
70. b. See e.g. Schmidt, U., Lieberknecht, A., Bökens, H., Griesser, H.: J. Org. Chem. *48*, 2680 (1983), and references cited therein
71. Cf. for instance: Glotzmann, C., Haslinger, E., Langer, E., Lehner, H.: Monatsh. Chem. *106*, 187 (1975)
72. Glotzmann, C., Langer, E., Lehner, H., Schlögl, K.: ibid. *105*, 907 (1974)
73. Krois, D., Lehner, H.: Tetrahedron *38*, 3319 (1982)
74. Krois, D., Lehner, H.: J. Chem. Soc., Perkin Trans. 1, *1982*, 477 and 2369
75. Vögtle, F., Meurer, K., Mannschreck, A., Stühler, G., Puff, H., Roloff, A., Sievers, R.: Chem. Ber. *116*, 2630 (1983)
76. Gault, I., Price, B. J., Sutherland, I. O.: J. Chem. Soc., Chem. Commun. *1967*, 540
77. Kainradl, B., Langer, E., Lehner, H., Schlögl, K.: Liebigs Ann. Chem. *766*, 16 (1972)
78. Krieger, C., Lehner, H., Schlögl, K.: Monatsh. Chem. *107*, 195 (1976)
79. Glotzmann, C., Langer, E., Lehner, H., Schlögl, K.: Tetrahedron Lett. *1975*, 675; Monatsh. Chem. *106*, 763 (1975)
80. Gschwend, H. W.: J. Am. Chem. Soc. *94*, 8430 (1972)
81. a. Wittek, M., Vögtle, F., Stühler, G., Mannschreck, A., Lang, B. M., Irngartinger, H.: Chem. Ber. *116*, 207 (1983)
81. b. Vögtle, F., Palmer, M., Fritz, E., Lehmann, U., Meurer, K., Mannschreck, A., Kastner, F., Irngartinger, H., Huber-Patz, U., Puff, H., Friedrichs, E.: Chem. Ber. *116*, 3112 (1983)
82. Hesse, G., Hagel, R.: Chromatographia *6*, 277 (1973) and *9*, 62 (1976); see also: Blaschke, G.: Angew. Chem. *92*, 14 (1980); Int. Ed. Engl. *19*, 13 (1980)
83. Schlögl, K., Widhalm, M.: Chem. Ber. *115*, 3042 (1982)
84. Nakazaki, M., Hirose, Y., Shimizu, T., Suzuki, T., Ishii, A., Makimura, M.: J. Org. Chem. *45*, 1428 (1980)
84. a. Anet, F. A. L., Miura, S. S., Siegel, J., Mislow, K.: J. Am. Chem. Soc. *105*, 1419 (1983)
85. Krois, D., Lehner, H.: unpublished results
86. Griffin, jr., R. W., Orr, N.: Tetrahedron Lett. *1969*, 4567

87. Ramey, C. E., Boekelheide, V.: J. Am. Chem. Soc. *92*, 3681 (1970)
88. Parham, W. E., Johnson, D. R., Hughes, C. T., Meilhahn, M. K., Rinehart, J. K.: J. Org. Chem. *35*, 1048 (1970)
89. Zipplies, M. F., Krieger, C., Staab, H. A.: Tetrahedron Lett. *1983*, 1925
90. Lehner, H.: Monatsh. Chem. *107*, 565 (1976); Keller, H., Krieger, C., Langer, E., Lehner, H.: ibid. *107*, 1281 (1976)
91. a. Otsubo, T., Stusche, D., Boekelheide, V.: J. Org. Chem. *43*, 3466 (1978)
b. Kamp, D., Boekelheide, V.: J. Org. Chem. *43*, 3470 (1978)
92. Schuster, J. C., Völlenkle, H.: Monatsh. Chem. *109*, 1155 (1978)
93. Umemoto, T., Otsubo, T., Misumi, S.: Tetrahedron Lett. *1974*, 1573
94. Hefelfinger, D. T., Cram, D. J.: J. Am. Chem. Soc. *93*, 4754 (1971)
95. Delton, M. H., Gilman, R. E., Cram, D. J.: ibid. *93*, 2329 (1971)
96. a. Akabori, S., Hayashi, S., Nawa, M., Shiomi, S.: Tetrahedron Lett. *1969*, 3727
b. Hefelfinger, D. T., Cram, D. J.: J. Am. Chem. Soc. *92*, 1073 (1970)
97. Horita, H., Kannen, N., Otsubo, T., Misumi, S.: Tetrahedron Lett. *1974*, 501
98. Kannen, N., Otsubo, T., Misumi, S.: Bull. Chem. Soc. Jpn. *49*, 3208, 3307 (1976)
99. Kannen, N., Otsubo, T., Sakata, Y., Misumi, S.: Bull. Chem. Soc. Jpn. *49*, 3203 (1976)
100. Schlögl, K.: in Methodicum Chimicum (Ed. Korte, F.), Vol. I/1, S. 235, Thieme, Stuttgart, 1972
101. Tribout, J., Martin, R. H., Doyle, M., Wynberg, H.: Tetrahedron Lett. *1972*, 2839
102. Lightner, D. A., Hefelfinger, D. T., Powers, T. W., Frank, G. W., Trueblood, K. N.: J. Am. Chem. Soc. *94*, 3492 (1972)
103. Nakazaki, M., Yamamoto, K., Maeda, M.: Chemistry Lett. *1980*, 1553
104. Falk, H., Schlögl, K.: Angew. Chem. *80*, 405 (1968); Int. Ed. Engl. *7*, 383 (1968)
105. Mislow, K., Brzechffa, M., Gschwend, H. W., Puckett, R. T.: J. Am. Chem. Soc. *95*, 621 (1973)
106. Lehner, H., Paulus, H., Schlögl, K.: Monatsh. Chem. *112*, 511 (1981)
107. Keller, H., Krieger, C., Langer, E., Lehner, H., Derflinger, G.: Tetrahedron *34*, 871 (1978)
108. Helmchen, G.: personal communication
109. Falk, H., Schlögl, K.: Monatsh. Chem. *102*, 33 (1971), and papers cited therein
110. Langer, E., Lehner, H.: Monatsh. Chem. *104*, 644 (1973); Keller, H., Lehner, H.: ibid. *106*, 1117 (1975)
111. a. Keller, H., Krieger, C., Langer, E., Lehner, H., Derflinger, G.: J. Mol. Struc. *40*, 279 (1977)
b. Keller, H., Krieger, C., Langer, E., Lehner, H., Derflinger, G.: Liebigs Ann. Chem. *1977*, 1296
112. See for instance: Ruch, E.: Acc. Chem. Res. *5*, 49 (1972); Ruch, E., Runge, W., Kresze, G.: Angew. Chem. *85*, 10 (1973); Int. Ed. Engl. *12*, 20 (1973); Richter, W. J., Richter, B., Ruch, E.: ibid. *85*, 21 (1973); Int. Ed. Engl. *12*, 30 (1973); Neudeck, H., Schlögl, K.: Chem. Ber. *110*, 2624 (1977); Neudeck, H., Schlögl, K., Richter, B.: Monatsh. Chem. *110*, 931 (1979)
113. Nakazaki, M., Chikamatsu, H., Naemura, K., Hirose, Y., Shimizu, T., Asao, M.: J. Chem. Soc., Chem. Commun. *1978*, 668
114. Mizuno, H., Nishiguchi, K., Otsubo, T., Misumi, S., Morimoto, N.: Tetrahedron Lett. *1972*, 4981
115. Vogel, E., Roth, D.: Angew. Chem. *76*, 145 (1964); Int. Ed. Engl. *3*, 228 (1964)
116. Dobler, M., Dunitz, J. D.: Helv. Chim. Acta *48*, 1429 (1965); Grammaccioli, C. M., Simonetta, M.: Acta Cryst. *B 27*, 2231 (1971)
117. Kuffner, U., Schlögl, K.: Monatsh. Chem. *103*, 1320 (1972)
118. a. Vogel, E., Günther, H.: Angew. Chem. *79*, 429 (1967); Int. Ed. Engl. *6*, 429 (1967)
b. Vogel, E.: Pure Appl. Chem. *28*, 355 (1971)
c. Masamume, S., Darby, N.: Acc. Chem. Res. *5*, 272 (1972)
d. Vogel, E.: Pure Appl. Chem. *54*, 1015 (1982)
e. Marshall, J. A., Conrow, R. E.: J. Am. Chem. Soc. *105*, 5679 (1983)
119. Kuffner, U., Schlögl, K.: Tetrahedron Lett. *1971*, 1773
120. LaBar, R. A., Jones, W. M.: J. Am. Chem. Soc. *96*, 3645 (1974)
121. Schäfer-Ridder, M., Wagner, A., Schwamborn, M., Schreiner, H., Devrout, E., Vogel, E.: Angew. Chem. *90*, 894 (1978); Int. Ed. Engl. *17*, 853 (1978)
122. Schlögl, K., Widhalm, M., Vogel, E., Schwamborn, M.: Monatsh. Chem. *114*, 605 (1983)
123. Bentley, P. H., Todd, M., McCrae, W., Maddox, M. L., Edwards, J. A.: Tetrahedron *28*, 1411 (1972)

124. Vogel, E., Tückmantel, W., Schlögl, K., Widhalm, M.: Tetrahedron Lett., in preparation.
125. Cremer, D., Kraka, E.: personal communication
126. Anet, F. A. L.: J. Am. Chem. Soc. *84*, 671 (1962)
127. Mislow, K., Perlmutter, H. D.: ibid. *84*, 3591 (1962)
128. Salisbury, L. E.: J. Org. Chem. *43*, 4991 (1978)
129. Gardlik, J. M., Johnson, L. K., Paquette, L. A., Solheim, B. A., Springer, J. P., Clardy, J.: J. Am. Chem. Soc. *101*, 1615 (1979)
130. Gardlik, J. M., Paquette, L. A.: Tetrahedron Lett. *1979*, 3597
131. Gardlik, J. M., Paquette, L. A., Gleiter, R.: J. Am. Chem. Soc. *101*, 1617 (1979)
132. Krebs, A.: Tetrahedron Lett. *1968*, 4511
133. Krebs, A., Odenthal, J., Kimling, H.: ibid. *1975*, 4663
134. Eyer, M., Schlögl, K., Schölm, R.: Tetrahedron *37*, 4239 (1981)
135. Tratteberg, M.: Acta Chem. Scand *B 29*, 29 (1975); Ermer, O.: Angew. Chem. *86*, 672 (1974); Int. Ed. Engl. *13*, 604 (1974)
136. Levin, C. C., Hoffmann, R.: J. Am. Chem. Soc. *94*, 3446 (1972)
137. Cope, A. C., Ganellin, C. R., Johnson, jr., H. W., van Auken, T. V., Winkler, H. S.: ibid. *85*, 3276 (1963)
138. Cope, A. C., Mehta, A. S.: ibid. *86*, 5626 (1964)
139. Manor, P. C., Shoemaker, D. P., Parkes, A. S.: ibid. *92*, 5260 (1970)
140. Moscowitz, A., Mislow, K.: ibid. *84*, 4606 (1962)
140. a. Binsch, G., Roberts, J. D.: ibid. 5157 (1965)
141. Cope, A. C., Pawson, B. A.: ibid. *90*, 636 (1968)
142. Thulin, B., Wennerström, O., Somfai, I.: Acta Chem. Scand. *B 32*, 109 (1978)
142. a. Marshall, J. A., Konicek, T. R., Flynn, K. E.: J. Am. Chem. Soc. *102*, 3287 (1980)
143. Cope, A. C., Moore, W. R., Bach, R. D., Winkler, H. J. S.: J. Am. Chem. Soc. *92*, 1243 (1970)
144. Bach, R. D., Mazur, U., Brummel, R. N., Lin, L.-H.: ibid. *93*, 7120 (1971)
145. Moore, W. R., Anderson, H. W., Clark, S. D., Ozretich, T. M.: ibid. *93*, 4932 (1971)
146. See for instance: Rösner, M., Köbrich, G.: Angew. Chem. *87*, 715 (1975); Int. Ed. Engl. *14*, 708 (1975), and ibid. *86*, 775 (1974); Int. Ed. Engl. *13*, 741 (1974)
147. Feringa, B., Wynberg, H.: J. Am. Chem. Soc. *99*, 602 (1977)
148. Cf.: Marshall, J. A., Black, T. H.: ibid. *102*, 7581 (1980)
149. Marshall, J. A., Flynn, K. E.: ibid. *105*, 3360 (1983)
150. Nakazaki, M., Yamamoto, K., Maeda, M.: J. Chem. Soc., Chem. Commun. *1980*, 294

Carbohelicenes and Heterohelicenes

Wim H. Laarhoven and Wim J. C. Prinsen

Laboratory for Organic Chemistry, Catholic University of Nijmegen, Toernooiveld, 6525 ED Nijmegen, The Netherlands

Table of Contents

1 Introduction, Terminology

The first synthesis of phenanthro[3,4-c]phenanthrene accomplished via a twelve steps route was reported by Newman in 1956 [1]. He proposed the name *hexahelicene* referring to the non-planar model of the molecule due to the overcrowding of terminal rings. Though the photoconversion of stilbene into phenanthrene became known in 1960 [2] it took several years before the reaction was used for higher aromatics. Martin used the method in 1967 for the preparation of heptahelicene [3], and Wijnberg for the synthesis of sulfur containing heterohelicenes in 1968 [4]. The simplified accessibility stimulated the study of the chemistry and physical properties of the helicenes, especially because these molecules deviate from what aromatic hydrocarbons 'ought' to be: planar and rigid.

In the middle of this period of intensive research two reviews have appeared. Wijnberg published in 1971 a review on the chemical, photochemical and spectral properties of thiophenes [5] in which the sulfur-containing heterohelicenes take a major part. Martin gave a review, mainly devoted to carbohelicenes, in 1974 [6]. In the present review a part of the fundamental earlier work is shortly compiled and some studies, only shortly mentioned in the previous reviews without references to the literature, have also been included, but the larger part concerns more recent work.

In general, only helicenes with six or more benzene or other rings will be treated. The lower homologs, dibenzo[c,g]phenanthrene (pentahelicene) and benzo[c]phenanthrene will only be mentioned for comparison in some instances.

The main progress in the chemistry of helicenes in the last ten years came from:

i) the large increase in x-ray analyses of various helicenes
ii) the resolution of enantiomers by HPLC techniques, mainly based on charge — transfer complexation
iii) the asymmetric syntheses by different methods
iv) the use of force-field calculations
v) the refinement of the interpretation of NMR-spectra.

In this review the nomenclature and numbering originally given by Newman [1] will be used. As an example the numbering of hexahelicene is given. In the example the quaternary carbon atoms are not numbered, but on addition of a group at these carbon atoms they are numbered as usual in polycondensed aromatic systems. For the x-ray analyses (and also ^{13}C-NMR etc). the quarternary carbon atoms are numbered separately.

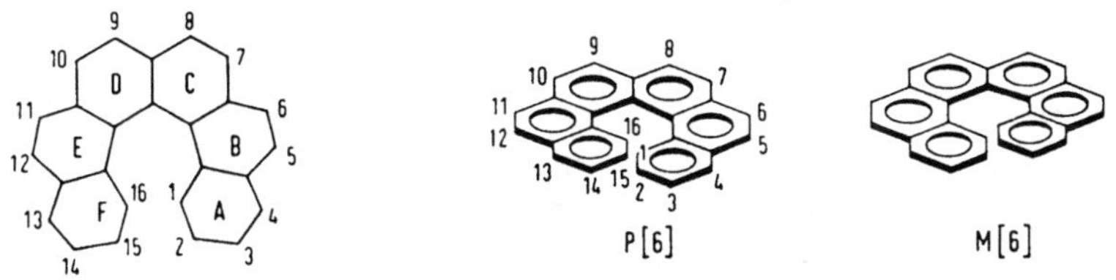

The benzene rings in the helicenes are indexed by letters as shown. Sometimes helicenes are indicated by short nomenclature symbols, via the number of the constituting benzene (or other) rings between brackets. So, hexahelicene is [6], decahelicene is [10]. In principle this short hand notation is consequently used in Tables, not in the text.

For the heterohelicenes, where isomers are possible, a new way for abbreviation is introduced by giving after the number in brackets the correct sequence of benzene and hetero rings, using symbols B = benzene, S = thiophene, NH = pyrrole, N = pyridine (the position of the hetero atom is indicated by the position number), O = furan etc.

Examples are:

[6] BOBBOB

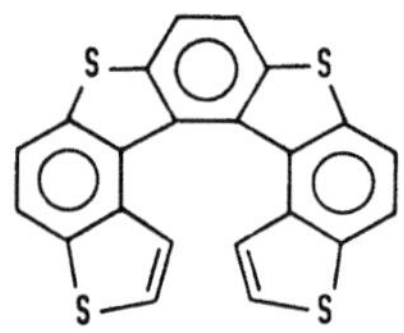

[7] SBSBSBS

[6] (4N) BBBBB

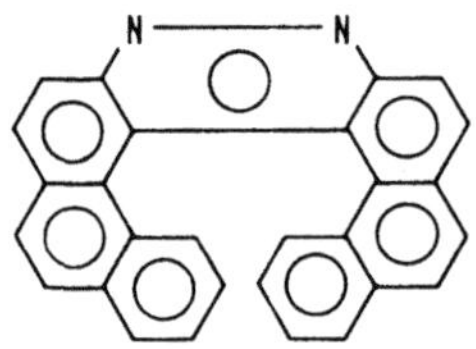

[7] BBB(NN)BBB

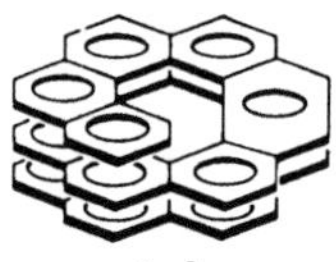

[14]

[15] SBSBSBSBSBSBSBS

2 Syntheses

2.1 Photochemical Syntheses

2.1.1 Unsubstituted Carbohelicenes

The photochemical synthesis of helicenes by irradiation of 1,2-diarylethylenes in a dilute solution and in the presence of an oxidizing agent is based on the well known photocyclodehydrogenation of stilbene into phenanthrene. There is an overwhelming amount of literature on this type of photoreaction. Details about scope and limitation can be found in previous reviews [7,8]. Therefore, only a short survey will be given of the mechanism of the reaction.

Irradiation of stilbene gives rise to a rapid interconversion of the Z- and E-isomers (Z1 and E1).

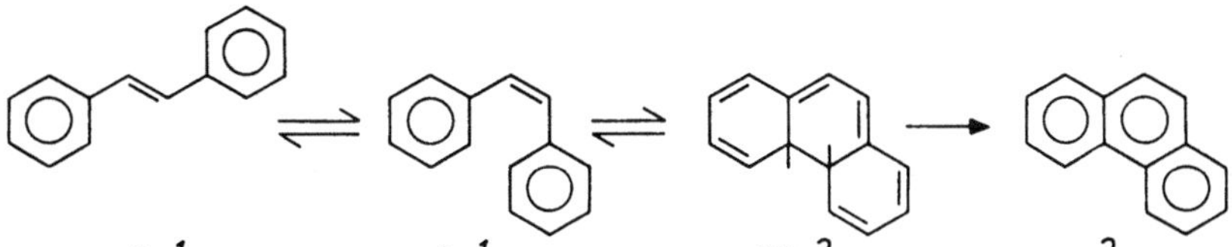

E-1 Z-1 DHP-2 3

Scheme 1

Cis-stilbene (Z1) also undergoes a conrotatory cyclization reaction into trans-4a,4b-dihydrophenanthrene (DHP, 2), a short living, not isolated product with an absorption in the visible spectrum at 450 nm. In the absence of an oxidizing agent DHP will return to the starting material, by both a thermal and a photochemical ring opening reaction.

In the presence of an oxidizing species (oxygen, iodine, tetracyanoethylene and others) DHP is dehydrogenated into phenanthrene. This oxidation can also be thermal or photochemical. In general, meta substituted stilbenes give rise to two isomeric substituted phenanthrenes; ortho-substituted stilbenes can lose the substituent on cyclization.

Similar reactions occur also with other 1,2-diarylethylenes[1]. The aryl-olefins are usually prepared by a Wittig reaction [9] in one of its modifications, using an ylide and an arylaldehyde, or by a Siegrist [10] reaction using a methyl aromatic and an anilide. A more unconventional synthesis is given by Jutz [11] who starts with the reaction of 2-naphthylmethylcyanide and the bis-perchlorate salt of 2-dimethylamino-1,1-bis-(dimethyliminiomethyl)ethylene, which resulted in the presence of a base in 1,5-dicyano-1,5-dinaphthyl-3-(dimethylaminomethylene)-penta-1,4-diene, that on heating gives 1-(1′-cyanophenathr-3′yl)-2-naphthylacrylonitril. Irradiation of the latter compound in the presence of iodine gives rise to 7,10-dicyanohexahelicene. Because of the difficult availability of substituted 2-naphthylmethylcyanides this method might have a restricted applicability.

Ring closure of arylolefins into helicenes can be effected by irradiation of a dilute solution. In principle all solvents can be used that do not absorb at the wavelength of irradiation and do not react with the helicene precursor or the oxidizing agent. Methanol, hexane, cyclohexane, benzene are usual solvents.

To prevent dimerization the concentration is kept low (10^{-3} mol/l) but lower concentrations are allowed, when required by the bad solubility of the precursor. As an oxidizing agent 5.0 mol-% iodine in an atmosphere of air is frequently used. To prevent oxidation of the endproduct sometimes deaeration of the solvent and use of 100 mol-% iodine may be a better choice. Other useful agents are π-acceptors like tetracyanoethylene (TCNE) in dichloromethane as solvent. In most cases the helicenes are well separated from the irradiation mixture by evaporating the solvent and chromatography of the residue. In cases where the separation of the helicene and cis-olefin is difficult it is of advantage to irradiate until the olefin has completely reacted.

---

1 In the following the short notation for 1,2-diarylethylenes of Martin [29] will be used. The molecules are referred to by figures in parentheses indicating the number of annelated benzene rings in each aryl moiety. So, stilbene is (1 + 1) and styrylbenzo[c]phenathrene is (1 + 4).

In all cases studied the E,Z-isomerization has the higher quantum yield, so it is of no importance from which isomer the photoreaction starts. In analogy to the primary product from stilbene the cyclization products from other diaryl olefins will be denoted by DHP. The lifetime of such DHP's can vary from some milliseconds to several hours [12].

Depending on the symmetry of the diaryl ethylenes one to four cyclization products may be formed. The photoreaction is, however, very selective; e.g. on irradiation of 2-styrylbenzo[c]phenanthrene (4) hexahelicene (6) is obtained in 85% yield, whereas no trace of benzo[a]naphtho[1,2-h]-anthracene (9) is formed.

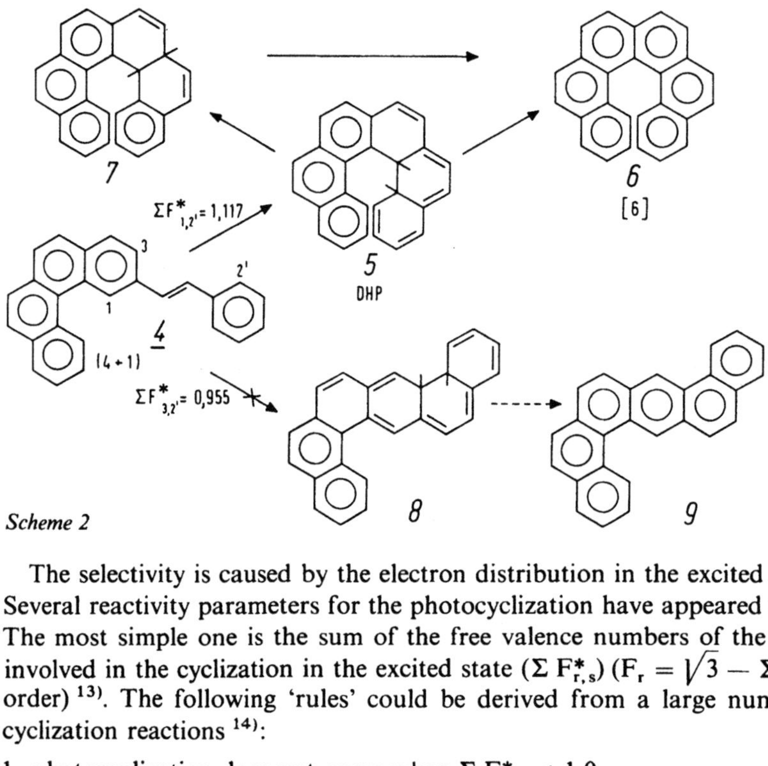

Scheme 2

The selectivity is caused by the electron distribution in the excited diarylethylene. Several reactivity parameters for the photocyclization have appeared to be valuable. The most simple one is the sum of the free valence numbers of the atoms r and s involved in the cyclization in the excited state ($\Sigma F^*_{r,s}$) ($F_r = \sqrt{3} - \Sigma P_r$, P = bond order) [13]. The following 'rules' could be derived from a large number of photocyclization reactions [14]:

1. photocyclization does not occur when $\Sigma F^*_{r,s} < 1.0$
2. when two or more cyclizations are possible according to rule 1 only one product arises if $\Delta(\Sigma F^*_{r,s}) > 0.1$; more products are formed if the difference is smaller
3. the second rule holds only when only planar or only non planar products (penta or higher helicenes) can arise. When planar as well as non planar products can be formed the planar aromatic is, in general, the main product, provided that $\Sigma F^*_{r,s} > 1.0$ for its formation.

Illustrative examples are given in Schemes 2 and 3, e.g. (1+4) 4 gives 6 (not 9), but (3+2) gives mainly 9 (+6).

A more simple way to determine the preferred route for the photocyclization of a diarylethylene is to count the number of benzene rings in the DHP's. From Scheme 3 as an example, it can be deduced that *11* (1-naphthyl (n) + 1-benzene (b)) is preferred to 12 (2b), to 13 (1n) and to 14 (1b).

The kinetics of the photoreaction of 1,2-diarylethylenes (1+4), (2+3), (2+4), (3+3), (3+4), (4+4) leading to the helicenes [6], [7], [8] and [9] are thoroughly investigated by E. Fisher et al. [15] using static methods at temperatures as low as −180 °C and flash methods in a temperature range down to −100 °C. The results fully confirm the qualitative expectations of the calculations. The quantum yield of

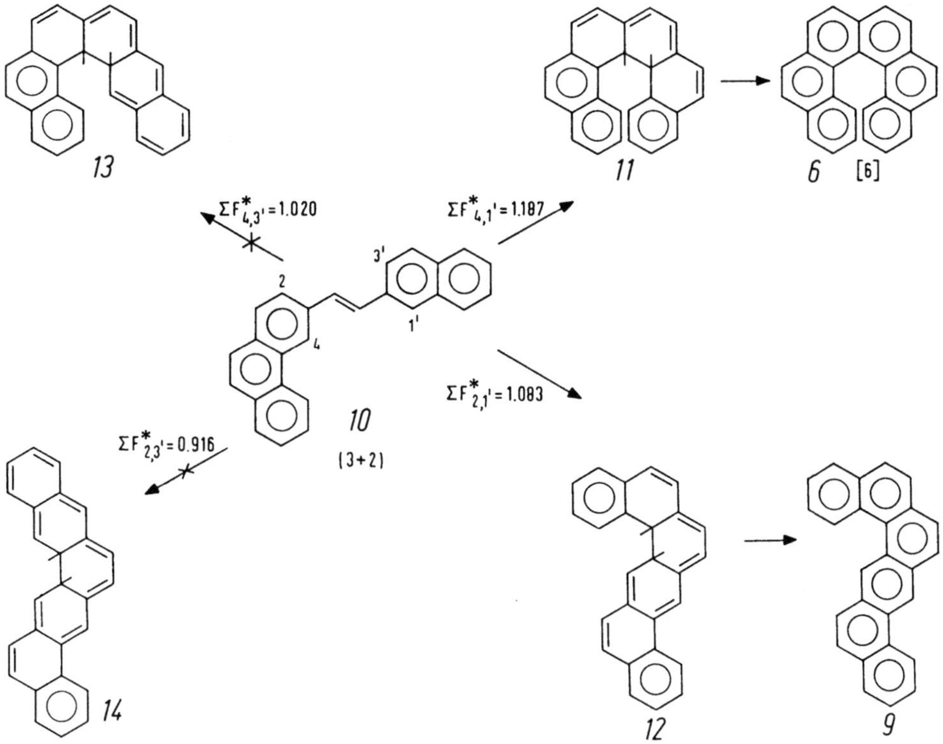

Scheme 3

cyclization which is 0.03 for (4+1) at room temperature decreases when moving to higher homologues ((3+3) > (3+4) > (4+4)). There is a pronounced variation of the quantum yield of cyclization with the temperature, indicating that there is a substantial energy barrier on the pathway from the singlet-excited cis-compound to the DHP corresponding to a helicene and a smaller barrier to the DHP corresponding to a non-helicenic product.

In the photocyclization of 2-styrylbenzo[c]phenanthrene under anaerobic conditions another more stable dihydrohexahelicene is formed [15a] from DHP 5 (Scheme 2). Its structure was elucidated as trans-6a,16d-dihydrohexahelicene 7 [16]. It is converted into [6], both thermally and photochemically.

Besides from 1,2-diarylethylenes it is possible to prepare helicenes from proper bis-arylvinyl aromatics by two subsequent photocyclization reactions; e.g. hexahelicene can be prepared from 2,7-distyrylnaphthalene (1+2+1) (60%), and tridecahelicene from 3,6-bis(2-benzo[c]phenanthrylvinyl)-phenanthrene (4+3+4) (52%).

In this context it is important to keep in mind, however, that di(2-naphthyl)ethylene (2+2) and 3-styrylphenanthrene (3+1) form pentahelicene on irradiation, but in the presence of iodine it cyclizes into benzo[g,h,i]-perylene [17]. The second cyclization takes place even when a substituent is present at a carbon atom involved in the reaction as shown in the photoreaction of 1-(2-naphthyl)-2-(7-phenyl-2-naphthyl)ethylene (15), which forms benzo[a]coronene (18) [17,18,19] (Scheme 4; see also Scheme 33).

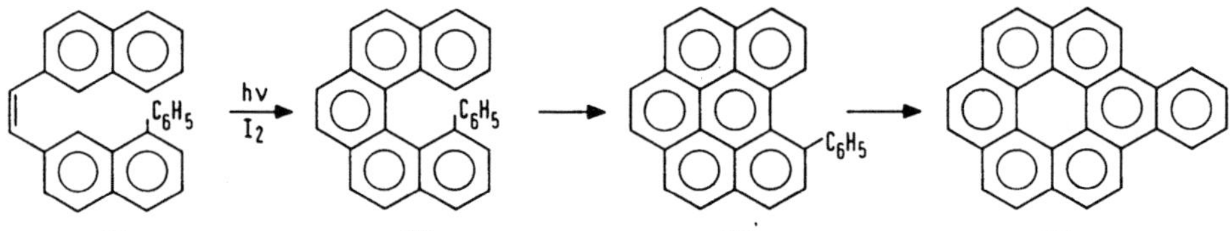

15 16 17 18

Scheme 4

Therefore, photoreactions via pentahelicenic intermediates may lead to undesired products. In the synthesis of a 3,15-disubstituted heptahelicene (20) starting from a (3+1+1) derivative (19) the coronene derivative (21) was formed as a side product [20].

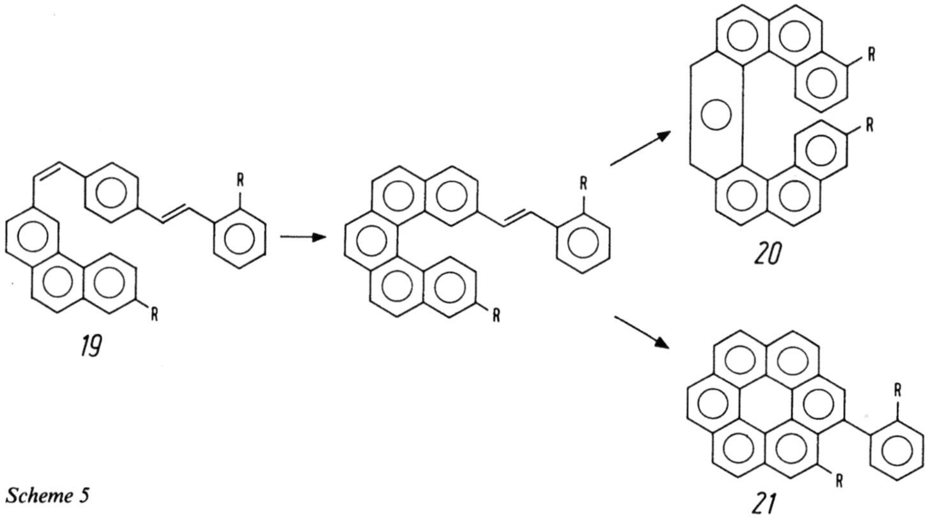

19 20 21

Scheme 5

In Table 1 all the unsubstituted carbohelicenes synthesized are tabulated together with their photoprecursors, the chemical yields and some physical properties.

2.1.2 Substituted Carbohelicenes

Substituted carbohelicenes can be synthesized in the same way as unsubstituted helicenes, when a substituted 1,2-diarylethylene is used as the precursor, e.g. substituted styrylbenzo[c]phenanthrenes lead to substituted hexahelicenes. With a meta

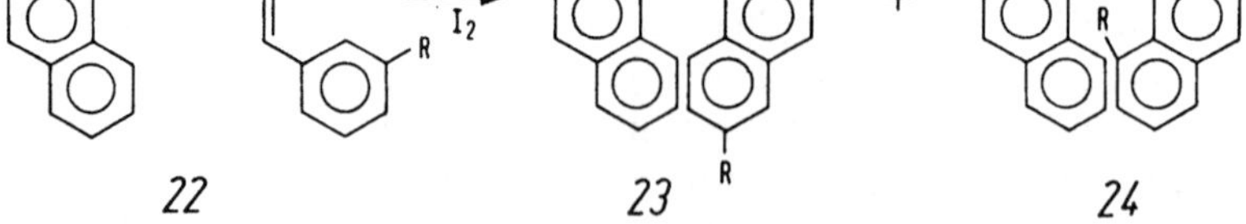

22 23 24

Scheme 6

Table 1. Photosynthesis of Carbohelicenes from Arylethylenes

Carbo-helicenes	Precursors[a]	Yield (%)	m.p. (°C) racemate	m.p. (°C) enantiom.	Ref.
[6]	(4+1)	87	240–242	265–267	[26,27] [63,117]
	(3+2)	25			[27,61]
	(2+1+1)	55			[28]
	(1+2+1)	60			[28]
[7]	(4+2)	20	245–255	309–310 [80]	[29,27] [117]
	(3+3)	61			[3,117] [27]
	(2+1+2)	20			[28]
	(1+3+1)	41			[24]
[8]	(6+1)[b]	40	330–331	364–365 [80]	[30,31] [29,27]
	(4+3)	68			[63,117]
	(4+1+1)	30			[63]
[9]	(6+2)[b]	74	379–380	398–399 [80]	[31] [29,27]
	(4+4)	48			[117]
	(4+1+2)	50			[63]
[10]	(6+3)[b]	79			[31]
	(4+3+1)	30			[63]
	(4+1+3)	20			[63]
[11]	(6+4)[b]	45			[31]
	(4+3+2)	80			[63]
	(4+1+4)	84			[32,63]
	(3+3+3)	54			[32]
[12]	(4+3+3)	30			[63]
	(3+4+3)	32			[32]
	(2+6+2)	42			[32]
[13]	(6+3+2)[b]	2	414–415		[31]
	(4+3+4)	52			[33,63]
[14]	(4+4+4)	10			[32]
	(3+6+3)	45			[32]

[a] 1 = phenyl
+1+ = 1,4-phenylene
2 = 2-naphthyl
+2+ = 2,7-naphthylene
3 = 3-phenanthryl
+3+ = 3,6-phenanthrylene

[a] 4 = 2-benzo[c]phenanthryl
+4+ = 2,11-benzo[c]phenanthrylene
6 = 2-hexahelicyl
+6+ = 2,15-hexahelicylene

[b] synthesis starting with optically pure 1-formylhexahelicene

substituent in the styryl group two helicenes are often formed: the 1- and 3-substituted isomer (23 and 24). Irradiation of 2,7-bis(m.methylstyryl)naphthalene (25) gives rise to a mixture of three products [21]: 1,16-dimethyl, 1,14-dimethyl and 3,14-dimethylhexahelicene (26, 27, 28). The ratio between the isomers depends on the lifetime of

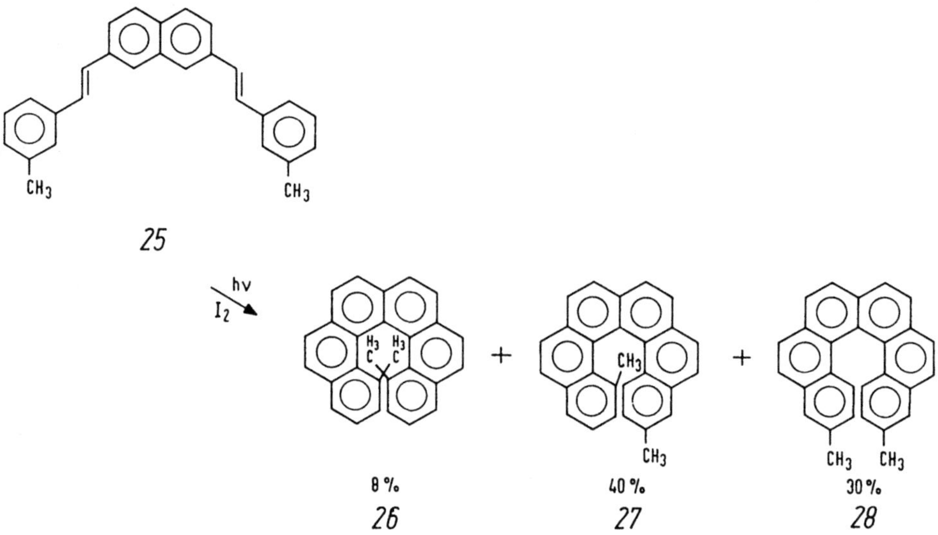

Scheme 7

the corresponding DHP's and the rate of oxidation [22]. As the lifetime of the more hindered DHP is usually shorter the amount of the 1-substituted product tends to be lower than that of the 3-substituted one. 3-Methylstyrylbenzo[c]phenanthrene irradiated in benzene in the presence of 5 mol-% iodine gives rise to 72% 3-methyl-[6] and 8% 1-methyl-[6] [23]. When the same compound is irradiated in hexane in the presence of 100 mol-% iodine both isomers are formed in nearly the same amount [21].

To enhance the formation of the 1-substituted helicene Martin [6] introduced in the meta substituted styrylbenzo[c]phenanthrene an additional ortho-bromo substituent. In good yield 1-alkyl-4-bromohelicenes (30) were formed, from which the bromo substituent could be withdrawn with $LiAlH_4$ (Scheme 8).

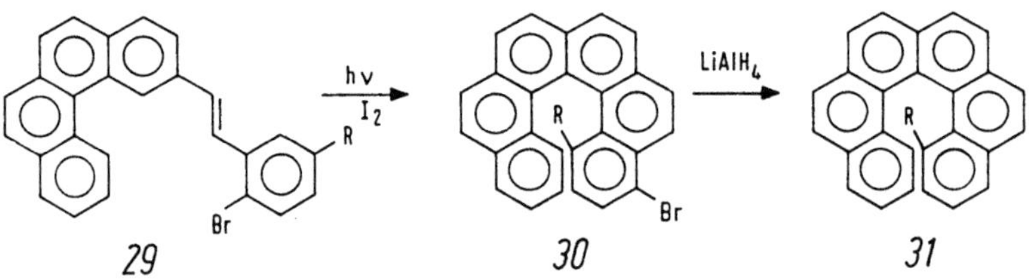

Scheme 8

The high flexibility of the helix structure is well demonstrated by the fact that even 1,3-di-tert.butylhexahelicene can be synthesized. The very low yield (1%) in this case is in part also due to the photoinstability of the product [23].

2-Substituted hexahelicenes are synthesized irrespective of the bulkyness of the substituent in high yields; 2-tert.butylhexahelicene has been obtained in 80% yield [23] and the mixture of meso- and racemic 2,2'-bis-hexahelicyl (33) in 50% yield [25].

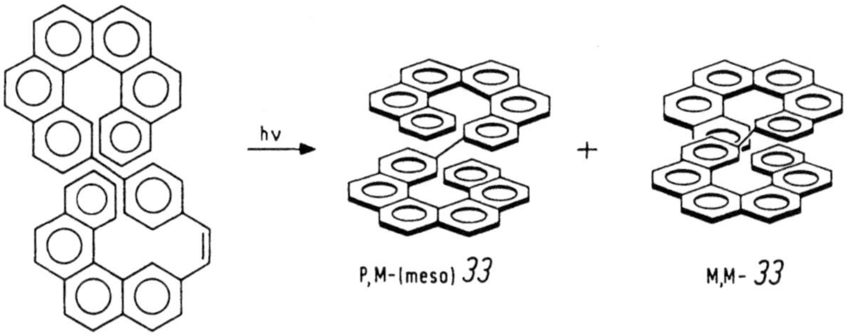

32

Scheme 9

A 3,15-disubstituted heptahelicene (35) was prepared as the starting compound for two products which possess a bridge connecting the terminal rings: 3,15-ethano- and 3,15-(oxa-2′-propano)heptahelicene [20] (37 and 38).

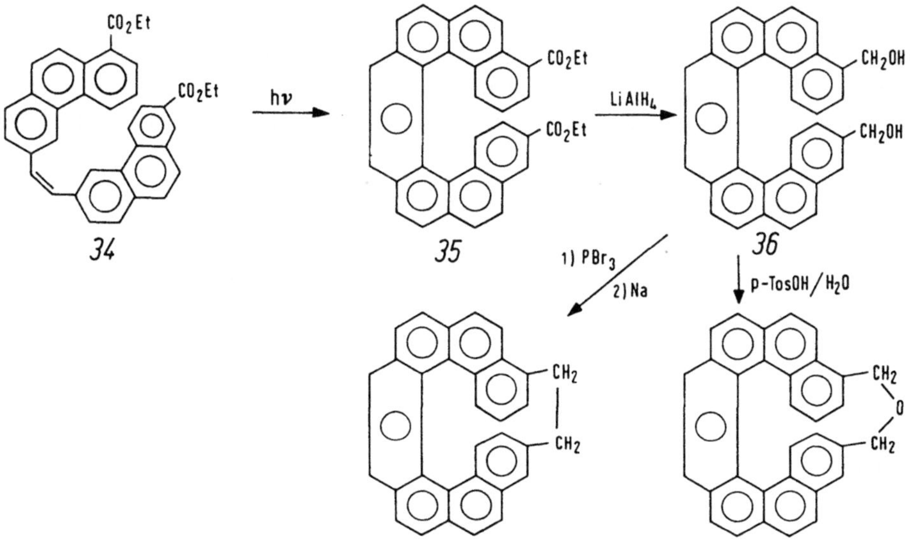

Scheme 10 37 38

Martin has described the synthesis of hexahelicene-2-carbaldehyde by irradiation of a polymer supported 1,2-diarylethylene containing a masked aldehyde function [26] suspended in benzene for 4 h. After prolonged hydrolysis (60 h) the product was obtained in low yield (5%).

Other substituted helicenes have been prepared by a direct substitution reaction of hexahelicene (see Sect. 8).

2.1.3 Benzo- and Double Helicenes

To examine the influence of annelated benzene rings on the helicity of helicenes several benzohexa- and benzoheptahelicenes have been synthesized.

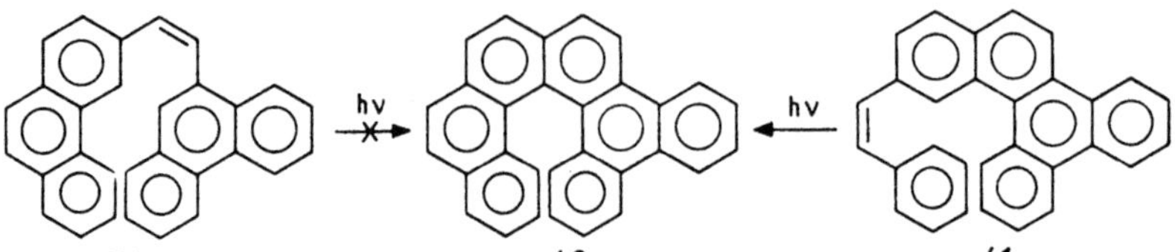

39 40 41

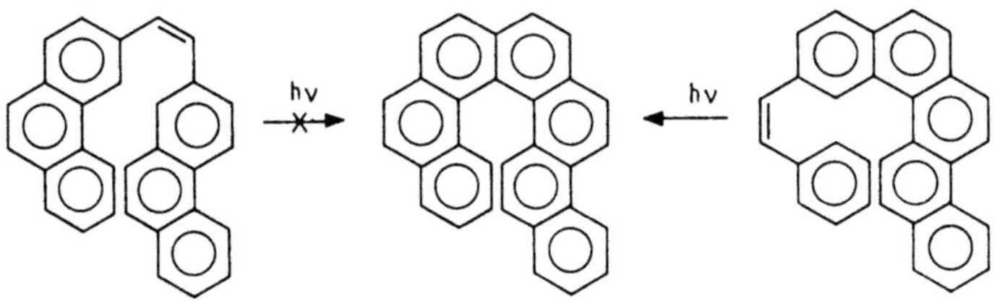

42 43 44

Scheme 11

In Table 2 the precursors, yields and melting points of these compounds are given, whereas in Scheme 11 the preparations of benzo[f]- and benzo[c]-hexahelicene are formulated more completely, showing that the products are formed via the styryl-substituted compounds 41 and 44, but not via the diphenanthryl ethylenes 39 and 42.

Table 2. Synthesis of Benzo- and Double Helicenes from Aryl Olefinic Precursors

Helicenes	Precursors		Yield (%)	m.p. (°C)	Ref.
Benzohexahelicenes					
benzo[f] *40*	(5+1)		60	247–253	27)
benzo[i] *45*	(4+2)		90	212–214	27)
	(5+1)		66		
benzo[c] *43*	(5+1)		65	252–253	27)
benzo[b] *46*	(3+3)		22	272–277	27)
	(5+1)		10		
Benzoheptahelicenes					
benzo[l] *47*	(5+2)		45	358–359	34)
dibenzo[i.o.] *48*	(4+4)		50	338–340	34)
dibenzo[f,r] *49*	(5+3)		20	340–341	34)
tribenzo[f,l,r] *50*	(6+3)		90	375–380	34)
Double helicenes					
diphenanthro[3,4-c;		dl	58	about 320	35, 36, 37)
3′,4′-l]chrysene *52*	(3+2+3)	meso	12	400–402	
dl hexaheliceno[3,4-c]-hexahelicene *54*	(7+4)			410°	38)
dl diphenanthro[4,3-a; 3′,4′-o]picene *55*	(6+4)		73	452–454	39, 40)
dl benzo[s]diphenanthro-[4,3-a; 3′,4′-o]picene *56*	(7+4)		26	454–459	39)

From the latter compounds only planar molecules are formed in good yields, in accordance with the $\Sigma\, F^{*}_{r,s}$-rule.

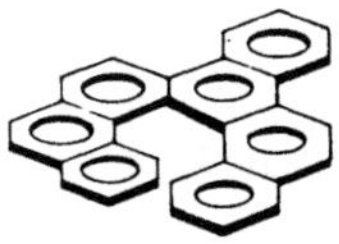
45

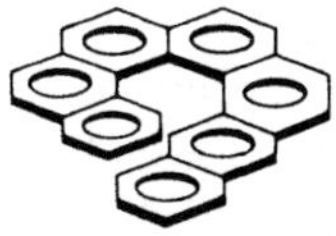
46

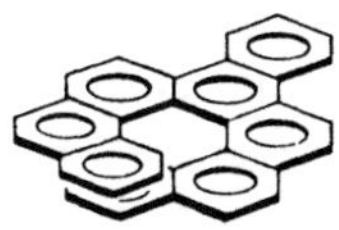
47

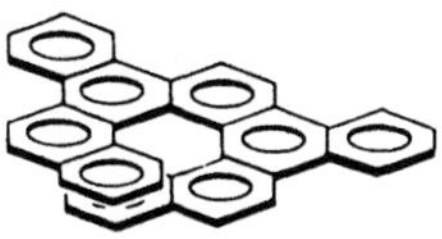
48

49

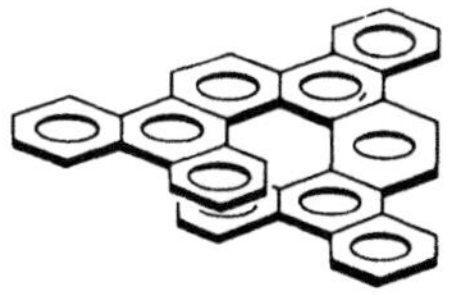
50

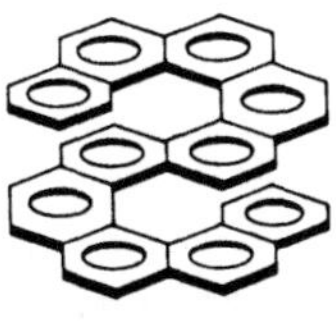
P,M (meso) 52

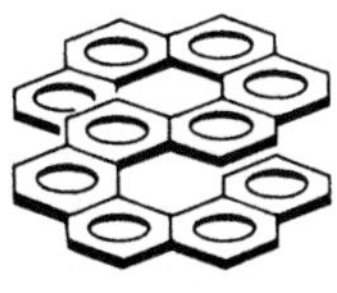
M,M-52

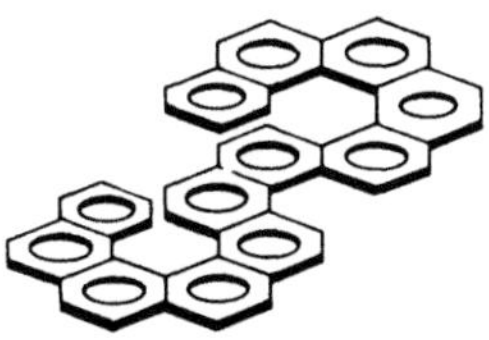
P,P-54

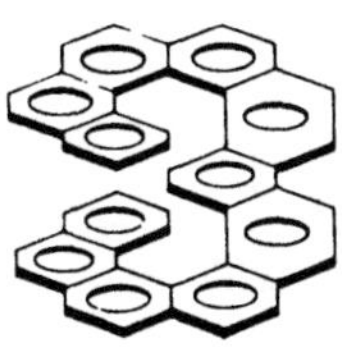
P,P-55

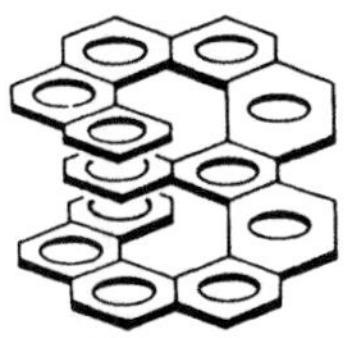
P,P-56

The double helicenes 52, 54 and 56 have been synthesized because they may occur in a racemic and meso form and to study possible isomerization between stereoisomers.

Only from 52 two diastereomeric forms could be isolated.

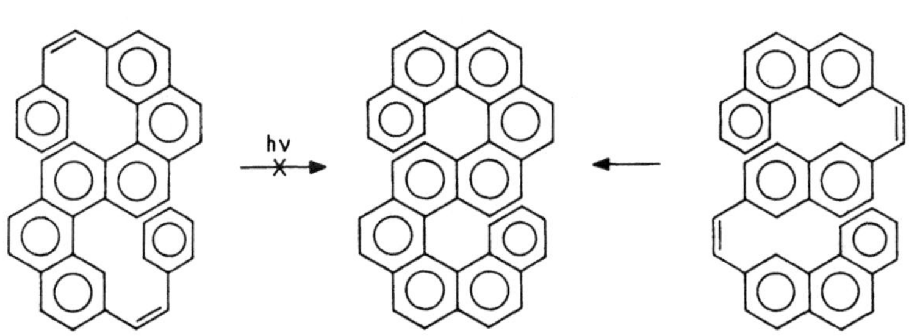

51 52 53

Scheme 12

It is noted that in the synthesis of 52 the precursor 53 gave the product in good yield (70%) (only small amounts of other photocyclization products were isolated), whereas the distyryl aromatic 51 did not yield the required product, in accordance with the relevant free valence numbers.

By chromatography on Al_2O_3 the racemic mixture and the meso form could be separated. The racemic form melted at about 320°, resolidified and melted again at the melting point of the meso form [35,37]. From the other double helicenes only one isomer could be isolated.

2.1.4 Photosyntheses of Helical Compounds others than Helicenes

Irradiation of 2-styrylbenzo[c]phenanthrene (1+4) in a primary amine gives rise to the formation of 5,6-dihydrohexahelicene 57.

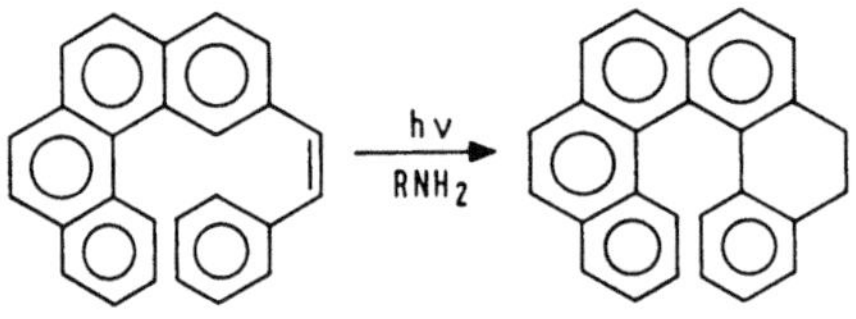

4 57

The 5,6-dihydrohexahelicene is formed via the primary DHP [22]. It is also formed when (1+4) is irradiated in deaerated benzene in the presence of only 1% of iodine [41]. The mechanism of its formation via the latter method is not yet elucidated.

A special type of photosynthesis of a helical compound is the irradiation of 1,1'-spirobi[benzindene] substituted with two tert. butyl groups [42] 58.

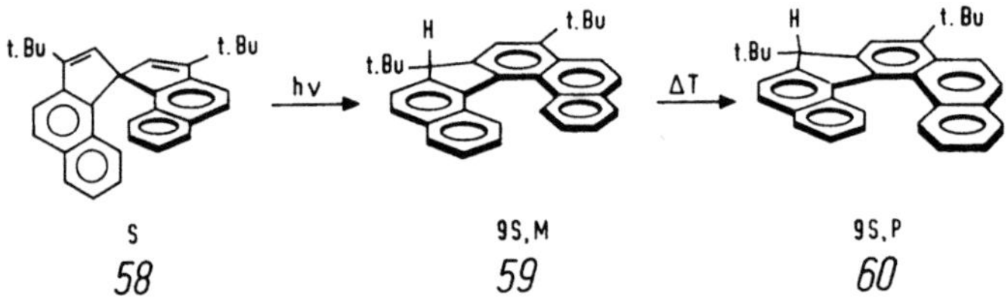

S 9S,M 9S,P
58 59 60

Scheme 13

By the irradiation of optically active 58 the helical compound 59 is formed stereospecifically. However, this product (with the tert. butylgroups at a short distance) is thermolabile. It converts spontaneously into 60, already at 0 °C, thus showing mutarotation. The process was followed by ORD, which showed a complete reversal of the spectrum.

Another helicene containing a five membered ring (61) was formed from 5,6-diketohexahelicene as will be described in Section 8. Treatment of *61* with Li gives rise to the 'aromatic' helicene 62 [43].

1-Phenylbenzo[c]phenanthrene *64* may be conceived as a hexahelicene, lacking one of the rings (ring B), but showing similar overcrowding of the terminal group. It is easily synthesized [44] by irradiation of 1 (3-phenanthryl)-4-phenyl but-1-en-3-yne

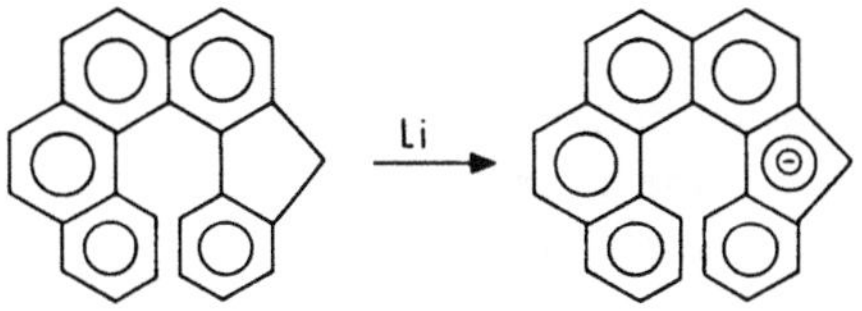

61 62

Scheme 14

(63). A similar procedure can be used to obtain 4,5-diphenylphenanthrene (66) and several related compounds. The compound *66* is a heptahelicene lacking rings B and E and has interesting photochemical and chiroptical properties [45,46].

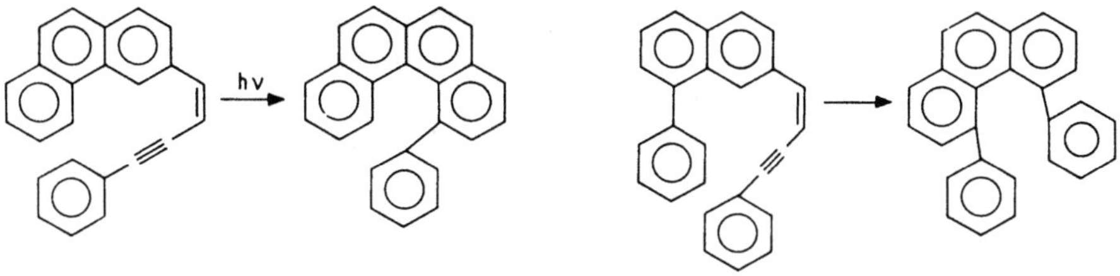

63 64 65 66

Scheme 15

A helical compound, lacking even three rings is the o.quaterphenylophan 67 [47]. It looks like the overbridged heptahelicene 37.

Reviews of compounds having a helical structure, others than helicenes, have been given by Mislow [48] and by Voegtle [49].

Bis-2,13-pentahelicenylene (69), consisting of two pentahelicene moieties and because of its propellor shape also named propellicene is obtained via a photocyclodehydrogenation reaction [50].

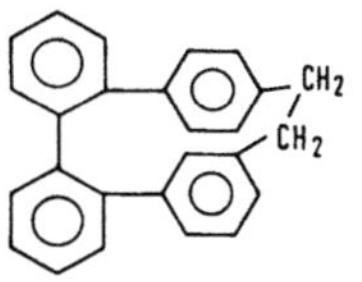

67

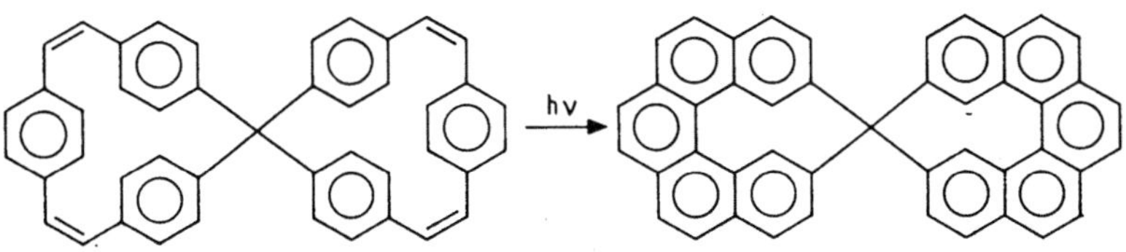

68 69

Scheme 16

2.1.5 Heterohelicenes

In general, the photosynthesis of heterohelicenes is similar to that of carbohelicenes. Because the nature, the number and the positions of the heterocyclic rings in a helicene can vary, the diversity of heterohelicenes is much larger than that of carbohelicenes. More than 30 different thiahexahelicenes are possible, but only a limited number is synthesized until now. Reports on heterohelicenes containing pyridine, pyrole and furane rings are even very few.

The photocyclodehydrogenation of thienyl ethylenes is well-defined when both thiophene rings are bound via a C(2) atom to the ethylenic bond as in (70). In other cases, however, more cyclization products are possible. To predict the photocyclization mode for heterohelicenes the $\Sigma F^*_{r,s}$ rule fails in many cases, because correction factors for the hetero atoms in the Hückel MO calculation have to be introduced and the systems are not well comparable with carbocyclic diaryl ethylenes. A better reaction parameter in these cases is the Mulliken overlap population (n_{rs}) [51], introduced by Muszkat [52] for these cases. The overlap populations of the atoms r and s in ground and excited state ($n_{r,s}$ and $n^*_{r,s}$), are calculated using the extended Hückel method. Cyclizations should not occur when $n_{r,s}$ and $\Delta n^*_{r,s}$ ($= n_{r,s} - n^*_{r,s}$) have negative values. (This method can also be used for diaryl olefins, but in these cases calculation of $\Sigma F^*_{r,s}$ is more simple.).

In Table 3 some examples of syntheses of heterohelicenes are given. Similar to the synthesis of carbohelicenes heterohelicenes can be obtained by the photoreaction of 1,2-diheteroaryl ethylenes as well as from bis(heteroarylvinyl) aromatics (diolefins).

Table 3. Synthesis of Thiaheterohelicenes

Helicene	Precursor	Yield (%)	Melting point (°C)	Ref.
[6] SSBSBS	(3 · SBS + 2 · SS)	67	210–211.5	[53]
[6] SSBSBB	(3 · BBS + 2 · SS)	60	220–222	[53]
[6] SBSSBS	(1 · S + 2 · SS + 1 · S)	51	279–280	[53]
[6] BBSSBB	(1 · B + 2 · SS + 1 · B)	50	289–291.5	[53]
[6] SBBBBS	(1 · S + 2 · BB + 1 · S)	50	240–241	[53]
[6] SBSBSB	(2 · BS + 3 · SBS)	40	210–212	[54]
[7] SBSBSBS	(3 · SBS + 3 · SBS)	50	269–278	[55]
[9] SBSBSBSBS	(5 · SBSBS + 3 · SBS)	38		[55]
[11] SBSBSBSBSBS	(3 · SBS + 3 · SBS + 3 · SBS)	8		[55]
[13] SBSBSBSBSBSBS	(7 · SBSBSBS + 5 · SBSBS)	17		[55]
[15] SBSBSBSBSBSBSBS	(5 · SBSBS + 3 · SBS + 5 · SBSBS)	5		[55]

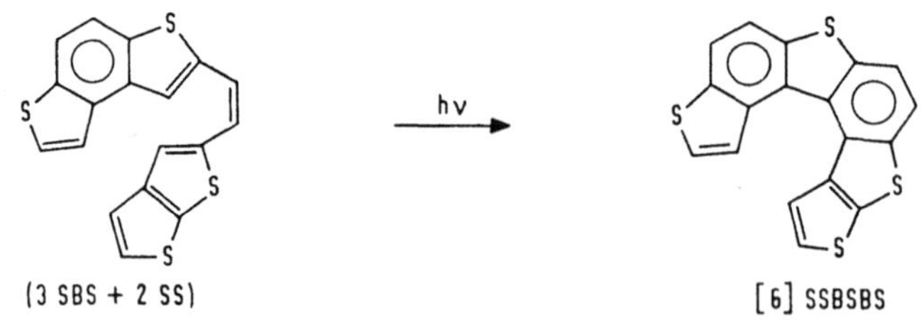

70 71

Scheme 17

According to Kawazura [55] product formation by the latter method affords lower yields. The reaction conditions are completely similar as given and discussed for carbohelicenes.

2.2 Non-Photochemical Syntheses

Carbohelicenes cannot be made via a groud state chemical method that can compete with the photochemical procedure in simplicity and yield. Some heterohelicenes, however, can be synthesized better by a non-photochemical way.

The first heterohelicene (72) ever made was obtained by Fuchs and Niszel [56] in 1927 by heating 2,7-dihydroxynaphthalene with phenylhydrazine and sodium-hydrogen sulphite. The reaction was also investigated by Zander [57].

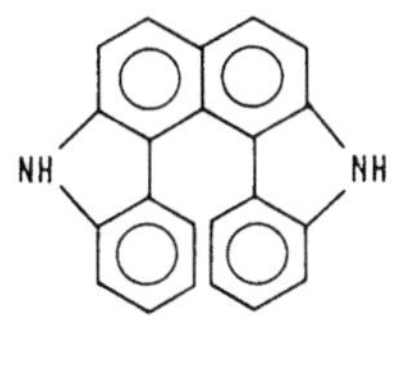

72

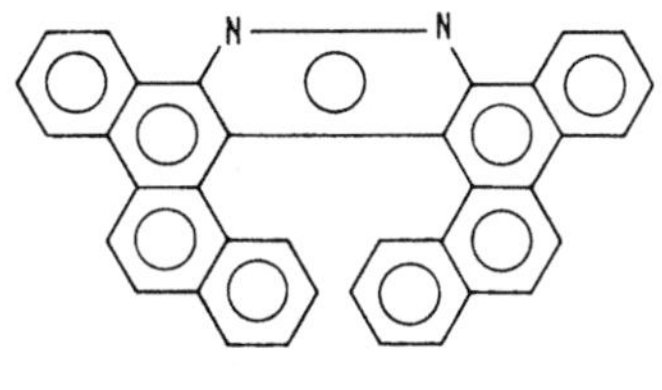

73

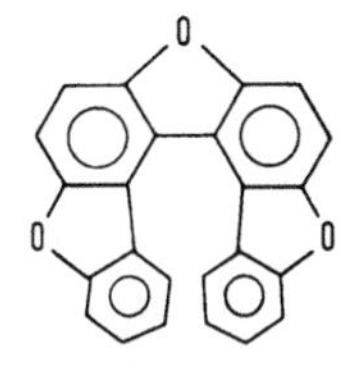

74

Small yields of heterohelicenes were obtained by treating bis phenylhydrazone of p.quinone with sulfuric acid [82].

Another nitrogen-containing helicene (73) was prepared in high yield by treating the diazonium salt of 6-aminochrysene with sodium sulphite and cyclization of the resulting 6,6'-diazochrysene with sulphuric acid [58]. Non-photochemical preparations of helicenes containing furan rings were performed by Chatterjea [59].

A heptahelicene containing three furan rings (74) was made starting with the treatment of p-benzoquinone with strong acids. The resulting 2-hydroxybenzofurane was treated successively with iodine chloride and copper giving a biaryl derivative which was cyclized with HBr [60].

2.3 Asymmetric Syntheses of Helicenes

Several studies have been undertaken to get optically active helicenes by means of an asymmetric photocyclization. Four types of asymmetric photosynthesis can be distinguished for helicenes:

- asymmetric synthesis with circularly polarized light (Sect. 2.3.1),
- asymmetric synthesis in chiral solvents (Sect. 2.3.2),
- asymmetric synthesis in cholesteric liquid crystals (Sect. 2.3.3),
- chemically induced asymmetric synthesis (Sect. 2.3.4).

2.3.1 Asymmetric Synthesis with Circularly Polarized Light

Photocyclodehydrogenation of 1,2-diarylethylenes by irradiation with circularly polarized light (CPL) has given helicenes, which show an enantiomeric excess up to

0.5%, depending on the precursor and the wavelength used. Results for varying precursors are compared in Table 4, which also shows wavelength-dependence in the case of [7]- and [8]helicene. In all cases studied right-handed CPL forms M-helicenes and left-handed CPL P-helicenes. The optical yield is calculated by multiplying [α] of product/[α] of pure enantiomer by a factor 100.

This asymmetric photosynthesis might be explained by three possible mechanisms:

1. primary formation of a racemic helicene, followed by light-induced assymetric destruction.
2. primary light-induced formation of a racemic mixture of a trans-dihydrohelicene (DHP), with subsequent light-induced asymmetric ring opening to the starting olefin, i.e. partial photoresolution of the intermediate DHP.
3. the diarylethylene precursor, existing as a mixture of P- and M-forms in rapid equilibrium at room temperature gives different amounts of excited P- and M-forms on irradiation with CPL. In the excited state cyclization is faster than racemization.

The first possibility is ruled out by the observation that asymmetric photodestruction by irradiation with CPL is much slower than asymmetric photocyclization. Moreover, it results in an optical activity of opposite sign in comparison with the optical activity observed in asymmetric photocyclization of the corresponding 1,2-diarylethylene. Finally, the optical yield and the wavelength of the CPL required for different precursors of [8]-helicene are different and the latter does not show any connection with the CD-Spectrum of [8]-helicene [65].

The second mechanism seems highly unlikely as the ring opening of DHP must be much slower than the oxidation into helicenes with iodine [66]. Moreover, the wavelength-dependence of the optical yield of [8]-helicene does not correlate with that expected from the proposed CD-spectrum of the intermediate DHP [62,65,67]. This mechanism can also not explain the difference in optical yield of [6] from (1+4) and (2+3).

The remaining mechanism involves the preferential photocyclization of a special conformer of the cis-1,2-diarylethylene. The parent compound cannot be planar and must be considered as a mixture of several interconverting conformers. In solution, there is a slight preference for the cis-syn-conformations over the cis--anti-conformations [68].

The preferential excitation of the P- or M-cis-syn-conformer by CPL is related to the g-factors of the conformers. This optical anisotropy factor $g = (\varepsilon_L(P.S) - \varepsilon_R(P.S))/\varepsilon$ in which $\varepsilon_L(P.S)$ and $\varepsilon_R(P.S)$ are the extinction coefficients of the P-cis-syn-conformer for left- and right-handed circularly polarized light, respectively. ($\varepsilon_L(P.S) = \varepsilon_R(M.S)$.)

The P- and M-cis-syn-conformers give photoreactions, whose rates are proportional to ε_L when left-handed CPL is used. Therefore, a high optical yield requires a high g-factor [69]. After the excitation cyclization occurs leading to a DHP, which is readily oxidized to helicene. The precursor remains racemic as racemization in the ground state, is easy. Racemization in the excited state, however, may be slow, compared to cyclization. According to Kagan [69] this mechanism explains the asymmetric synthesis well.

From Table 4 it appears that photocyclization of 1-(2-naphthyl)-2-(3-phenanthryl)-ethylene (2+3) with CPL leads to a higher optical yield (0.2%) than that of 2-styryl-

Table 4. Asymmetric Syntheses of Helicenes by Irradiation with Circularly Polarized Light (CPL)

Prescursor	Helicene	Wavelength in nm	Optical yield	Ref.
(1+4)	[6]	370 ± 6	0.05	61,62,63)
(2+3)	[6]	370 ± 6	0.21	61)
(1+2+1)	[6]	290–370	0.05	63)
(1+1+2)	[6]	290–370	0.05	63)
(o.Br-1+4)	4-Br-[6]	370 ± 6	0.26	62)
(o.Cl-1+4)	4-Cl-[6]	370 ± 6	0.33	62)
(p.F-1+4)	2-F-[6]	370 ± 6	0.08	62)
(p.Br-1+4)	2-Br-[6]	370 ± 6	0.17	62,70)
(2+4)	[7]	370 ± 6	0.75[a]	62,65)
(2+4)	[7]	400 ± 6	1.29[a]	62,65
(3+3)	[7]	370 ± 6	0.6	62)
(1+1+4)	[8]	290–370	0.15	63)
(3+4)	[8]	290–370	0.3	63,64)
(3+4)	[8]	290 ± 6	0.42[b]	65)
(3+4)	[8]	310 ± 6	0.21[b]	65)
(3+4)	[8]	350 ± 6	1.37[b]	65)
(3+4)	[8]	400 ± 6	2.00[b]	65)
(4+4)	[9]	290–370	0.36	63)
(3+1+4)	[10]	290–370	0.06	63)
(4+1+4)	[11]	290–370	0	63)
(4+3+3)	[12]	290–370	0	63)
(4+3+4)	[13]	290–370	0	63)

[a, b] These values are not calculated from the [α] of pure enantiomers and are relative
[a] for [7] and [b] for [8]helicene

benzo[c]phenanthrene (1+4) which gives an optical yield of 0.05%. Calvin [62, 65, 70] ascribed this difference to partial equilibration of the M- and P-conformers in the excited state, which should be easier for cis-syn 2-styrylbenzo[c]phenanthrene than for cis-syn 1-(2-naphthyl)-2-(3-phenanthryl)ethylene. According to Calvin such equilibrations are caused by a rotation of an aryl residue, which is easier when it concerns a smaller residue (easier in (1+4) than in (2+3). (Scheme 18.))

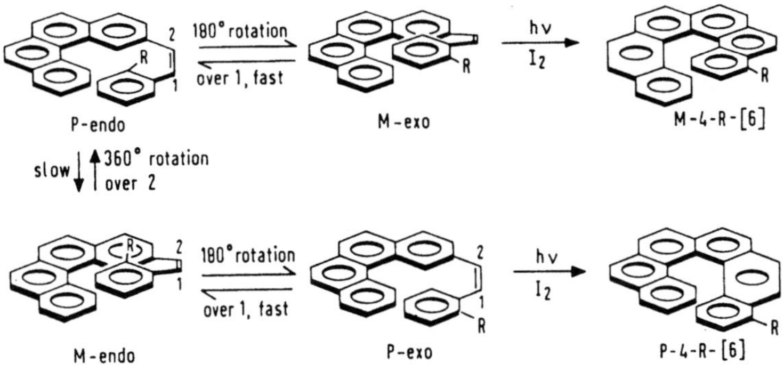

Scheme 18

To exclude the possibility that the difference in optical yield is caused by a different g-value of the precursors several ortho substituted (1+4) precursors were irradiated.

The ortho substituents used were supposed to have no influcence on the absorption spectrum and to possess the same circular dichroism as the parent compound. The ortho substituted compound can only photocyclize in its exo-conformations, but equilibration by a rotation over the phenylethylene bond converts an M-exo into a P-endo form (or P-exo into M-endo) which cannot cyclize. Indeed, the optical yields of the 4-substituted [6]helicenes from the ortho substituted (1+4) precursors are not lowered in comparison with the unsubstituted (2+3) precursor. On the other hand (4+1) precursors having a para substituent give again a low optical yield. The values found suggest that racemization in the excited state is slightly hindered, what may be due to steric hindrance.

According to this concept the low optical yield of hexahelicene from the (1+1+2) precursor suggests that the cyclization occurs via the (1+4) intermediate, what is in accordance with the prediction from the Σ F*-rule [63]. The asymmetric synthesis of [10] via [4+3+1) has a very low optical yield (0.03%). No asymmetric synthesis is observed with (4+1+4) and (4+3+2) for [11], (4+3+3) for [12] and (4+3+4) for [13] helicene. There is no clear explanation for this behaviour [63].

2.3.2 Asymmetric Synthesis in Chiral Solvents

Laarhoven et al. [71] have studied the influence of chiral solvents on the optical yield of [6]helicene, synthesized by photodehydrocyclization of various precursors. Most thoroughly investigated was the photochemical ring closure of 2-styrylbenzo[c]-phenanthrene in optically active fluids (Table 5).

Table 5. 2-Styrylbenzo[c]phenanthrene Irradiated in Chiral Solvents

	$[\alpha]_D$ (°)	$[\alpha]_D$ neat (°)	Optical yield %[a]
a. (+)-α-Pinene		− 7	0.21
b. (S)-(−)-Ethyl lactate (20 °C)		+ 15	0.42
c. (S)-(−)-Ethyl lactate (100 °C)		+ 6	0.18
d. (S)-(−)-Ethyl mandelate		+ 71	2.1
e. (S)-(+)-Ethyl O-benzoylmandelate		+ 2	0.04
f. (±)-Ethyl mandelate		0	0.0
g. (RR)-(+)-Diethyl tartrate		− 40	1.1
h. (RR)-(+)-Diethyl O,O'-dibenzoyltartrate-benzene (1:3)	+28.2	+113	(3.0)
i. (RR)-(+)-Diethyl O,O'-dibenzoyltartrate-benzene (1:5)	+18.5	+111	(2.9)
j. (S)-(+)-Ethyl O-benzoyllactate		−30	0.84
k. (S)-(+)-Ethyl O-(α-naphthoyl)lactate		− 57	1.6
l. (4-phenylbenzoyl)lactate benzene (1:5)	−17	−102	(2.7)
m. (S)-(+)-Ethyl O-(2-phenylbenzoyl)lactate		− 30	0.84

[a] Optical yield = ([α] of product/[α] of pure enantiomer) × 100%

The influence of the chiral solvent is ascribed to the unequal diastereomeric interactions between the chiral medium and enantiomeric forms of solutes. This concerns not only the cis-syn conformers (P and M) of the parent compound in the ground-state, but also in the excited-state, and even the enantiomeric DHP's belonging to them. These different interactions with the solvent in several phases of the photocyclization cause that the chiral solvent is more cooperative in the formation of one of the enantiomers of the product than in that of the other.

The results collected in Table 5 suggest that hydrocarbon residues, especially aromatic groups, in the solvent are strongly responsible for the interaction with cis-(1+4). The position of the largest hydrocarbon residue apparently determines whether P- or M-[6]-helicene will be formed in excess. Replacement of the methyl group in (S)-ethyl lactate (b) by a phenyl group giving (S)-ethyl mandelate (d), increases the optical yield fivefold.

Introduction of a benzoyl residue in the polar hydroxy function of (S)-ethyl lactate gives, however, [6]-helicene in which the antipodal, laevorotatory enantiomer predominates.

Enlargement of the aromatic residue in (S)-ethyl O-benzoyllactate by annelation with a benzo group (k) or introduction of a phenyl substituent (m) leads to a significant increase of the optical yield. Remarkable is that the optical yield also increases when the photosynthesis starts from (1+2+1) instead of (1+4). Irradiation of (1+2+1) in S-(+)-ethyl O-benzoyllactate gives an optical yield of 1.7%.

The latter result is explained by assuming that the chiral solvent acts as a matrix in which the product of the first photocyclization step is already formed in a preferential conformation.

2.3.3 Asymmetric Synthesis in Cholesteric Liquid Crystals

Macroscopical helical structures formed by cholesteric liquid crystals have been used as chiral helical media for the asymmetric synthesis of helicenes.

Several cholesteric mesophases have been studied as media for the photodehydrocyclization of (4+1). The results are collected in Table 6.

The experiments of Table 6 indicate that a cholesteric mesophase with a right-handed helix gives an excess of (+)-[6]-helicene, which is known to have also a right-handed helix. The optical yield is small but significant. Nakazaki et al. [72] did not find induction in an isotropic mesophase (Table 6, c), while Hibert [73] got still an enantiomeric excess in an isotropic and in several compensated nematic mesophases (Table 6, f, i, j, k, l). Hibert explained these findings by discerning two effects:

a. in compensated nematic and isotropic phases the asymmetric induction results only from diastereomeric solute-solvent interactions, as is the case in other chiral solvents [71] (see 2.3.2). In the compensated nematic phase the induction is increased by a factor 10 compared with the isotropic phase (Table 6, f, i, j), because the higher local ordering intensifies the solute-solvent interactions;

b. the helical order of the cholesteric mesophase causes a clear additional effect. The effect of a right-handed helix is additive (Table 6, g, h), that of a left-handed helix opposite (Table 6, e). The magnitude of this additional effect seems inversely proportional to the pitch of the cholesteric helix (see Table 6, g and h). This additive effect of the macroscopical helix on the optical yield has also been demonstrated in an achiral, mechanically twisted mesophase [74], where the induction solely depends upon the handedness of the pitch (Table 6, m–q). A right-handed helix gives right-handed (+)-[6]-helicene in excess, while a left-handed helix yields mainly (—)-[6]-helicene.

Table 6. Photocyclodehydrogenation of (4+1) in Cholesteric Liquid Crystals

	Liquid crystal	Mesophase	Temp. °C	Pitch and handedness	$[\alpha]_D$ (°)	Optical yield	Ref.
a	cholesteryl nonanoate-cholesteryl chloride (3:2)	cholesteric	23	R	+40	1.1	72)
b		cholesteric		0.5μ; L	+10.5	0.29	73)
c		isotropic	55–60		0	0	72)
d	cholesteryl benzoate	cholesteric	145–150	R	+37	1.0	72)
e	cholesteryl chloride-cholesteryl myristate (7:4)	cholesteric	31 ± 1	−3.5μ; L	+2.0 ± 0.3	0.05	73)
f		compensated nematic	41 ± 1		+7.0 ± 0.5	0.19	73)
g		cholesteric	51 ± 1	−4.9μ; R	+10.8 ± 0.7	0.29	73)
h		cholesteric	61 ± 1	−1.6μ; R	+15.5 ± 1	0.43	73)
i		isotropic	70 ± 1		+1.0 ± 0.2	0.03	73)
j		isotropic	80 ± 1		+0.8 ± 0.2	0.02	73)
k	cholesteryl chloride-cholesteryl palmitate (68:32)	compensated nematic	65 ± 1		+4.7 ± 0.4	0.13	73)
l	cholesteryl chloride-cholesteryl nonanoate (68:32)	compensated nematic	32 ± 1		+6.1 ± 0.5	0.17	73)

		Mesophase	conc. (WT %)	Handedness	$[\alpha]_{325}$ (°)	Optical yield	Ref.
m	p.cyanophenyl p-butyl-benzoate p-cyanophenyl p-heptyl-benzoate (1:1)	mechanically	1.0	R	+260 ± 60	0.03 ± 0.04	74)
n		twisted nematic mesophase	1.0	L	−250 ± 60	0.04 ± 0.01	74)
o			0.5ᵃ	L	−520 ± 60	0.08 ± 0.01	74)
p			0.5ᵃ	R	+580 ± 60	0.09 ± 0.04	74)
q		no twist	0.5ᵃ	—	ca. 0	ca. 0	

ᵃ With 2.0 wt % of anthracene

The helix of the mesophase has probably a steric influence upon the conformational equilibrium of cis-syn (4+1). It seems unlikelv that this contribution is caused by CPL, generated in the cholesteric mesophase [75].

2.3.4 Chemically Induced Asymmetric Synthesis

The stereoselective formation of helicene skeletons by photocyclodehydrogenation of 1,2-diarylethylenes carrying a chiral group had been studied systematically by Martin et al.

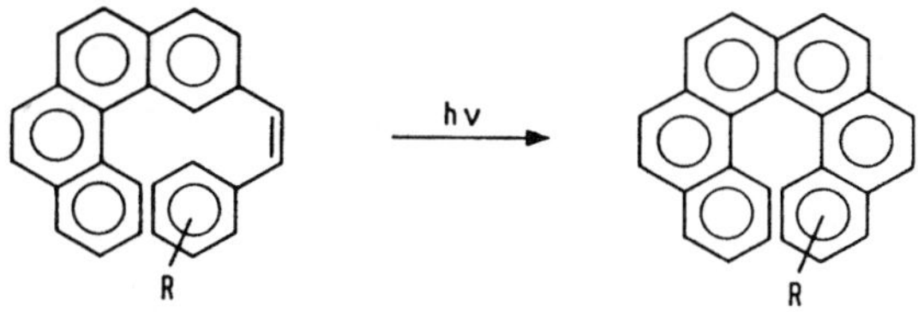

75 76

Scheme 19

In order to study the influence of a given chiral group at different positions in the precursor styrylbenzo[c]phenanthrene several carboxy styrylbenzo[c]phenanthrenes were synthesized and esterified with chiral alcohols, and the products were used to prepare hexahelicenes *76* substituted at various positions. The bromo-esters (29) were used as precursors for [6]-helicenes, substituted at position 1.

Table 7. Chiral Groups as Inductor for the Asymmetric Synthesis of Hexahelicene

Chiral group R in 75	Position in helicene 76	Ratio of diastereomeric helicenes	Ref.
-$CO_2CH(CH_3)Bu^t$	1	64:36	[75]
-$CO_2CH(CH_3)Bu^t$	2	51:49	[75]
-$CO_2CH(CH_3)Bu^t$	3	51:49	[75]
-$CO_2CH(CH_3)Bu^t$	4	52:48	[75]
-$CO_2CH(CH_3)Bu^t$	5	54:48	[75]
-(—)CO_2-menthyl (+80 °C)	1	(—)80:(+)20	[76]
-(—)CO_2-menthyl (+25 °C)	1	(—)76:(+)24	[76]
-(—)CO_2-menthyl (0 °C)	1	(—)62:(+)38	[76]
-(—)CO_2-menthyl (—78 °C)	1	(—) 2:(+)98	[76]
-(—)CO_2-menthyl	2	(—)47.5:(+)52.5	[77]
-(—)CO_2-menthyl	3	(—)47.5:(+)52.5	[77]
-$CCH_3(OCH_3)Bu^t$	2	51:49	[75]
-$CO_2CH(CH_3)C_6H_5$	2	52:48	[75]
-CO_2CH_2-2(+)-(P) [6]	2	(+)58:(—)42	[75]

The results, collected in Table 7 show an appreciable enantiomeric excess for 1-substituted helicene, a lower excess for 5-substituted helicenes and low selectivity in the formation of 2-, 3- or 4-substituted products.

The chemically induced asymmetric photocyclization of the 1-(—)-menthyl ester shows a striking temperature dependence [76], a completely reversed ratio is obtained at low temperature. When the chiral group is placed at C(11) of the benzo[c]phenanthryl group (77) the effect is of the same order (about 5% diastereomeric excess) as when the same group is placed at the paraposition of the phenyl group, or when a properly substituted naphthyl, phenanthryl ethylene (78) is used as the starting compound.

Optically pure (M)-(—)-2-formylhexahelicene (76, R = 2-formyl), prepared from the corresponding resolved (—)-carboxy-menthyl ester, was used as a common precursor for the photochemical synthesis of optically pure (M)-[8]-, (M)-[9]-, (M)-[10]-, (M)-[11]-, and (M)-[13]-helicene [31]. Wittig reactions between (M)-(—)-formylhexahelicene and suitable phosphonium salts led to 1,2-disubstituted ethylenes, all containing the same chiral moiety, which could be subjected to photocyclodehydrogenation (Table 1).

In a similar way Wijnberg et al. [4, 54] used an optically active heptaheterohelicene viz. [7]BBSBSBS, partially resolved from (—)-α-pinene (optical purity 7.5%) as a starting compound for the preparation of an optically active undecaheterohelicene [11]BBSBSBSBSBB.

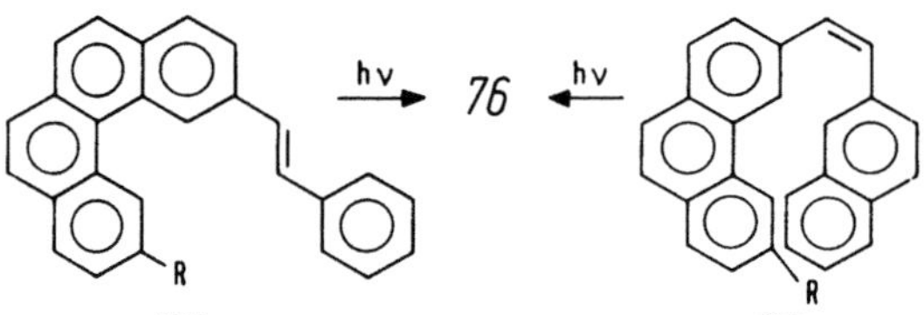

77 78

Scheme 20

3 Chiroptical Properties

3.1 Optical Resolution of Enantiomers

The first resolution of [6]-helicene was achieved by Newman and Lednicer [78] by crystallization with the aid of a chiral complexing agent, 2-(2,4,5,7-tetranitro-9-fluorenylidene aminooxy)propionic acid (TAPA), which was especially designed for this purpose. (R)- and (S)-TAPA (79a) form diastereomeric charge-transfer (CT) complexes with the enantiomers of hexahelicene. Several other helicenes could also be resolved using this reagent.

Resolution of helicenes has also been performed by the very laborious method of picking single crystals [4,5,29,79,80]. After recrystallization of the partially resolved mixture the procedure can be repeated, until no more variation in optical rotation occurs. Because some helicenes crystallize into racemic crystals by lamellar intergrowth of pure P and pure M forms (see Sect. 6) the crystal picking method is not always applicable, however. A [7]-heterohelicene was partially resolved by crystallization from the chiral solvent [54]: (—)α-pinene.

Conventional chromatography on a silica gel column coated with optically active TAPA results in partial resolution of some helicenes, but the technique is slow and inefficient [81,40]. For chromatographical resolution columns of optically active polymers, triacetyl cellulose [82] and more recently (1)-poly(triphenylmethyl methacrylate) [83, 84] have been employed.

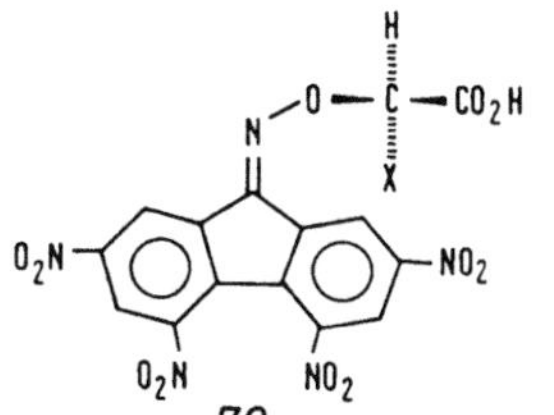

79

High Performance Liquid Chromatography (HPLC) on columns coated with or bonded to TAPA can resolve the enantiomers of both carbo- and heterohelicenes completely. The method is not only more efficient [85-87] but also less time-consuming than those given above.

It turns out the TAPA has a better resolving capacity than derivatives containing larger alkyl groups at the chiral centre (TABA (79b), TAIVA (79c) and TAHA (79d) [85b)]. The bulkiness of the group of the CT-acceptor is crucial for the ease of resolution of the helicenes. In Fig. 1 the resolution factors[2] (r) for carbohelicenes with some acceptors are plotted vs. the number of benzene rings. With R(—)-TAHA no resolution was detected for any of these helicenes. Pentahelicene could only be resolved by R(—)-TAPA bonded to silicagel and by R(—)-TABA coated on silicagel

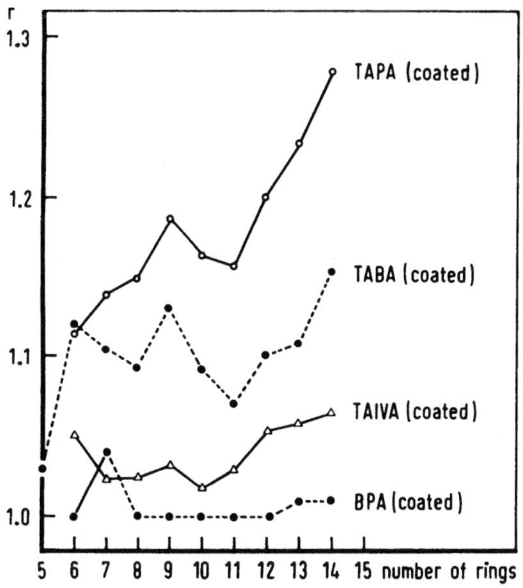

Fig. 1. Resolution factor (r) for carbohelicenes on different chiral selectors

2 The parameters used in HPLC are: capacity factor $k' = \frac{t_R - t_0}{t_0}$ with t_R = retention time, t_0 = non-sorbed time; resolution factor $r = k_1'/k_2'$; Resolution $R = 2\left(\frac{t_{R_1} - t_{R_2}}{w_1 + w_2}\right)$, with w = bandwidth.

(see Fig. 1). For pentahelicene the resolution factor with R(—)-TAPA is 1.049, with R(—)-TABA 1.033. The resolutions are R_s(TAPA) = 0.18 and R_s(TABA) = 0.33.

A suggested explanation [85b)] for the diminished resolving capability of the chiral selectors[3] with larger alkyl substituents supposes that in the CT-complex this alkyl group must be accommodated to the semi-cavities of the helicenes. The decrease of the r-value with increasing size of the alkyl substituent is then acceptable as a larger alkyl group at the asymmetric carbon will be accommodated less readily. Space-filling models of the diastereomeric associates agree with this explanation. These models indicate that the CT-complex of P(+)-[6]-helicene and R(—)-TAPA is more stable than the complex of M(—)-[6]-helicene and R(—)-TAPA. This agrees with the observation that P(+)-helicenes are retained longer on columns with R(—)-TAPA than M(—)-helicenes [85b)].

Calculation [89)] of the association constants (K) of the diastereomeric CT-complexes of S(+)-TAPA and [6]-helicene from ORD and CD spectra revealed a small difference between K+ (K for (+)(+)[4] = 4.6 (at 20 °C)) and K— (K for (—)(+)[4] = 5.5 (at 20 °C)). These values fit into the deduction from HPLC experiments [85)] that (+)(—) and (—)(+) are more stable than (—)(—) and (+)(+).

From Table 21 (Sect. 4.4) it follows that the CT-complex of the thiaheterohepta-helicene P(+) [7]-SBSBSBS and S(+)-TAPA in tetrachlorethane is more stable than the complex of the same helicene with R(—)-TAPA [90)].

In the separation, however, of the enantiomers of this helicene by HPLC on a column, composed of S(+)-TAPA bonded to aminated silicagel, M(—)-[7] has the longer retention time, pointing to a higher association constant for the complex with this enantiomer [87)].

This may suggest that the CT-complexes in solution are different from those formed during resolution of helicenes by HPLC on TAPA modified columns. Interestingly, when the three alternating thiahelicenes [3], [5] and [7] were subjected to HPLC on silica bonded to TAPA, the order of elution ([3]>[7]>[5]) was equal to that of the complexation constants with another π-acceptor (TCNQ) [91)] (see Sect. 4.4).

In Fig. 2 the internal angle[5] of helicenes is plotted vs. the resolution factor; for thiahelicenes on TAPA-bonded silicagel and for the carbohelicenes on TAPA-bonded- and on TAPA-coated silicagel. It appears that helicenes with internal angles between 540° and 720° are less well resolved. Because, the band width for columns containing covalently bonded TAPA is larger, the resolution on this column is worse.

Wijnberg [86)] separated several heterohelicenes on aluminium oxide coated with TAPA. The resolution factors are comparable with those of TAPA on silicagel; they ranged from r = 1.12 for [7]-BOBBBOB to 1.20 for hexahelicene and some thiahexahelicenes. Other chiral selectors have been tested as to their usefulness for the resolution of helicenes. Binaphthyl-2,2′-diylhydrogenphosphate (BPA) (*80*) raised great expectations [88)] because its chirality is of the same kind as that of

3 Selector and selectant are cybernetic terms for separating agent and sample input, respectively.

4 The first symbol stands for the rotation of hexahelicene, the second one for that of TAPA.

5 The internal angle is found by summation of contributions of the individual rings: for a benzene ring 60°, for a thiophene ring 45° [5)].

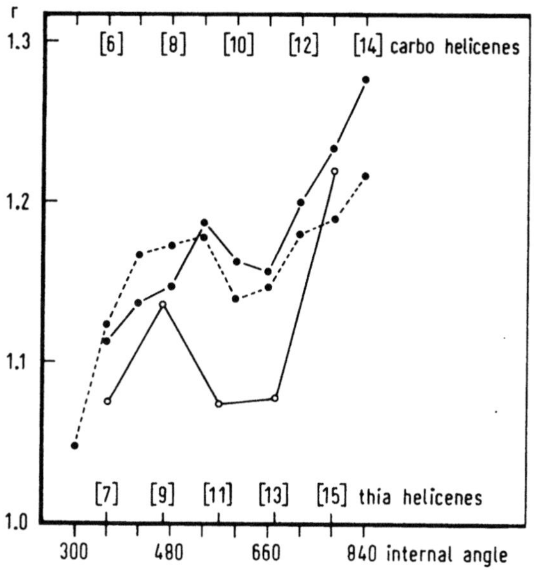

Fig. 2. Resolution factors (r) of some selectors vs the internal angle of carbo- and thiaheterohelicenes
○—○—○ Thiaheterohelicenes on TAPA bonded on silicagel
●—●—● Carbohelicenes on TAPA coated on silicagel
●- -●- -● Carbohelicenes on TAPA bonded on silicagel

the helicenes. Carbohelicenes show, however, a poor resolution on BPA (see Fig. 1), indicating that this selector forms only a weak or no charge transfer complex. Thiahelicenes or helicenes substituted with bromine have a greater electron-donating capacity what results in a stronger interaction with BPA and a much better resolution.

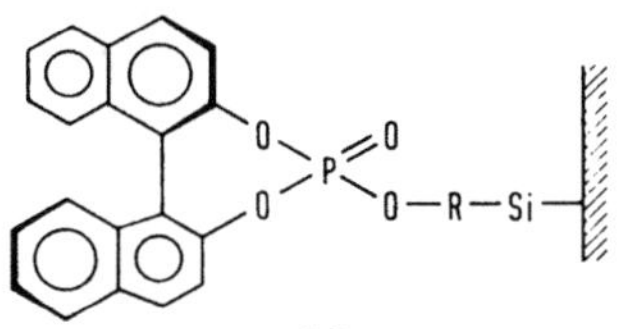

80

Among the naturally occurring chiral substances, tested in the resolution of helicenes, riboflavine, and several nucleotides and nucleosides have appeared successful. Some results are given in Table 8.

From Table 8 it is obvious that the resolution always increases with an increase of the number of benzene rings and that riboflavine is a more powerful selector than the nucleotides, but not as good as TAPA. An interesting experiment shows that it is not always necessary to have the selector coated or bound to the solid phase but that it can sometimes be used as well, dissolved in the mobile phase. The n-dodecyl ester of N-(2,4-dinitrophenyl)-L-alanine is able to discriminate between the enantiomers of 1-aza-[6]-helicene, when used as a chiral dopant in the mobile phase in HPLC on a reversed phase column [93] (see Table 9). The usefulness of this dopant must be due to the known ability of a dinitrophenyl moiety to form CT-complexes with polycyclic aromatic hydrocarbons; the presence of a chiral 'site' near this group causes resolution of helicenes, because the steric interactions in diastereomeric complexes will be quite different.

Table 8. Resolution of Helicenes by TAPA, Riboflavine, Nucleosides and Nucleotides Coated on Silicagel

Coated material		Capacity and resolution of helicene									Ref.
		[6]	[7]	[8]	[9]	[10]	[11]	[12]	[13]	[14]	
TAPA	K'_1	4.08	6.15	6.58	8.57	12.81	13.67	17.53	21.44	23.58	[85]
	K'_2	4.54	7.00	7.55	10.17	14.89	15.80	21.04	26.45	30.12	
	r	1.113	1.138	1.147	1.187	1.162	1.156	1.200	1.234	1.277	
Riboflavine	K'_1	2.4	3.24	3.62	4.83	6.46	7.52	9.59	11.48	11.85	[92a]
	K'_2	2.4	3.41	3.87	5.20	7.02	8.23	10.46	12.70	13.39	
	r	1.00	1.052	1.069	1.077	1.087	1.044	1.091	1.106	1.130	
Adenosine	K'_1					2.04	2.88	3.68	4.53	5.26	[92b]
	K'_2					2.13	3.04	3.88	4.76	5.5	
	r					1.044	1.056	1.054	1.051	1.046	
2′-Deoxy-adenosine	K'_1					3.39	4.38	5.82	7.62		[92b]
	K'_2					3.52	4.52	6.06	7.89		
	r					1.038	1.032	1.041	1.035		
3′-AMP	K'_1							~6.6	~8.2		[92b]
	K'_2							~6.6	~8.2		
	r							<1.03	<1.03		
5′-AMP	K'_1					4.72	5.78	7.22	9.81	10.40	[92b]
	K'_2					4.97	6.11	7.64	10.44	11.17	
	r					1.053	1.057	1.058	1.064	1.074	
3′,5′-Cyclic AMP	K'_1							10.66	12.31	12.37	[92b]
	K'_2							11.0	12.71	12.89	
	r							1.032	1.032	1.042	
Guanosine	K'_1								5.61	6.28	[92b]
	K'_2								5.78	6.44	
	r								1.030	1.030	
Uridine	K'_1								5.71	6.36	[92b]
	K'_2								5.71	6.36	
	r								1.0	1.0	

Table 9. Resolution of 1-Aza-6-Helicene by the n-Dodecyl Ester of N-(2,4-Dinitrophenyl)-L-Alanine as a Chiral Dopant in the Mobile Phase[a] [93]

Mobile phase	Vr_1[b]	Vr_2[b]	r
85% MeOH/15% H_2O; 0.80 mM dopant	12.15	12.54	1.039
83% MeOH/17% H_2O; 0.70 mM dopant	15.16	15.94	1.060
82% MeOH/18% H_2O; 0.60 mM dopant	19.29	20.87	1.091

[a] Column: Ultrasphere ODS 5; 25 × 0.46 cm;
[b] Vr: Retention volume

Finally, it is noted that suitable enantiomers of helicenes have been applied as selector for non-helical compounds. The disodium salt of P-(+)-7,10-dicarboxy hexahelicene coated on silicagel was successfully used in resolving the N-(2,4-dinitrophenyl)-α-amino acid esters. Good resolutions were found for alanine, isoleucine, valine, phenylalanine and phenylglycine [94].

3.2 Optical Rotations

The absolute configurations of helicenes have been firmly established by calculations [95–98], by X-ray analysis [99] and by synthesis [100].[6] The dextrorotatory enantiomers of hexahelicene and a thiahexahelicene have been shown to possess the right-handed (P) helicity. From the correspondence of the ORD- and CD-curves of these hexahelicenes and higher helicenes positive rotations correspond in all cases with P-helicenes.

ORD-spectra have been reported for [6] [101], [7] [29,80], [8] [80] and [9] [80], and thiaheterohelicenes [96]. CD-spectra were reported for [6] [101], [7] [80], [8] [80], [9] [80] and for thiaheterohelicenes [54,96,102].

Table 10. Specific $[\alpha]^{a}$ and Molecular (M) Rotations of Carbo- and Heterohelicenes in Chloroform

Helicene	$[\alpha]^{25}_{5790}$ (°)	Ref.	M	$[\alpha]^{23}_{6328}$ ^{b}	Helicene	$[\alpha]^{23}$ (λnm)	M	Ref.
[5]				1970				
[6]	3640 ± 20	1)	11950	3570	[6]SBSBSB	3640 (436)	12600	96)
[7]	5900 ± 100	80)	22330	4570	[7]SBSBSBS	2990 (500)	12030	87)
[8]	7170 ± 100	80)	30730	6010	[9]SBSBSBSBS	3760 (500)	19115	87)
[9]	8150 ± 100	80)	39000	7380	[11]SBSBSBSBSBS	4550 (500)	27960	87)
[10]	8940 ± 100	31)	47260	6100	[13]SBSBSBSBSBSBS	8290 (500)	59730	87)
[11]	9310 ± 100	31)	53880	7200	[7]BOBBBOB	1784 (589)		86)
[12]				6850	[7]BBSBSBB	2177 (589)		86)
[13]	9620 ± 100	31)	65300	6870				
[14]				7620				

[a] When two different [α]-values are reported for the enantiomers, the largest value is given here, without + or −.

[b] These values from Ref. [85] are less accurate due to the very small concentrations used.

In Table 10 the optical rotations of helicenes are given.

The specific rotations of carbohelicenes increase with increasing number of benzene rings, but the increments become continuously smaller. In Fig 3 the molecular rotations $\left(M = \frac{Mw}{100}[\alpha]\right)$ are plotted vs. the internal angle of the helicenes. It appears that the molecular rotations of the carbohelicenes are on a smooth concave curve ($M = -370n + 14500n - 62500$) ($5 < n < 15$), what makes that subsequent additions of a benzene ring are accompanied with decreasing contributions to the total rotation.

For thiahelicenes, all having alternating thiophene and benzene rings, this regular course is not observed. This may be an indication that these helicenes are constructed

6 The absolute configuration of the (−)-[2,2]-paracyclophane carbaldehyde, whose unique stereochemistry was used by Martin to force the hexahelicene helix in the M-configuration, was according to Nakazaki [83] not yet firmly established. Nakazaki regarded Martins work rather as a most reliable method to assign the (R)-configuration to (−)-[2,2]-paracyclophanecarbaldehyde, since the (M)-helicity of (−)-[6]-helicene had been well established by various methods.

more irregularly than carbohelicenes, as can also be concluded from the analysis of the NMR-spectra (Sect. 4.1).

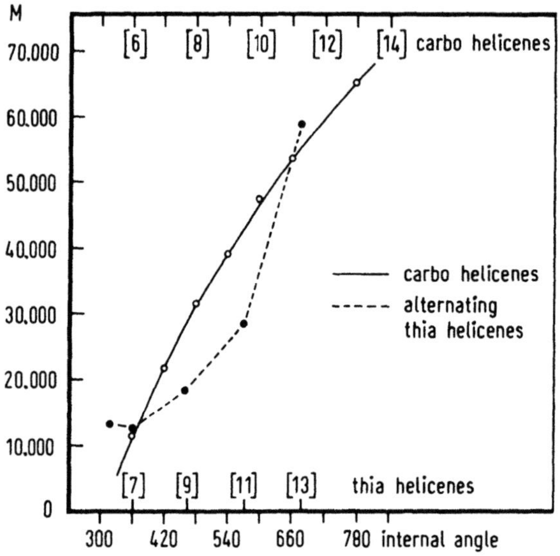

Fig. 3. Molecular rotation (M) plotted vs the internal angle of helicenes

3.3 Thermal Racemization

Perhaps the most intriguing observation of helicenes is the unexpected ease with which these compounds racemize thermally. According to Martin [103] three pathways for the thermal racemization can be considered: 1) via bond breaking; 2) via an internal double Diels-Alder adduct, and 3) via a direct inversion. Martin rejected the first possibility because it is not in accordance with kinetic data for the racemization. He could exclude the second possibility in an elegant way by using appropriately substituted derivatives [103].

For direct inversion the terminal rings have to pass each other which process can only occur by molecular deformations, spread over a large number of bonds and angles. It was supposed that the racemization occurs via an intermediate state in which the terminal rings are parallel. Calculations by a force-field method [104] (see Section 7) using this model gave reasonably good values for the energy of activation for hexahelicene (Ea = 171.7 kJ $\cdot$ mol^{-1} (exp. 150.5) and for heptahelicene (Ea = 191.0 kJ mol^{-1} (exp. 171.7)). According to Lindner [104] the racemization of pentahelicene proceeds via a planar conformation.

Determination of the racemization parameters for some methyl-substituted hexahelicenes showed that the ΔH-values are in general between those of hexa- and heptahelicene (Table 11). The activation entropies were, however, much larger for helicenes having substituents at C(1) and C(16). When the terminal rings should be parallel in the transition state, 1-methyl groups should bring about only a small effect. The large increase of the activation entropy points, therefore, to a transition state which has the terminal rings in an orthogonal position to each other. Such a conformation

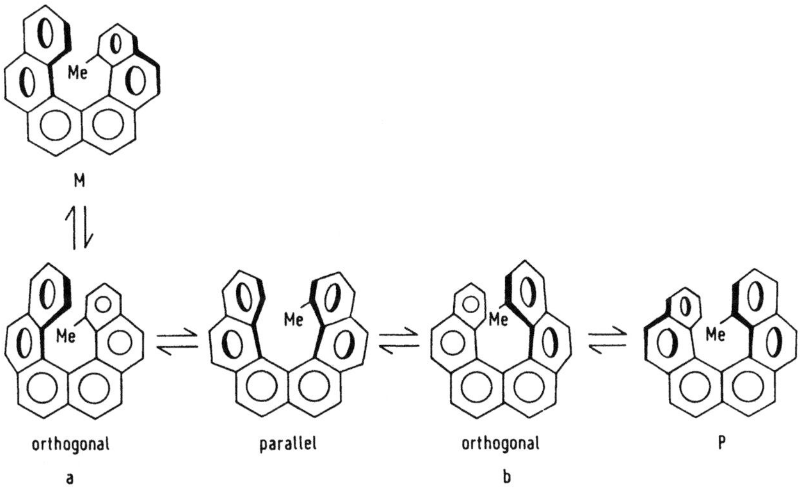

Scheme 21

is passed twice during racemization [105] (*a* and *b* in Scheme 21) and in the 1-methyl substituted helicenes one of them (*a*) is more strained than in the corresponding unsubstituted helicene.

According to the Scheme and in accordance with the experimental data substituents at C(3) and C(14) will have no effect on the $\Delta G^{\neq}$-value, whereas methyl groups at C(2) and C(15) cause a small but significant increase in $\Delta G^{\neq}$ (9.6 kJ · mol^{-1} at 240°).

These results can be compared with those of quasi-helical molecules like 1-substituted benzo[c]phenanthrenes (*81* and *82*) and 4,5-diphenyltriphenylenes like *83* and *84* (Scheme 22). Replacing the methyl group in *81* by a phenyl group lowers the activation force-energy of racemization by about 25 kJ · mol^{-1}, whereas the phenyl group is certainly not smaller than the methyl group [45]. The transition state, having orthogonal terminal rings which is required for racemisation can relatively easily be

Table 11. Thermodynamic Parameters for the Racemization of Helicenes

Helicene	ΔG (kJ · mol^{-1})	(T °C)	ΔH (kJ · mol^{-1})	ΔS (J · mol^{-1} × K^{-1})	$t_{1/2}$ (min)	(T °C)	Ref.
[5]	103.9	(196°)	95.3	−17.1	62.7	(57°)	106)
[6]	154.3	(196°)	146.2	−17.6	13.4	(221.7)	107)
[7]	178.1	(269°)	169.4	−16.9	13.4	(295)	103)
[8]	181.9	(270°)	171.5	−19.2	3.1	(293.2)	103)
[9]	188.3	(270°)	174.5	−29.0	12.3	(293.5)	103)
3,14-diMe [6]	153.9	(196°)			87	(196)	105)
2,15-diMe [6]	165.2	(240°)			222	(240)	105)
1-Me [6]	183.2	(269°)	161 ± 2	−41 ± 2	231	(269)	105)
1,16-diMe [6]	186.5	(270°)	158 ± 2	−54 ± 2	444	(270)	105)
double helicene (*55*)	136.1	(210°)					40)
[6] BSBSBS			92.0		13	(25)	128)
[6] BSBSBB			106.3		241	(25)	128)

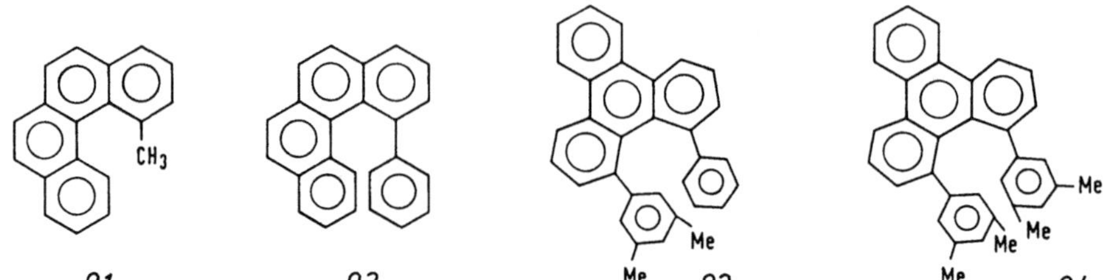

81	82	83	84
$\Delta G^{\ddagger}rac = 92\ KJ \cdot mol^{-1}$	$\Delta G^{\ddagger}rot = 56\ KJ \cdot mol^{-1}$	$\Delta G^{\ddagger}rot = 60\ KJ \cdot mol^{-1}$	$\Delta G^{\ddagger}rot = 62\ KJ \cdot mol^{-1}$
	$\Delta G^{\ddagger}rac = 67\ KJ \cdot mol^{-1}$	$\Delta G^{\ddagger}rac = 64\ KJ \cdot mol^{-1}$	$\Delta G^{\ddagger}rac = > 88\ KJ \cdot mol^{-1}$

Scheme 22

reached as a consequence of the relatively easy rotation of the phenyl group which does not induce distortion in the benzo[c]phenanthrene moiety.

In the quasi-heptahelicene 83 the activation energy of rotation of the phenyl groups ($\Delta G^{\ddagger}_{rot}$) is about equal to $\Delta G^{\ddagger}_{rac}$ [46]. Introduction of two methyl groups at the second phenyl ring (84) does not change the activation energy of rotation but $\Delta G^{\ddagger}_{rac}$ is strongly increased (not measurable with the NMR-method). The large influence of the additional methyl groups can only be explained by assuming that also in this compound the two phenyl rings are in an orthogonal position in the transition state of the racemization.

Another demonstration of the remarkable flexibility of helicenes is found with the double helicene 55. Dissolved in naphthalene it racemizes at 210 °C (more than 230 °C below its melting point) with about the same rate as hexahelicene at this temperature. From this observation it must be concluded that the racemization does not occur via the meso-form which has the terminal rings at one side of the central ring, but, due to the large number of bonds, via a high-vibrational state in which the terminal rings are at a non-hindering distance. Not only the racemizations mentioned in Table 11, but also the thermal racemization of [10] and [11] helicenes has been carried out successfully.

4 Spectral Properties

4.1 NMR-Spectroscopy

^{1}H NMR-spectroscopy has appeared a powerful method for the recognition of helicenes and for an analysis of their conformations in solution. This is especially due to the appearance of quasi first order spectra for the protons in the terminal rings, even at 90 MHz.

In Fig. 4 the NMR-spectra of hexahelicene at 90 and 500 MHz are shown. In the latter spectrum the signals of the protons H_5 and H_6, and of H_7 and H_8 appear as AB-patterns. The four proton pattern of the terminal rings can easily be assigned by the rules of Martin and coll [107], implying that $^3J_{H_1,H_2} > {}^3J_{H_3,H_4}$ and $^4J_{H_1,H^3} < 4J_{H_2,H_4}$. In general, the coupling constants of protons in helicenes do not deviate from those in planar aromatic compounds [107, 108, 109, 110]. In heptahelicene [110] and several benzohelicenes [109] long-range couplings are present, viz. epicoupling between

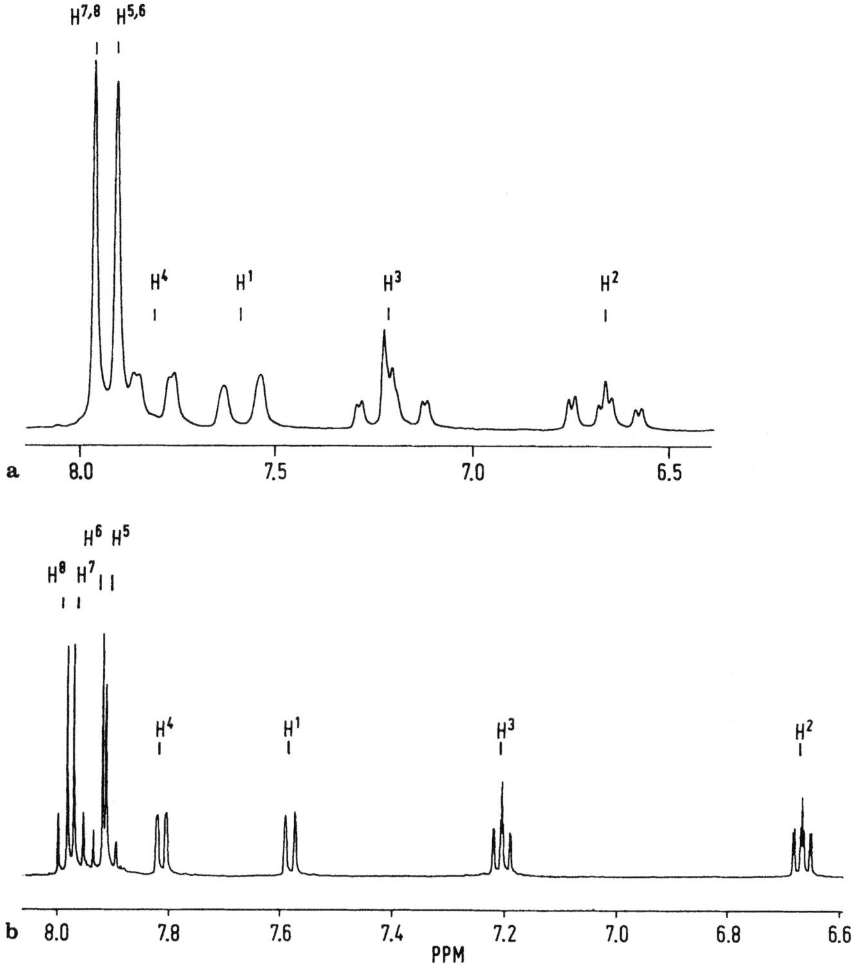

Fig. 4. NMR spectra of hexahelicene a. at 90 MHz, b. at 500 MHz

H(1) and H(6) and pericouplings between H(4) and H(5), and between H(6) and H(7).

There are marked solvent effects as illustrated in Table 12. Especially the ASIS (aromatic solvent induced shift, $\delta_{CDCl_3} - \delta_{C_6D_6}$) for H(1) is large. Besides, concen-

Table 12. Solvent Effects on the Chemical Shifts of Hexahelicene

	$CDCl_3$	C_6D_6	CS_2	CCl_4	$AsCl_3$
H(1)	7.58	7.83	7.47	7.52	7.57
H(2)	6.65	6.52	6.53	6.71	6.70
H(3)	7.18	7.00	7.08	7.13	7.26
H(4)	7.78	7.63	7.67	7.72	7.86
H(5, 6)	7.87	7.69	7.77		7.96
H(7, 8)	7.92	7.72	7.82		8.02

Table 13. Chemical Shifts (in PPM) of some Modified and Substituted Hexahelicenes

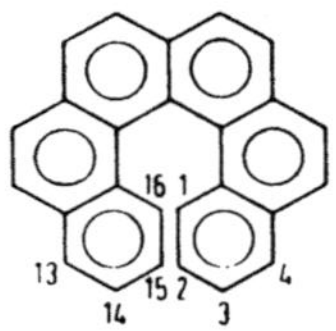

[6]

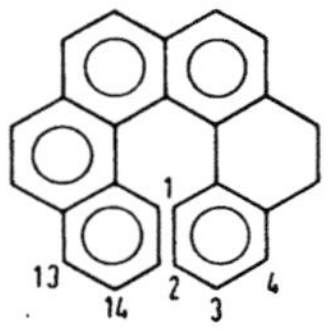

57

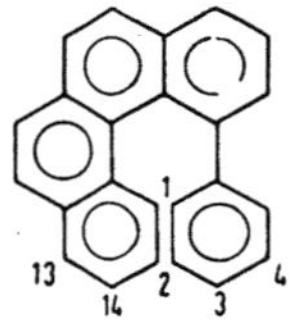

64

61

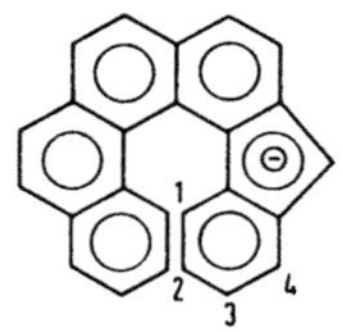

62

1-Me [6] 1,3-dit. Bu [6]

	[6]	57	64	61	62	1-Me [6]	1,3-di-t.Bu [6]
H(1)	7.58	6.26	6.04[a]	6.38	6.73	—	—
H(2)	6.65	6.31	6.44[a]	6.77	6.19	7.06 (6.61)[b]	6.81
H(3)	7.18	6.80	6.94[a]	7.12	6.73	7.06 (7.17)[b]	—
H(4)	7.78	7.18	7.46[a]	7.55	7.47	7.58 (7.72)[b]	7.47
H(13)	7.78	7.16	7.65	8.02	7.73	7.63	7.50
H(14)	7.18	7.16	7.19	7.50	7.17	6.97	6.90
H(15)	6.65	6.85	6.88	7.17	6.80	6.32	6.22
H(16)	7.58	7.94	7.88	8.08	8.80	6.71	6.43
Ref.	22)	41)	45)	113)	43)	22)	22)

[a] at −40 °C, at higher temperature the signals coalesce; [b] the values between parentheses are the δ-values corrected for the influence of the substituent

tration effects are observed, especially leading to upfield shifts of the non-overcrowded protons at higher concentrations; this effect is negligible in benzene [107,109,110].

The theoretical work of Haigh and Mallion [108,111] has given a sound base for the interpretation of the spectra. The authors argued that the chemical shift of a strongly overcrowded proton like H(1) of hexahelicene depends on three different effects:

i) deshielding due to the ring currents of the benzene ring to which the proton belongs and the neighbouring rings
ii) shielding due to the ring current of overlapping rings
iii) deshielding due to Van der Waals interaction with nearby H- and especially C-atoms.

The δ-value of H(1) in hexahelicene is 7.58 ppm in $CDCl_3$. Depending on the contribution of each of these factors the chemical shift of H(1) and H(16) of a hexahelicene derivative or analogous molecule can vary, however, between 6.0 and 8.0 ppm. Until now the largest difference between H(1) and H(16) ($\delta\Delta = 1.7$ ppm) in one molecule is found in 5,6-dihydro- and 6a,16d-dihydrohexahelicene. By modifying the helicenes in a number of different ways (a: introducing a non-aromatic ring; b: introducing a substituent at C(1); c. bridging the terminal rings; d: changing the bond length of a peripheral bond by annelation of a benzene ring) it has appeared to be possible to observe the influence of the three effects almost separately.

a) The protons of ring A of 5,6-dihydrohexahelicene 57 absorb at higher field than those of hexahelicene (Table 13) by the absence of a ring current in ring B. The opposite shift of H(16) has to be ascribed to the same cause. Analysis of the ABCD-pattern of the C(5)–C(6) protons of 57 and comparison with the corresponding δ-values of 9,10-dihydrophenanthrene revealed that 57 occurs in only one conformation, very similar to that of hexahelicene [41,112]. The analogy of the NMR-spectra of 57 and of 1-phenylbenzo[c]phenanthrene (*64*) which can be considered as hexahelicene lacking ring B is striking.

When ring B in 57 is exchanged for a cyclopentadiene ring as in *61* the distance between the terminal rings will increase, what will result in a decrease of the Van der Waals interaction and less shielding. The experimental data follow the expected trend, even after correction for the influence of the five-membered ring structure [113].

By aromatization of the five-membered ring with Li the helicene 62 is obtained. In this compound the ring current in ring B is restored, but the distance between the terminal rings is enlarged compared to hexahelicene. By an accurate analysis of the shifts and comparison with the δ-values of the fluorenyl-anion the shifts of all protons except that of H(16) could be ascribed to the effects of i, ii and iii. The large downfield value of H(16) must be ascribed to the neighbourhood of the Li^+-ion in the contact ionpair.

b) Introduction of a substituent at C(1) of hexahelicene will result in an enlargement of the pitch of the helix and in a reduction of the deshielding Van der Waals interaction. In Table 13 the effect of substituents at C(1) on the δ-value of H(16) is given. For 1,3-di-tert-butylhexahelicene an upfield shift of 1.15 ppm is observed. The influence of the position of alkyl substituents can be deduced from Table 14 where the chemical shifts of the substituent protons are given. As expected the δ-value of a 1-methyl substituent is always at higher field than that of the methyl groups at other positions, but its exact value varies (0.54–0.82) in dependence on the presence of substituents at other hindering positions. From the spectrum of 1,16-dimethylhexahelicene it was concluded that the overlap of the terminal rings is increased compared to the parent compound and that the angle between the terminal rings is smaller. This conclusion was fully consolidated by the X-ray analysis [143].

Table 14. Chemical Shifts in PPM of Methyl Groups at Different Positions of Hexa- and Heptahelicene

	1-(16)-Me	2-(15)-Me	3-(14)-Me	4-(13)-Me
1-Me [6]	0.88[a]; 0.80[b]			
2-Me [6]		1.71[a]; 1.70[b]		
3-Me [6]			2.31[a]; 2.27[b]	
4-Me [6]				2.70[b]
1,3-diMe [6]	0.77[b]		2.25[b]	
1,14-diMe [6]	0.89[a]; 0.81[b]		2.24[a]; 2.18[b]	
1,16-diMe [6]	0.62[a]; 0.54[b]			
1,3,14,16-tetraMe [6]	0.60[a]		2.26[a]	
1,2,3,4-tetraMe [6]	0.76[a]	1.60[a]	2.25[a]	2.72[a]
2,15-diMe [6]		1.76[a]		
3,14-diMe [6]			2.29[a]; 2.26[b]	
4,13-diMe [6]				2.73[a]
1,15-diMe [7]	0.72[a]		2.36[a]	
3,15-diMe [7]		2.13[a]	2.36[a]	
toluene	2.25[a]			

[a] in $CDCl_3$; [b] in CS_2

c) The presence of an ethano bridge from C(3) to C(15) in the terminal rings of heptahelicene does not change the shifts of the protons of the central rings, but those of H(2), H(4), H(16) and H(17) differ considerably when compared with 3,15-dimethylheptahelicene (see Table 15). These effects can only be caused by a change in the helix conformation. The $-CH_2OCH_2$-bridge from C(3) to C(15)

Table 15. Chemical Shifts (in PPM) of Heptahelicene and Some Derivatives

	[7]	3,15-diMe- [7]	3,15-ethano- [7] (37)	3,15-(2′-oxapropano)- [7] (38)	benzo[1]- [7] (47)
H(1)	7.02	6.86	6.91	7.05	6.69
H(2)	6.26	6.06	5.85	6.27	6.26
H(3)	6.75	—	—	—	6.75
H(4)	7.15	6.91	6.61	6.93	7.07
H(5)	7.36	7.25	7.06	7.20	7.23
H(6)	7.60	7.53	7.42	7.51	7.49
H(14)	7.36	7.57	7.43	7.86	7.23
H(15)	7.15	—	—	—	7.07
H(16)	6.75	6.58	6.47	6.58	6.75
H(17)	6.26	6.12	6.03	6.01	6.26
H(18)	7.02	6.89	6.85	6.96	6.69
Ref.	[20]	[20]	[20]	[20]	[34]

in 38 has a clearly less pronounced effect. The decrease of the distance between the terminal rings was affirmed by the X-ray analysis [20].

d) The effect of an additional benzo group is illustrated with the δ-values of benzo[1]-heptahelicene (47) (Table 15). Compared with heptahelicene H(1), H(5) and H(6) are considerably shifted, whereas these protons cannot have a direct influence of the newly attached benzo group [34].

The effect was ascribed to an alteration of the conformation of the helix in such a way that the Van der Waals interaction is diminished, resulting in more shielding of H(5) and H(6). The introduction of the benzo group causes enlargement of the C(9)–C(10) bond and consequently some stretching of the helix, extending the overlap of the terminal rings further to ring B and F. Also in this case the X-ray analysis [114] of tribenzo[f,l,r]heptahelicene has consolidated this conclusion from the NMR-analysis.

The influence of the increase of the overlapping region is more clearly observed in the series of carbohelicenes and thiahelicenes. In the former series (Table 16) the upfield shifts of protons under the influence of increasing overlap is gradually extended to protons in the rings B, C and even D.

Table 16. Chemical Shift (in PPM) of the Carbohelicenes[a]

	[6]	[7]	[8]	[9]	[13]
H(1)	7.58	7.02	7.02	7.08	6.10
H(2)	6.65	6.26	6.38	6.34	5.82
H(3)	7.18	6.75	6.96	6.96	6.55
H(4)	7.78	7.15	7.29	7.31	
H(5)	7.87	7.59	7.15	7.03	
H(7)	7.92	7.78	7.41	7.13	
H(8)	7.92	7.85	7.77	7.25	
H(9)		7.87		7.42	
H(10)			7.17	7.78	
H(11)				7.94	

[a] in $CDCl_3$

In the thiahelicene series (Table 17) the helicity depends on the relative number of thiophene and benzene rings in the helix. A benzene ring contributes 60° to the internal angle of the helix and a thiophene ring 45° [5]; yet thiahelicenes with the same number of thiophene and benzene rings do not show equal chemical shifts as appears from the data for the hexahelicenes in Table 17. In the series of uneven thiahelicenes all having alternating thiophene and benzene rings it is striking, that δH(2) becomes gradually smaller in a similar way as in the carbohelicene series. The upfield shift of H(1) is much larger; the δ-value falls eventually to 4.20 ppm, and such a low value for an aromatic proton has only been found in extremely constraint metacyclophanes [115]. This is visualized graphically in Fig. 5 where the relevant data of the carbo- and thiahelicenes are related to the variation of the internal angle in both series.

Table 17. Chemical Shifts (in PPM) of some Thiaheterohelicenes

	[6] SSBSBS	[6] SBSSBS	[7] SBSBSBS	[9] SBSBSBSBS	[11] SBSBSBSBSBS	[13] SBSBSBSBSBSBS	[15] SBSBSBSBSBSBSBS
H(1)	7.55	7.74	6.75	5.60	5.21	4.71	4.20
H(2)	7.32	7.40	6.91	6.65	6.62	6.45	6.23
H(3)	7.85	7.83	8.02	7.47	7.52	7.66	7.4
H(4)	7.71	7.71	8.02	7.49	7.54	7.67	
H(5)	7.63		8.00	8.18	7.60	7.58	
H(6)	7.73			8.18	7.60	7.58	
H(7)	7.11				8.13	7.59	8.0
H(8)	7.20					7.79	
H(9)							7.00

ᵃ in $CDCl_3$

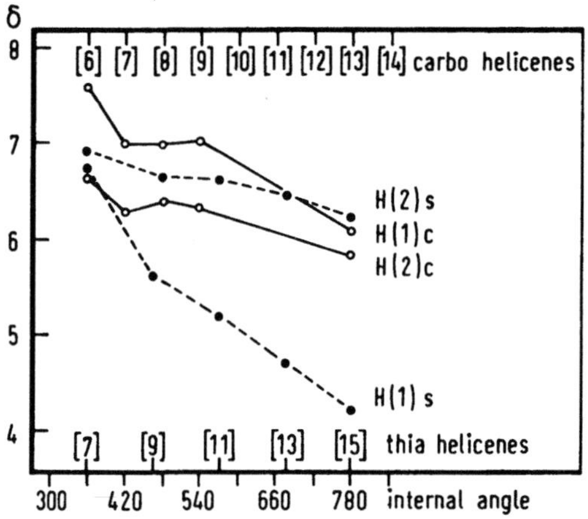

Fig. 5. Chemical shifts (δ) in ppm of the protons H(1) and H(2) of carbo- and alternating thiaheterohelicenes vs the internal angle of the helicenes.
c = carbohelicenes, s = thiahelicenes

Y. Yamada [102] ascribes the large shift to the presence of the bulky sulfur atoms. In these thiahelicenes the sulfur atoms from two parallel thiophene rings lie close together. By repulsive forces they are kept as far away as possible. This brings about a distortion of the molecules, resulting in a shorter distance between H(1) and the opposite ring. By an accumulation of these distortions in the higher members of the series the distance becomes smaller and smaller resulting in an increasing upfield shift. A consequence of this concept is that the protons of the rings B, C, D etc. compared to those in the carbohelicenes remain at relatively low field, as is found indeed.

Large differences are observed between the chemical shifts of some corresponding protons in diastereomers of the double helicene 52 and of hexahelicylhexahelicene 33 (Table 18). These differences are comprehensible because in the racemic double helicene the terminal rings are at the same side of the central naphthalene moiety leading to deshielding of H(1) and H(2) and shielding of H(15). In the mesoform with a terminal ring at each side of the central part of the molecule the interaction between the terminal rings is completely absent. This explanation is corroborated by the observation of a Nuclear Overhauser effect in H(2′) when H(1) is irradiated [36].

Table 18. Chemical Shift (in PPM) of Some Protons of Double Helicenes and Bis-2,2′-Hexahelicyl

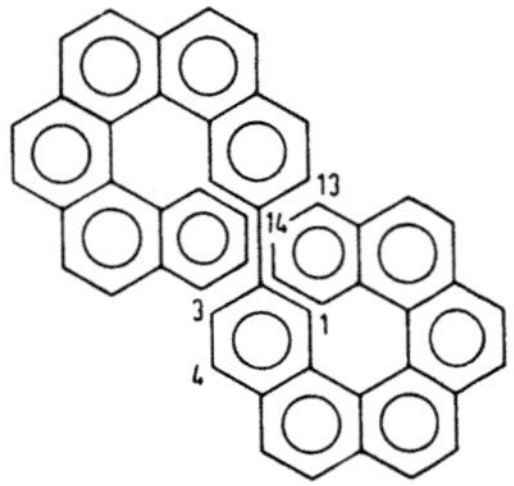

33

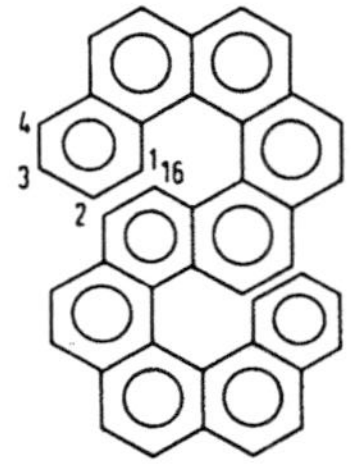

52

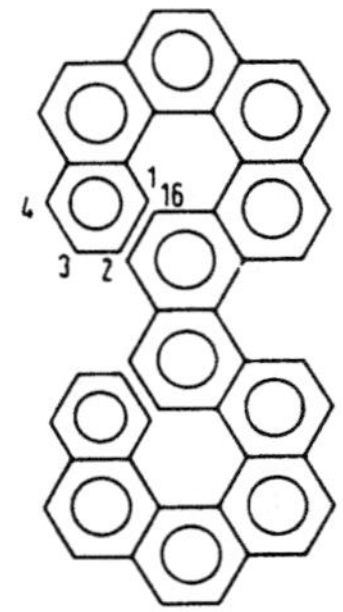

54

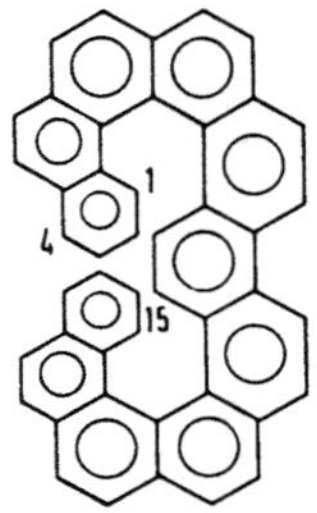

55

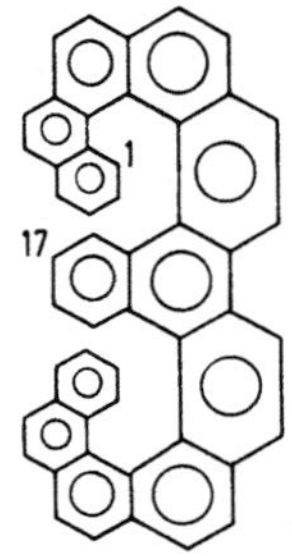

56

	33		52		54	55	56	
	dl	meso	dl	meso				
H(1)	7.40	7.62	8.59	7.02	7.51	7.22		6.84
H(2)	—	—	7.12	6.40	6.24	6.15		5.85
H(3)	5.92	6.20	7.29	7.30	6.82	6.74		6.37
H(4)	7.45	7.42	7.92	7.82	7.68	7.42		6.88
H(13)	7.87	7.26	—	—	—	—	(H(15))	6.38
H(14)	7.22	6.67	—	—	—	—	(H(16))	5.45
H(15)	6.65	6.34	7.10	7.21	6.58	6.70	(H(17))	5.45
H(16)	7.57	7.40	7.75	7.78	7.41	6.70	(H(18))	6.38
Ref.	[25]	[25]	[36,37]	[35,37]	[38]	[39,40]		[39]

From the NMR-data for dl 33 and meso 33 it can be concluded that in the racemic form both hexahelicenyl moieties are almost perpendicular, whereas in the mesoform the central biphenyl moiety is planar [25]. The three double helicenes 54, 55 and 56 could only be isolated in one form. For 55 this was the racemic form having the terminal rings at opposite sides of the central part, as could be deduced from the isolation of optically active samples. From racemization experiments (vide infra) it could be deduced that the mesoform can hardly exist. Comparison of the NMR spectra of 54 and 56 with that of 55 suggests that 54 and 56 are also racemic forms. The chemical shift of H(16) and H(17) in 56 has the lowest value found for protons in carbohelicenes.

Besides ^{13}C-spectra of penta-, hexa- and heptahelicene in which all the protonated carbon atoms could be assigned [116,179] no other reports on ^{13}C-NMR analysis of helicenes have been published.

4.2 Photoelectron Spectra

Photoelectron spectra (PES) of helicenes have been measured to trace whether interactions between the orbitals of overlapping benzene rings occur. Evidence for such interactions has been found for cyclophanes [118,119]; bands from symmetry-equivalent π-orbitals of the benzene rings in these compounds were split due to transannular π–π-overlap.

Obenland and Schmidt [120] measured the PES of benzene, naphthalene, phenanthrene and all other orthoannelated hydrocarbons up to [14]. In Table 19 the first two bands of the vertical π-ionization potential of helicenes are given.

Table 19. Vertical π-π-Ionization Potentials in eV [120]

Helicene	IP_1	IP_2	Helicene	IP_1	IP_2
[5]	7.51	7.7	[10]	6.99	6.99
[6]	7.37	7.5	[11]	6.95	7.1
[7]	7.25	7.4	[12]	6.93	6.93
[8]	7.15	7.15	[13]	6.91	6.91
[9]	7.07	7.07	[14]	6.88	6.88

The effect of progressive annelation in the helicene series is a suppression of the intensities of all π-bonds relative to the σ-bond system. Furthermore, the π-bonds become more symmetrical and the vibrational structure is blurred out. This indicates a substantial geometry change upon ionization.

The observed π-IP's were compared with calculated values, obtained from four π-electron models. On the whole the agreement between calculated and observed π-IP's was good, indicating that transannular effects are not present. The authors ascribe the absence of such an effect in helicenes to the greater flexibility of these molecules, compared with cyclophanes. From and INDO-MO-calculation on hexahelicene [121] it was concluded that the four HOMO's have predominantly π-character, in contrast to the other occupied MO's. A linear combination of the four MO'S correlates well with the PE-spectrum. Apparently transannular interaction in hexahelicene is unfavourable.

The PES have appeared of great value for the interpretation of UV-spectra (see next section).

4.3 UV-Spectra

The longest wavelength absorption band of carbohelicenes shifts with increasing number of benzene rings to higher wavelengths. The structure of the absorption curve becomes more and more diffuse, and differentiation between the α, p and β-bands[7] cannot readily be made for helicenes higher than octahelicene [123]. In Fig. 6 the spectra of [6], [14] and the double helicene 55 are pictured.

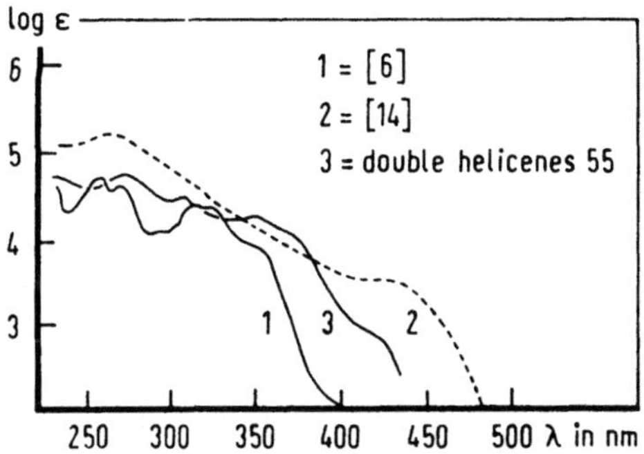

Fig. 6. Ultra violet spectra of [6] and [14] and the double helicene *55*

From the photoelectron spectra of planar aromatics [124] it was found that the position of the α-band relative to the p-band can be predicted from the first two IP's. When the difference $\Delta IP = IP_2 - IP_1 > 0.7$ eV, then the p-band is the first band in the optical spectrum; when $\Delta IP \leqslant 0.5$ eV the α-band is at longer wavelength.

From Table 19 it can be derived that ΔIP is small for all helicenes, so that the α-band is expected to precede the p-band. (This is observed experimentally for [6], [7] and [8].) From the study of planar aromatic compounds [124] it was also found that the p-band correlates linearly with the first IP, whereas for the α- and β-band energies a linear correlation with the mean of the first two IP's was established.

Since both the IP_1- and IP_2-values lie on smoothly descending lines, when plotted against the number of benzene rings, also the energies of α-, p- and β-band (E_α, E_p and E_β) must be smoothly descending. On this basis the identification of the transitions in the spectra of the higher helicenes is straightforward. For heptahelicene the identification in this way is different from the older assignment [97], which could be shown to be inadequate. For octa- and nonahelicene calculated values for E_α, E_p, E_β and E_β/E_α agree well with experimental data.

---

7 Nomenclature according to Clar [122].

Schmidt concluded [120] that the optical spectra of the carbohelicenes do not show abnormalities compared to planar aromatics; all effects due to the helicity of these molecules are incorporated in the ionization potentials.

In Fig. 7 a plot of the wavelengths of α-, p- and β-bands against the number of rings is given for carbo- and alternating thiahelicenes, together with some values for substituted helicenes. It shows the increase of λ_{max} for all the bands.

A substituent at C(1) of hexahelicene [22] and heptahelicene [20] results in a bathochromic shift of the α- and p-band and a loss of fine structure. A bridge over the terminal rings of heptahelicene (*37*) has the same effect. The spectra of the double helicenes are more structured than the carbohelicenes with the same number of benzene rings (see Fig. 6.).

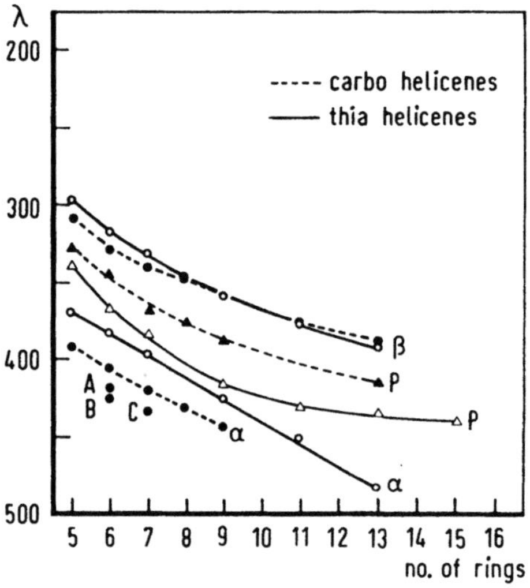

Fig. 7. Wavelengths of the α-, p- and β-bands of the UV-spectra of the carbo (------) and thiaheterohelicenes. The position of the α-bands of 1,16-dimethylhexahelicene, 1,3-di-tert.butylhexahelicene and of 1,15-dimethylheptahelicene are given by A, B and C, respectively

In contrast to the UV-spectra of carbohelicenes, the spectra of thiahelicenes possess well resolved bands. The α-bands have a much higher intensity compared with those of the carbohelicenes whereas the p-bands have lower intensities [96,102]. As a consequence the α- and β-bands can be observed separately, even for the higher members of the series of thiahelicenes. In comparing the carbo- and the alternating thiahelicenes UV-data in Figure 7 reveal that the α-bands of the carbohelicenes absorb at longer wavelengths whereas the β-bands in both seires are at almost the same position. It seems as if the E_α/E_β-ratio in these thiahelicenes increases with increasing number of rings.

In general, the incorporation of a thiophene group in hexahelicene results in a blue shift of all the absorption bands. The shift is small for the β-band and may be 10–30 nm in the α-band depending on the place of the thiophene ring. The largest effect is found when this group is in the middle of the helix, whereas when two thiophene rings as terminal rings ([6]-SBBBBS) are present the spectrum does not possess the typical heterohelicene feature [53].

4.4 Charge-Transfer Complexation

As mentioned before (Sect. 3.1) Newman developed tetranitrofluorenylidene oxaminopropionic acid (TAPA, 79) as a new agent for the resolution of hexahelicene. Its molecular structure, including a large moiety with strong electron acceptor properties, promised good complexing properties towards aromatic compounds like hexahelicene.

The complexation of TAPA with hexahelicene was studied in more detail by Brown et al. [126] with NMR-spectroscopy. By measuring the NMR-spectra at different concentrations of donor vs. acceptor the shift in the position of the proton signals can be used for determination of the association constant. The method is analogous to the Benesi-Hildebrand treatment of CT-complexation, which is based, however, on UV-data.

In Table 20 the limiting shifts are given for the complexation of hexahelicene with (±)TAPA.

Table 20. Limiting Shifts of the Protons of [6] and of 85 with TAPA [126]

	[6] + (±) TAPA (230 °K)	M85 + R-TAPA (248 °K)	P-85 + R-TAPA (248 °K)
H_1	.71	.74	.74
H_2	.37	—	—
H_3	.30	.33	.38
H_4	.38	.25	.25
$H_{5,6}$	.38	.34	.34
H_7	.52	.49	.44
H_8	.82	.79	.61

The protons most strongly affected in hexahelicene are H(1) and H(8) suggesting that these protons are most strongly exposed to the ring current of the acceptor. A more detailed study using 2,15-dimethoxyhexahelicene (85) as the donor and R-TAPA as acceptor demonstrates that two diastereomeric complexes were present both in equilibration with the respective enantiomers of the helicene. The association

Table 21. Thermodynamic Parameters for the Formation of Some Charge Transfer Complexes

		K ($dm^3 \cdot mol^{-1}$)	ΔH ($kJ \cdot mol^{-1}$)	ΔS ($Jmol^{-1} \cdot K^{-1}$)	Ref.
benzene	TCNQ[a]	0.13 (298 °K)	− 2.0 ± 0.8	−23.5 ± 3.2	91)
benzothiophene	TCNQ	0.41 (298 °K)	− 5.7 ± 0.5	−26.7 ± 2.0	91)
[3]SBS	TCNQ	1.73 (298 °K)	−11.2 ± 0.2	−33.1 ± 0.6	91)
[5]SBSBS	TNCQ	5.28 (298 °K)	−11.7 ± 0.5	−25.7 ± 1.6	91)
[7]SBSBSBS	TCNQ	4.85 (298 °K)	−13.0 ± 0.4	−30.5 ± 1.0	91)
(+)-[7]SBSBSBS	R(−)TAPA	5.04 (248 °K)	−14.6	−45.6 ± 0.8	90)
(+)-[7]SBSBSBS	S(+)TAPA	6.78 (248 °K)	−15.6 ± 0.2	−47.0 ± 0.7	90)
(+)-7,10-dicarbomethoxy [6]	R(−)TAPM	6.75 (293.5 °K)	−9.14 ± 0.06	−15.2 ± 0.2	127)
(−)-7,10-dicarbomethoxy [6]	R(−)TAPM	6.74 (293.5 °K)	−9.94 ± 0.7	−18.0 ± 0.3	127)

[a] 7,7,8,8-tetracyano-p.quinodimethane

constants of both complexes are equal: K(M-85) (R-TAPA) = 53 ± 3 and K(P-85) (R-TAPA) = 57 ± 3, the less well complexed (M-85) showing the greater H(8) shift.

The detailed structure of the complex depends on the enantiomer involved. In the (M-85) (R-TAPA) complex the preferred conformation of the side chain engenders greater steric hindrance than in the (P-85) (R-TAPA) complex. (In formula 85 the nitro groups have been omitted from TAPA.)

A similar analysis has been done [91] on CT-complexation of the series benzene, benzothiophene, benzo[1,2-b, 4,3-b′]dithiophene ([3]SBS), and the homologs [5]SBSBS and [7]SBSBSBS with TCNQ as the acceptor. Beside association constants, also ΔH- and ΔS-values were determined by measurements at different temperatures (Table 21). It is striking that the gradual increase of the association constant in the series is interrupted at [5]SBSBS; the next member has even a lower value.

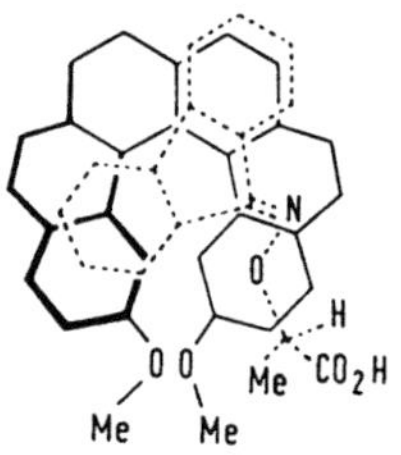

P(85) R-TAPA

85

It is apparent that this is not only due to a relatively small enthalpy change, but mainly to a strongly reduced entropy change. According to the authors this has to be ascribed to increased steric hindrance within the complex as a consequence of the appearance of a staggered conformation (See Sect. 6 on X-ray analysis).

Other such studies have been made on the complexation of (+)-[7]SBSBSBS with R- and S-TAPA [95] and of (+) and (−)-7,10-dicarbomethoxyhexahelicene and R-TAPM [127] (the methyl ester of TAPA).

In the former case the (+)-(+)-complex appeared to be more stable at 248 °K in the solvent used ($C_2H_2Cl_4$). The position of all protons in the NMR-spectrum were at higher field in the complexes; the shifts of TAPA ring protons were larger than that of helicene protons. Shifts of the protons of the propionic acid group were, however, small. Based on a detailed analysis of differences between the spectra of the diastereomeric complexes, the authors proposed the structure, in which the fluorenylidene plane of TAPA is located on the central region of [7] and the propionic side chain is directed toward the staggered terminals with the methine H closer to the helix. In (P[7]) (R-TAPA) the chiral group of TAPA is close to the upward terminal of P[7].

4.5 Photophysical Properties

All carbohelicenes fluoresce, and show a well defined fluorescence band which shifts progressively to longer wavelengths from hexahelicene to tetradecahelicene. In Fig. 8

the fluorescence spectra of some helicenes are pictured. The spectra include two bands of approximately similar intensity in hexahelicene. From hexa- to undecahelicene the bands at smaller wavelength correspond with the Franck-Condon maximum (0–0 transition) showing that there is little displacement of the nuclear configuration in the ground and excited state. A small change in the intensity of the 0- and 1-band indicates, however, a small displacement in the higher helicenes [123,125].

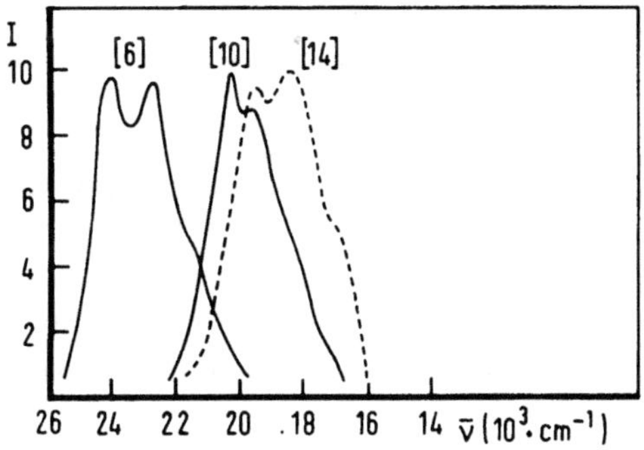

Fig. 8. Fluorescence spectra of [6], [10] and [14] in dioxane

In Table 22 the 0–0-fluorescence transition energy E_s is given together with other parameters of excited helicenes.

The triplet quantum yield φ_{isc} is for the helicenes nearly $(1 - \varphi_{fl})$ so that the rate of radiationless decay approximates k_{isc}. The low value of k_{fl} shows that the fluorescence of all the compounds occurs from the ^1L_b state. The fluorescence rate (k_{fl}) of the helicenes with an odd number of benzene rings decreases more slowly than of the helicenes with an even number of rings. This is explained by the difference in molecular rigidity of the two groups. The 'even' ring helicenes are symmetric about a naph-

Table 22. Excited State Parameters of the Helicenes[a]. Quantum yield of fluorescence (φ_{fl}), fluorescence lifetime (τ_{fl}), rate of fluorescence (k_{fl}), rate of intersystem crossing (k_{isc}), singlet state energy (E_S), triplet state energy (E_T) and lifetime of phosphorescence (τ_p)

Helicene	φ_{fl}	τ_{fl}(ns)	k_{fl} (10^6 s^{-1})[b]	k_{isc} (10^7 s^{-1})[c]	E_S (cm^{-1})	E_T (cm^{-1})	τ_p
[3]	0.12	6.3	1.90	1.40	28830	21730	3.5
[4]	0.18	35.5	5.07	2.31	26200	20000	2.5
[5]	0.04	25.5	1.57	3.76	24900	19700	2.2
[6]	0.041	14.5	2.83	6.61	24100	18800	1.8
[7]	0.021	13.8	1.52	7.09	22500	17100	0.67
[8]	0.014	10.0	1.40	9.86	22050	17200	0.51
[9]	0.014	9.6	1.46	10.27	21400	16900	0.18
[10]	0.0088	8.23	1.07	12.04	20650		
[11]	0.0088	7.24	1.22	13.69	20420		
[12]	0.0061	8.03	0.76	12.38	19600		
[13]	0.0097	8.55	1.13	11.58	19450		
[14]	0.065	8.71	0.75	11.41	19600		

[a] From Ref. [123], [125] and [130]; fluorescence values measured in dioxane at room temperature;
[b] $k_{fl} = \varphi_{fl}/\tau_{fl}$; [c] $k_{isc} = (1 - \varphi_{fl})/\tau_{fl}$

thalene nucleus and more rigid than the odd ring helicenes, which are symmetric about a benzene nucleus. That the rigidity is important follows also from the k_{fl}-value ($= 1.86 \times 10^6\ s^{-1}$) of the bridged helicene 3,15-ethanoheptahelicene (37), which fits better in the even helicenes curve (k_{fl} versus number of rings).

The difference between E_S and E_T is about constant for the helicenes with five to nine rings (Table 22), while the k_{isc} increase by a factor of about 3 [125]. This may suggest the participation of a quasi degenerate triplet state in the intersystem crossing. This most likely will be the ^3L_b state since a direct spin-orbit coupling between ^1L_b and ^3L_b is allowed in the helicenes because of their C_2-symmetry. It is proposed [125] that the increase in k_{isc} reflects the departure from coplanarity when the amount of benzene rings increases.

The maximum of the k_{isc}-values at undecahelicene may correspond to a maximum departure from coplanarity or to an optimization of the spin-orbit coupling of ^1L_b and ^3L_b due to degeneracy [125]. Later on it was shown with the aid of ESR-spectra in single crystals, in glass and in streched films that there is a transient magnetization in triplet hexahelicene [131], which demonstrates that the predominant path of intersystem-crossing leads to the population of the $|T_y\rangle$ state but that a small rate of population of the $|T_z\rangle$ state cannot be excluded.

4.6 ESR-Spectra

Because of the inherent non-planar structure of helicenes it seemed of interest to examine the spin distribution in helicene radical anions. For the mono anion of hexahelicene a set of 8 hyperfine splitting constants (hfsc's) and $3^8 = 6561$ ESR lines can be expected. Such a spectrum will be poorly resolved. Indeed, it was not possible to determine hfsc's from the ESR-spectrum of hexahelicene [132]. Using the ENDOR technique which reduces the amount of lines the eight hfsc's could be deduced, however, and the relative signs could be determined [133] by the triple resonance technique.

In Table 23 these experimental values for hexahelicene are compared with values, calculated by the HMO-McLachlan method. The agreement is unsufficient to assign couplings unambiguously to molecular positions.

Prolonged reduction of the mono anion resulted finally in a new paramagnetic species, having a different ESR- and ENDOR-spectrum. It was supposed to be a trianion [132], formation of a paramagnetic decomposition product could be excluded [133]. The ESR spectrum of hexahelicenes has also been studied on a single crystal, a glass, and a streched film [131].

The number of hfsc's of heterohelicenes is similar than that of carbohelicenes of comparible size, and for some alternating thiahelicenes assignment of hfsc's to distinct protons was possible by making use of several partly deuterated compounds [134]. The data in the Table concern a planar compound ([3]SBS), a compound having 'staggered' terminals ([5]SBSBS) and a compound having overlapping terminals ([7]SBSBSBS). The experimental and calculated spin distribution show the same trend for all three compounds and for two of them ([3]- and [5]-thiahelicene) experimental and calculated values agree rather well. This suggests that the spin distribution in these radical anions is not influenced by σ-π-mixing. The authors studied whether

Table 23. Proton Hyperfine Spin Couplings (a) of some Helicenes

[6]		[3]SBS[c]			[5]SBSBS[c]			[7]SBSBSBS[c]		
$a_{(exp)}$[a]	$a_{(calc)}$[b]	protons	$a_{(exp)}$	$a_{(calc)}$	protons	$a_{(exp)}$	$a_{(calc)}$	protons	$a_{(exp)}$	$a_{(calc)}$
−7.42	−7.84	1.6	0.028	0.045	1.8	0.051	0.038	1.10	0.039	−0.012
−5.67	−6.46	2.5	0.507	−0.520	2.7	0.293	−0.305	2.9	0.180	0.077
−3.35	−5.41	3.4	0.417	−0.420	3.6	0.473	−0.400	3.8	0.327	0.117
−2.28	−4.10				4.5	0.107	−1.25	4.7	0.02	0.019
−2.17	−3.05							5.6	0.287	0.094
−1.66	−0.86									
−0.59	−0.08									
+1.22	+1.10									

[a] Ref. [133]; [b] Ref. [135]; [c] Ref. [134]

better agreement between experimental and calculated data would be obtained for [7]SBSBSBS when transannular interaction was accepted. By introduction of a variable resonance integral (β_t) between C(1) and C(10) and between C(2) and C(9) in the calculation, the best fitting was obtained for $\beta_t = 0.2$; the improvement is only slight, but a weak transannular interaction could not be excluded.

5 Polarography

Plotting of the half wave reduction potentials ($\varepsilon_{1/2}$) of hexa- (—1.49 V), hepta- (—1.44), octa- (—1.43) and nonahelicene (—1.40), and benzo[i] (—1.44), benzo[l] (—1.43), benzo[c] (—1.48) and benzo[b]hexahelicene (—1.17) against the HMO electron affinities ($-m_{m+1}$) gives a linear relation ($\varepsilon_{1/2} = 2.48m_{m+1} - 0.16$) [136]. The slope of the line is equal to that given by Streitwieser [137] for a similar relation for planar aromatic molecules, but the intercept is different probably due to differences in entropy and solvation energy of molecules and anions which are incomparable for helicenes and planar aromatics.

The polarographic oxidation potentials of some methyl-substituted hexahelicenes were determined to investigate whether the more facile oxidation of 1,16-dimethylhexahelicene is due to the change in conformation compared to hexahelicene or to the electronic effect of the methyl groups [138]. It was found that both 3,14-dimethyl- and 1,14-dimethylhexahelicene have the same oxidation potentials ($\varepsilon_{1/2}$ (1st wave) = 1.23 V; $\varepsilon_{1/2}$ (2nd wave) = 1.45 V), whereas their conformations are different [24].

The decrease in oxidation potential of 1,3,14,16-tetramethylhexahelicene (1.13 and 1.32 V) compared to hexahelicene ($\varepsilon_{1/2} = 1.29$, no 2nd wave below 2 V) is small in comparison with that of naphthalene compared to its methyl derivatives. Therefore, it was concluded that the increased oxidizability of the methyl-substituted compounds is mainly due to the electronic effect of the methyl groups and not to conformational differences.

6 Crystallography

In the period under review the science and art of crystallography has made an extremely rapid progress in instrumentation and available software and as a consequence in insight and rate of analysis. This is well illustrated with examples from the X-ray analysis of helicenes.

When hexahelicene became known in 1955 its structure determination by X-ray was impossible because of the absence of a heavy atom, necessary to get good reflections. The first X-ray structure determination of hexahelicene was reported in 1969 for a complex of the compound with 4-bromo-2,5,7-trinitrofluorenone [139]. In the same year the X-ray structure of an uncomplexed helicene bearing sulfur atoms was published [140]. The structure of 2-bromohexahelicene was reported in 1971 [99] and two years later those of hexahelicene itself [141], 2-methylhexahelicene [142] and 1,16-dimethylhexahelicene [143].

In the next ten years eleven other helicene structures were resolved, most of them within a very short time.

Notwithstanding the sometimes relatively large differences between two individual helical structures it is possible to find some common features in the various X-ray analyses. To illustrate the differences in question two different projections of hexahelicene and 1,16-dimethylhexyhelicene are given in Fig. 9. In 9a the molecules have been projected on a plane perpendicular to the axis through C(21)–C(22) (see formula 86 for the numbering). In Fig. 9b the molecules have been projected on the least square plane through a terminal ring.

The apparent differences do not exclude the presence of a common feature which is generally found in all other helicenes studied, viz. the presence of a plane containing the atoms C(7), C(8), C(21), C(22), C(9) and C(10) (see Fig. 9a). It appears that other sets of atoms containing two C-atoms belonging to two rings and one or two adjacent pairs of outside C-atoms (e.g. C(19), C(20), C(9), C(10), C(11), C(12) also lie in a

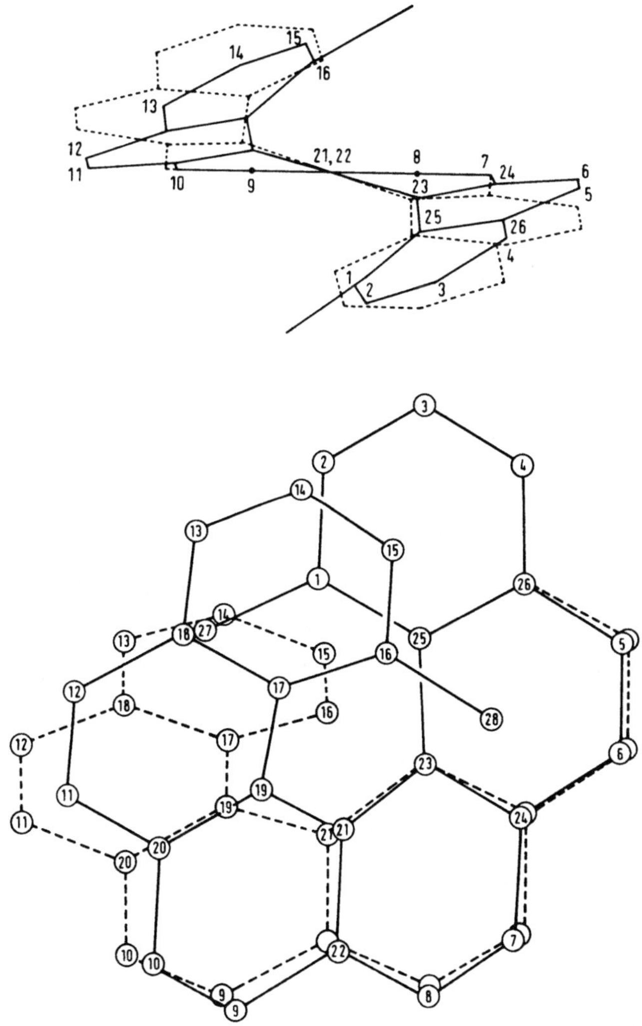

Fig. 9. Projection of hexahelicene (dotted lines) and 1,16-dimethylhexahelicene.
a. on a plane perpendicular to the axis through C(21)–C(22),
b. on the least square plane through a terminal ring

plane. Only at the end of the helix the planarity of such sets (C(17), C(18), C(11), C(12), C(13), C(14)) becomes slightly desturbed.

From Fig. 9b it appears that the terminal rings of both molecules deviate from planarity to the same extent. This also applies to other helicenes and even to other corresponding rings in different helicenes. Furthermore, the average bond length of corresponding bonds are equal in all helicenes. Compared to the bond length in benzene (0.139 nm) the C–C distances of the inner bonds in helicenes (C(1)–C(25), C(25)–C(23), C(23)–C(21) etc.) are lengthened to 0.143 $\pm$ 0.001 nm, the outer bond (C(5)–C(6), C(7)–C(8)), etc.) are shortened to 0.135 $\pm$ 0.001 nm, and the bonds belonging to two rings (radial bonds) are 0.142 $\pm$ 0.001 nm.

Differences between the geometrical structures of helicenes concern mainly the angles between the least square planes of subsequent benzene rings and the torsion angles of the bonds of the inner helix. In Table 24 some X-ray data are given. For heptahelicene two different crystals have been obtained. One of the crystals gave two

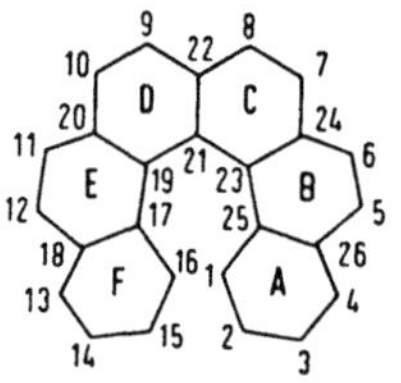

86

different sets of data, both of which are mentioned. From the Table it can be deduced that the angle between the mean planes of the terminal rings increases from benzo[c]-phenanthrene ([4], 27.4°) to hexahelicene ([6], 58°) but then decreases on further increase of the number of benzene rings. A decrease of this angle is also observed on substitution at carbon atoms in the overlapping region of the helix. In general, the average angle between two adjacent (internal) benzene planes is about 12.5°.

The torsion angles over the inner helix bonds are a good measure of the helicity of the molecules. In the Table the torsion angles over the penultimate bonds (dihedral angles C(1)–C(25)–C(23)–C(21) and C(21)–C(19)–C(17)–C(16)) are given. It appears that in most helicenes two different values are given. It appears that in most helicenes two different values are found pointing to a lack of C_2-symmetry (see also Fig. 9a). Differences between the two halves of the [11] double helicene (55) are also reflected in significant differences between the bond lengths of chemically equivalent bonds, which can amount to 0.0045 nm. These differences are not well understood, because the differences in the environment at both termini are not accompanied with unusually close intermolecular contacts [144].

From the regularities mentioned here de Rango et al. [145] designed two general models for helicenes, viz. the triple helical model and the stair case model. In the triple helical model all C-atoms are located on one of three helices: an inner helix with (n+1) C-atoms (n = number of benzene rings), a medium helix with (n+1) C-atoms and an outer helix with 2n C-atoms. In general, atoms of helicenes coincide very well with the 'best' helices, obtained by computation from crystallographic data.

Table 24. Some X-Ray Data of Helicenes

Helicene	Space group	Angle between terminal rings	Torsion angle over penultimate bond of inner helix	Shortest non-bonding C—C distance (in nm)	Special remarks	Ref.
[6]	$P_{2,2,2}$	58.5	11.2; 15.2	0.303	H(1) ... C(19) = 0.248 nm	141)
2-Me-[6]	$P_{2,2,2}$	54.5	12.0; 16.0		H(1) ... C(19) = 0.250 nm	142)
2-Br-[6]	$P_{2,2,2}$				Br at non-hindering position	99)
1-Me-[6]	$P_{2.1/c}$	42.4	24.1; 14.6	0.314		147)
1-CHO [6]	$P_{2.1/b}$	44.5		0.303	H of CHO directed to center of helix	148)
1-$COCH_3$ [6]	P_{bca}	42.5		0.309	O of $COCH_3$ directed to center of helix	148)
1,16-diMe [6]	$C_{2/c}$	29.6	31.4; 31.4	0.304	H(CH) ... C(19) = 0.266	143)
Ph = [6][a]	$P_{2.1/n}$		30.0; 31.0	0.303		176)
[6]NBBBBB		47.8				175)
[6]BSBSBB			11.8; 6.5	0.291		140)
DHI-[7][b]	P_{bcn}	69.1	4.8; 4.8		H(1) ... C(19) = 0.241 nm	149)
[7] A	P_{21}	30.7	16.8; 21.9	0.290	H(1) ... C(23) = 0.244 nm	150)
[7] B	$P_{21/c}$	33.8	20.7; 14.2		H(1) ... C(23) = 0.251 nm	151)
		32.3	20.8; 18.0		H(1) ... C(23) = 0.236 nm	
1,15-diMe-[7]	P_1	36.8		0.312		20)
3,15-CH_2OCH_2-[7]	P_{21}	10.4		0.295	C—O—C angle = 117°	20)
3,15-CH_2—CH_2-[7]	$P_{2na/c}$	15.2		0.295	-H_2C—CH_2- distance = 0.174 nm	20)
tribenzo[f,l,r] [7]	P_{bca}			0.293	H(1) ... C(23) = 0.240, bonds f, l and r not shortened	114)
[7]SBSBSBS-TCNQ	$C_{2/c}$	29.6	17.9; 17.9		Helicene and TCNQ stacked alternatively	154)
1-F-[7]	P_1	30.1 and 24.1				175)
[10]	P_{21}	A/H = 5.0; B/I = 4.7 C/J = 8.7	21.3; 17.8			152)
[11]	P_{21}	A/H = 4°; B/I = 1.1 C/J = 2°; D/K = 4.8	13.9;			153)
double [11][c]	$P_{21/c}$			0.298	two halves very similar to [6]	144)

[a] Ph = [6]: Biphenyleno[2.3-a]pentahelicene; [b] DHI-[7]: Diindeno[5.4-c; 4′,5′-g]phenanthrene (115); [c] double [11]: Diphenanthrenopicene (55).

Table 25. Helix Parameters (in nm) According to the Triple Helical Model

	[6]	1,16-diMe [6]	[7]	3,15-diMe [7]	3,15-CH_2OCH_2- [7]	3,15-CH_2CH_2- [7]	[10]	[11]
inner helix								
radius	0.130	0.132	0.133	0.133	0.136	0.136	0.133	0.135
pitch	0.326	0.324	0.317	0.311	0.306	0.306	0.322	0.322
median helix								
radius	0.263	0.269	0.271	0.270	0.277	0.277	0.273	0.275
pitch	0.459	0.394	0.388	0.394	0.339	0.327	0.367	0.364
outer helix								
radius	0.350	0.360	0.359	0.360	0.369	0.369	0.362	0.365
pitch	0.568	0.402	0.441	0.400	0.343	0.316	0.394	0.394

Table 25 gives values of the radius vector and the pitch of these three helices for several helicenes. It appears that the radius as well as the pitch of the inner helix are approximately constant. The radii of the median and outer helices increase with increasing number of benzene rings (n). The pitches of the median and outer helix decrease with increasing n and also by strain due to substituents or the presence of a bridge between the terminal rings. With the assumption that all C-atoms of a helicene are exactly located on three helices, characterized by a radius vector and a pitch value, only one additional parameter, the azimuthal angle φ is needed to define the atomic coordinates. As it was pointed out that for a given helicene the pitch of each helix is linearly related to the corresponding radius [20] a further reduction of the number of independent variables is possible. However, this triple helices model lacks a physical basis, as it does not take into account the occurrence of benzene rings or other planar groups.

The staircase model is based on the existence of planes through 6 (or 4) atoms, what points to a well regulated geometry. Accepting a regular arrangement of these planes a small number of independent parameters, related to the mutual positions of the planes should be sufficient for a full description of a helicene structure.

This regular staircase model corresponds to a four helices model in which all pitches are equal. It appears, however, that the radii, the pitch and the tilt angle are not sufficient to reproduce accurately all atomic positions.

In a refined staircase model the exact orientation of the planar groups is described using three angles. The authors [145] conclude that the triple helix model can be used to account for the optical rotation of helicenes, but that more sophisticated quantum mechanical calculations can better be based on the refined staircase model.

From the list of space groups in Table 24 it follows that hexahelicene crystals are apparently chiral. This is also true for crystals of hexahelicene grown from racemic solutions. However, dissolved single crystals of [6] display optical rotations which show only about 2% enantiomeric excess instead of 100%. Solid solutions are unlikely in this case (m.p. racemate 231–233°, optically pure [6] 265–267 °C) and ordinary twinning can also be rejected.

Green and Knossow [155] were able to show that the phenomenon is caused by 'lamellar twinning'. When crystals were allowed to dissolve slightly in hexane or carbontetrachloride a lamellar pattern was seen under the microscope. When layers of 10–30 nm were cleaved from the crystal they appeared to be 100% optical pure. The lamellar twinning of hexahelicene is not observed with X-ray because of the absence of observable anomalous dispersion with CuKα-radiation; the X-ray patterns of enantiomorphic crystals are identical. From the molecular packing of hexahelicene it can be deduced that planes perpendicular to the α-axis of the crystal can be drawn which divide the crystal but not induce disorder. Therefore, the [1.0.0] plane allows the coexistence of P- and M-molecules in a single crystal. No additional Van der Waals contacts shorter than normal are necessary. Calculations by Ramdas [156] demonstrate that the interfacial energy associated with twinning across the [1.0.0] plane is much lower than the energy for twinning across the planes [0.1.0], [0.0.1], [1.1.0], [1.0.1] or [0.1.1], in full agreement with the experiment.

7 Force-Field Calculations

Before any X-ray analysis of hexahelicene was available spatial models based on calculations were presented by Herraez [157] and by Kitaigorodsky [158]. An NMR analysis of alkylhexahelicenes [22] revealed that the former model was inferior to the second.

Lindner [104] used a π-SCF-force field method to calculate the energies of racemization of penta-, hexa- and heptahelicene. Two different pathways for the thermal racemization were compared. In one of them the terminal rings are brought in an almost parallel position to reach the intermediate state, as suggested by Martin (see Scheme 21). In the other way the terminal rings are in one plane whereas the remainder of the molecule is strongly bended.

The deformations of the benzene rings involved in the transition state of the former route are about the same for hexa- and heptahelicene. The bending angles of the C-atoms range from 0–10°. The angles between the benzene rings are 25, 22, 13, 22 and 25° for hexahelicene and 5, 37, 12, 12, 37 and 5° for heptahelicene. The transition state of the latter model has larger deformations and is less favorable.

In Table 26 the energies for the ground and transition state of hexa- and heptahelicene are given, together with the calculated and experimental values of the activation energy of racemization. The agreement between calculated and experimental data is satisfying for pathway 1.

Table 26. Calculated Energies of Hexa- and Heptahelicene [104]

	Atomization energy helicene in eV	Atomization energy of transition state in eV	Racemization energy (eV)	
			calculated	experimental[a]
[6]	223.91	222.13 (via pathway 1)	1.78	1.56
		220.87 (via pathway 2)	3.04	
[7]	257.07	255.09	1.98	1.78

[a] Ref. [103]

To investigate whether computational methods could be used for an accurate description of the conformation of helicenes a π-SCF-force field method was applied to hexahelicene, its 1-methyl- and 1,16-dimethyl derivatives and heptahelicene [112]. To avoid predestinate outcome of the calculations minimization of the energy was started from a nearly planar model, and the minima obtained were compared with those of a procedure starting with the geometry found by X-ray analyses.

The calculations starting with both models led to the same minima for hexahelicene and its 1-methyl derivative. For 1,16-dimethylhexahelicene the non-bonding distance between the CH_3 carbon at C(1) and C(16) had to be shortened to get identical minima (see Fig. 10). For heptahelicene small differences were found between the two models, but they were in the same order of magnitude as the accuracy which can be obtained by the program used.

The calculated structures were, as expected, more symmetrical than those obtained from X-ray analyses, but all the characteristics reported in the section on crystallo-

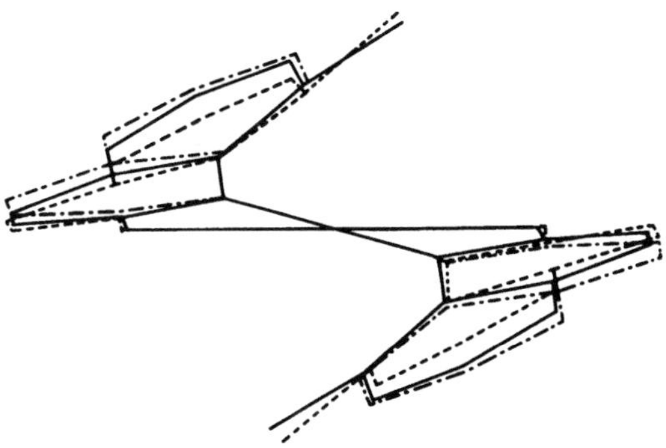

Fig. 10. Projections of 1,16-dimethylhexahelicene on a plane perpendicular to the axis through C(21)–C(22).
——— Crystallographic data
--------- Calculated starting from a planar model
-·-·-·-·- Calculated starting with the X-ray data and starting with the geometry obtained from the planar model (---------) using a shorter C(methyl)-C(1) distance

graphy were also present in the calculated structures. The differences between calculated and X-ray structures are generally small, e.g. differences between corresponding torsion angles 0.2–4.5°, and between corresponding non-bonding distances 0.0—0.2 mm. Angles between benzene rings do not differ more than 2.5°, but as a consequence of accumulation the angles between terminal rings show a greater difference.

The dihedral angles, calculated for the aliphatic moiety in 5,6-dihydrohexahelicene agree reasonably well [41,112] with the value obtained from NMR-data.

The potentials for the rotation of the 1-substituents in 1-formyl- and 1-acetylhexahelicene were calculated by Lindner [148]. For both helicenes two minima were present, differing 8 kJ/mol from each other, and separated by barriers of 29–60 kJ/mol.

The result suggests that rotations of the substituents are possible at room temperature. The lower minimum for the formyl group is at a dihedral angle C_2–C_1–C–O of 17°, for the acetyl group the minimum is found at a value of −158° for this angle. From X-ray data values of 31° and −139°, respectively, are found.

8 Chemical Reactions

In this section only reactions directly involving the helical skeleton are mentioned.

8.1 Intramolecular Reactions

Dougherty [159] reported in 1968 that the peak at m/e = 300 in the mass spectrum of hexahelicene ($C_{26}H_{16}$) is due to the ion of coronene ($C_{24}H_{12}^+$), formed by an internal Diels-Alder reaction, followed by a fragmentation yielding ethylene. Indeed, traces of coronene were detected on heating hexahelicene at 485 °C for 2 h in an evacuated tube.

Another (4+2)-cycloaddition, not accompanied with fragmentation, was observed by Martin et al. [160] when 1-formyl hexahelicene (87) was treated with the ylid of $(EtO)_2POCH_2CO_2Et$ in boiling benzene for 12 h. Two racemic stereoisomers (88a) and (88b) were formed in 80 and 12%, respectively. At room temperature only one isomer (88a) was obtained in 90%. The obvious intermediate β-(1-hexahelicycl)acrylic ester, could not be isolated (Scheme 23).

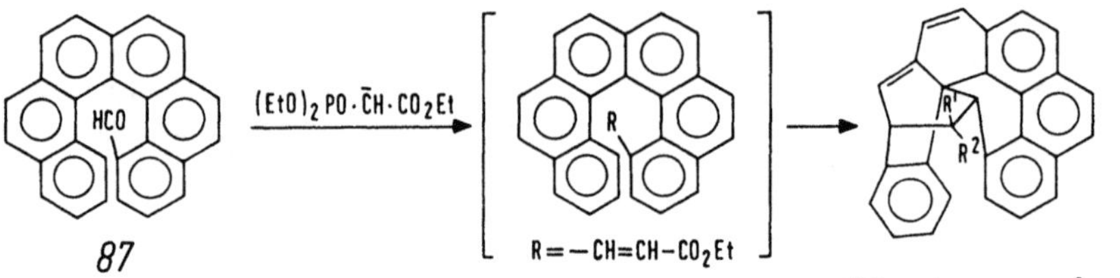

87

88 a R' = —CO₂Et, R² = H
b R' = H, R² = —CO₂Et

Scheme 23

The structure (88), indicated by spectroscopic methods was confimed for (88a) by X-ray analysis. Bridging of terminal rings by an internal Friedel-Crafts reaction was also reported by Martin [6]. Acid treatment of 1-hydroxymethylpentahelicene (89) resulted in the formation of (90). Wijnberg [162] reported the formation of methano-bridged heterohelicenes by treatment of the 1-methyl- and 1-ethylheteroheliccene (91) with N-bromosuccinimide (NBS)

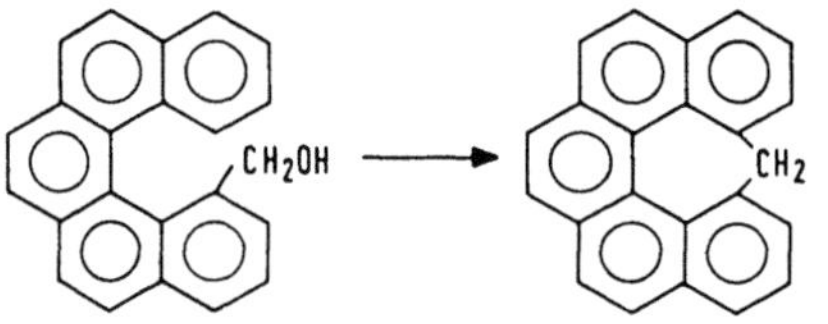

89 90

Scheme 24

in carbontetrachloride at 160° for 20 min. Product (92b) is formed in only one configuration, having the methyl group at the outer side of the core.

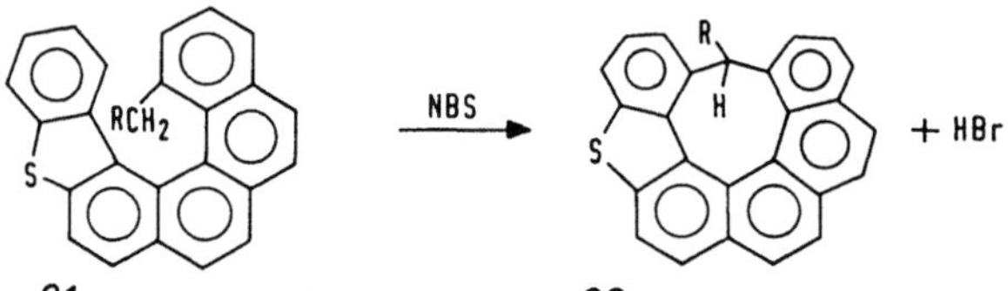

91 a R = H
b R = CH₃

92 a = R = H 35 %
b R = CH₃ 76 %

Scheme 25

A methanobridge is also formed when 1-hydroxymethylhexahelicene (93) is treated with an acid, but in this case the primary product (94) rearranges into the spiro compound (95). The same product is formed when (93) is heated in hexachloroethane at 200 °C for 30 min, either in the presence or absence of trifluoroacetic acid, but not in naphthalene as the solvent under the same conditions [163].

Even in the absence of a functionalized methyl group at C(1) a product like (95) can be formed. On heating 1,3,14,16-tetramethylhexahelicene (96), neat or as a concentrated solution in naphthalene, above 180 °C two spiro compounds (99) and (100)

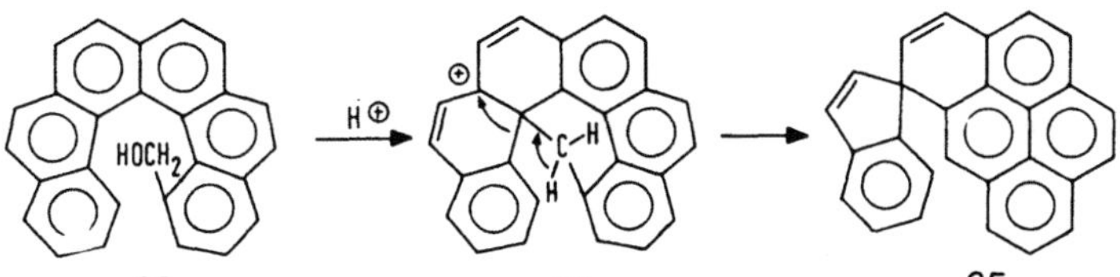

93 94 95

Scheme 26

are formed [164]. Because of the conditions of the reaction and the observation that the reaction is not accompanied with loss of optical activity a (partly) concerted pathway has to be involved. An initial [1.7] antarafacial hydrogen shift leads to (97), which gives (98) via a disrotatory electrocyclic reaction. The rearrangement of (98) into (99) may occur via a 1,2-phenyl radical shift or via a [π2s + σ2a] reaction followed by two 1,3-H shifts. Experimentally it was shown that (100) is formed from (99).

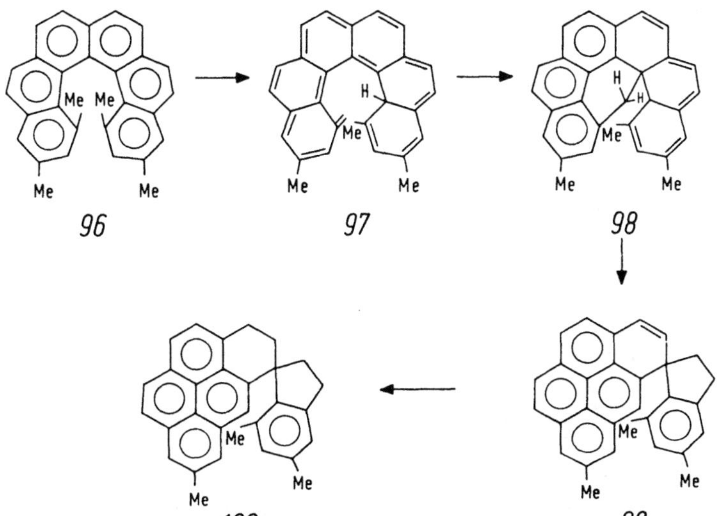

96 97 98 100 99

Scheme 27

The same reaction is observed with 1,16-dimethylhexahelicene but not with 1,3- and 1,14-dimethylhexahelicene, showing the strict steric requirements for this reaction (compare the differences in structure between 1,16-dimethyl- and 1-methylhexahelicene mentioned in the sections on NMR-spectra and crystallography). The different behaviour of (96) on heating in concentrated solution (rearrangement) and dilute solution (racemization) is ascribed to the smaller flexibility of the compound in the solid state.

The well-known intramolecular formation of an aryl-aryl bond under the influence of Friedel-Crafts catalysts (Scholl reaction [165]) is in the helicene series only observed with some penta- and hexaheterohelicenes. The products are named didehydroheterohelicenes. Some examples are given in Scheme 28.

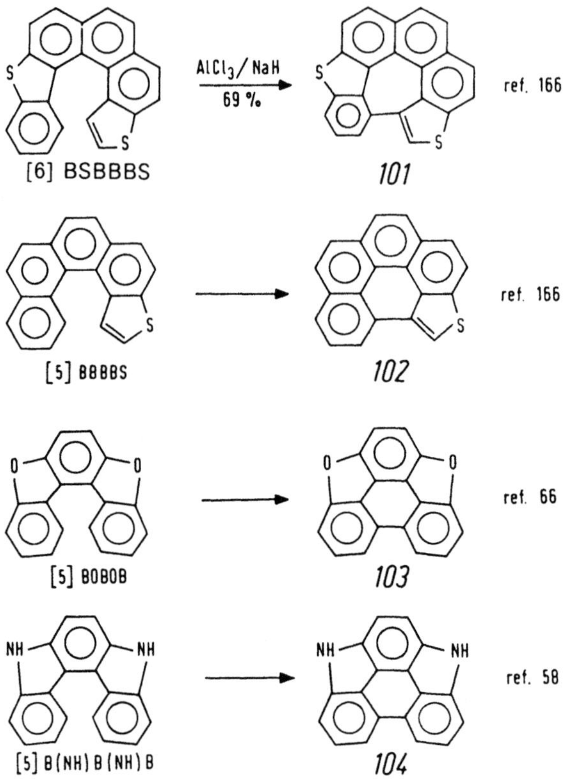

Scheme 28

Several other heterohelicenes resist this reaction even when the Friedel-Crafts catalyst ($AlCl_3$) is exchanged for FSO_3H, 90% H_2SO_4, 40% HF and other strong acids. Some dehydroheterohelicenes have been used to prepare heterocirculenes by a Diels-Alder reaction with maleic anhydride and subsequent hydrolysis and removal of the carboxylic acid groups [167].

Still another type of an intramolecular reaction is the ring enlargement by the insertion of a carbene. When 1-formylhexahelicene tosylhydrazone (105a) is heated in benzene, especially in the presence of sodium hydride, product (106) can be isolated [168]. Presumably, the reaction proceeds via the intermediates (105b) and (105c) (Scheme 29).

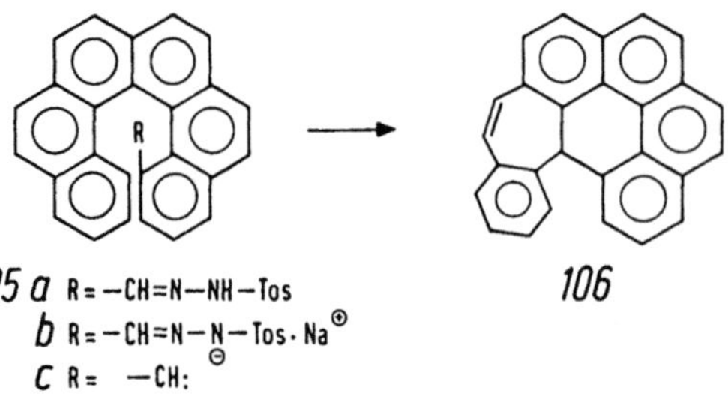

105 a R = –CH=N–NH–Tos
b R = –CH=N–$\overset{\ominus}{N}$–Tos · $Na^{\oplus}$
c R = –CH:

Scheme 29

Remarkably, (106) has also been isolated as a side product in the reduction of 4-bromo-1-methyloxycarbonyl hexahelicene with $LiAlH_4$ when followed by treatment with $LiAlH_4 + AlCl_3$ in ether at room temperature [168].

8.2 Intermolecular Reactions

8.2.1 Electrophilic Reactions

Several reactivity parameters adopted from theoretical chemistry have been used to predict the position at which electrophilic reagents will attack higher aromatics [137]. The applicability of such parameters to the non-planar hexahelicene was studied qualitatively in bromination, nitration and acetylation reactions [169], and in a quantitative way in the protiodetritiation of the eight monotritiohexahelicenes [170].

In Table 27 the simple Hückel indices, Nr (Dewar number), Fr (free valence number) and Lr (localization energy) are given, together with the Mulliken overlap population μr, calculated by the extended Hückel method and using the geometry of hexahelicene as determined by X-ray analysis [141] to account for the non-planarity.

Table 27. Reactivity Parameters of the Ring Atoms in Hexahelicene Nr (Dewar number), Fr (free valence number), μr (Mulliken overlap population) and Lr (localization energy), Rates of Protiodetritiation Relative to 9-T-Phenanthrene (k_{rel}) and Corresponding Partial Rate Factors (f)

Parameter	Position							
	C_1	C_2	C_3	C_4	C_5	C_6	C_7	C_8
Nr	1.88	2.05	2.11	1.83	1.66	1.85	1.81	1.82
Fr	0.444	0.407	0.404	0.451	0.452	0.449	0.449	0.448
Lr	2.338	2.456	2.479	2.311	2.2875	2.315	2.3065	2.314
μr	−0.131	−0.065	−0.117	−0.107	−0.117	−0.059	−0.87	−0.111
k_{rel}	5.76	2.56	0.563	4.963	13.55	3.74	11.75	10.0
f	10.250	4.400	0.905	8.770	25.000	6.525	21.600	18.200

Bromination of hexahelicene in CCl_4 leads to two bromine addition products 107 and 108. Differently from 9,10-dibromo-, 9,10-dihydrophenanthrene, which eliminates HBr in boiling CCl_4, 107 and 108 eliminate HBr only at higher temperatures (boiling toluene) leaving 5-bromo- and 5,12-dibromohexahelicene, respectively. Heating 107 or 108 in pyridine or triethylamine results in loss of bromine; hexahelicene is recovered almost quantitatively.

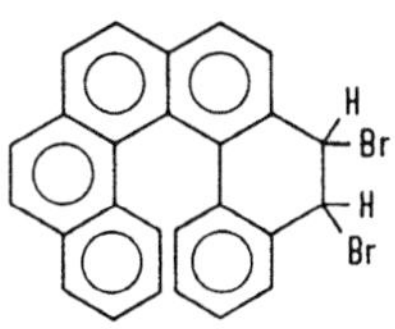

107

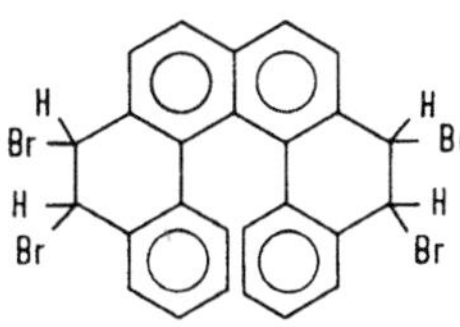

108

Scheme 30

Addition of 1 equivalent of bromine to hexahelicene in the presence of iodine gives rise to 70% 5-bromo- and 10% 5,12-dibromohexahelicene.

Under mild conditions nitration and acetylation of hexahelicene give the 5-nitro- and 5-acetyl substitution product as the main product in about 50% yield. In both cases another monosubstitution product is formed, which was identified tentatively by NMR as the corresponding 8-substituted hexahelicene. From the relative rates of detritiation (k_{rel}) or the partial rate factors (f) given in Table 27, it seems more probable, however, that the 7-isomers are formed as the side product, as the positional reactivity order of detritiation is C(5) > C(7) > C(8) > C(1) > C(4) > C(6) > C(2) > C(3). The preferred reactivity at C(5), found in electrophilic substitutions, is predicted by all the simple Hückel parameters, whereas the next two positions are correctly predicted by Nr and Lr. Judging from Nr-, Fr- and Lr-values the C(1) position does not experience much steric hindrance in the H-exchange. Relative to some other positions (C(4), C(6)) its reactivity is higher than expected. The Mulliken overlap population predicts, however, the highest reactivity for C(1) and leaves room for the supposition that this position is considerably masked.

The much higher reactivity of hexahelicene compared to phenanthrene as demonstrated in competition experiments [169] and apparent from the f-values [170] is in line with the regular increase in reactivity in the series: phenanthrene, benzo[c]phenanthrene, pentahelicene, hexahelicene. It demonstrates the effect of the increasing distortion of the aromatic rings, what destabilizes the ground state.

Another kind of electrophilic substitution, the lithiation following by exchange of lithium by a methyl group has been used for the introduction of methyl groups in the terminal thiophene ring of some thiahelicenes [171].

8.2.2 Oxidation and Subsequent Rearrangement

Hexahelicene is oxidized by chromic acid under very mild conditions (stirring for 24 h at room temperature in an acetic acid-water solution) giving hexahelicene-5,6-quinone (109) in 70% yield [138]. The increased oxidation rate of 1,3,14,16-tetramethylhexahelicene has to be the ascribed to the electronic effect of the multiple methyl substitution and not to the disturbance of the helical conformation as was concluded from the polarographic oxidation potentials (see Sect. 5) as well as from charge-transfer spectra [138].

On treatment of the quinone (109) with NaOH in a dioxane-water mixture (1:1) at 100° for several days a benzylic acid rearrangement takes place giving phenanthro-[3,4-c]fluorene (110) in 52% yield [113].

The Wolff-Kischner reduction of 110 gave the product 61 with a five-membered ring in 65% yield.

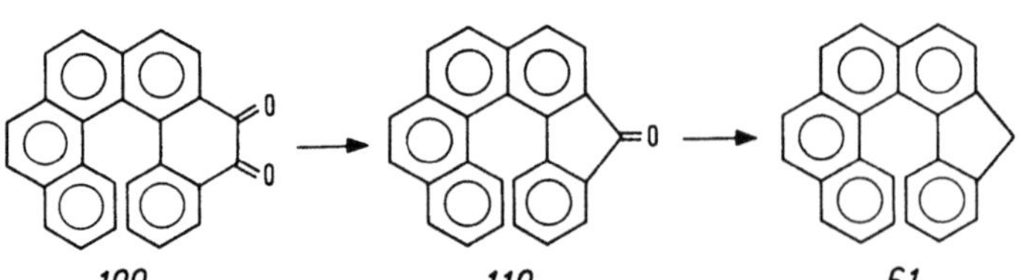

109 110 61

Scheme 31

8.3 Photoreactions

No photoreactions, except incidental photodestruction are known for hexa- and higher helicenes. As already mentioned in 2.1.1. pentahelicene forms benzo[g,h,i]perylene on irradiation in the presence of iodine as an oxidizing agent. Remarkably, the symmetrical benzoderivative (111) does not yield a similar cyclization product. This is ascribed to the antibonding character of the π-orbitals at C(1) and C(14) involved in the expected photoreaction as appeared from EHMO calculation [17b].

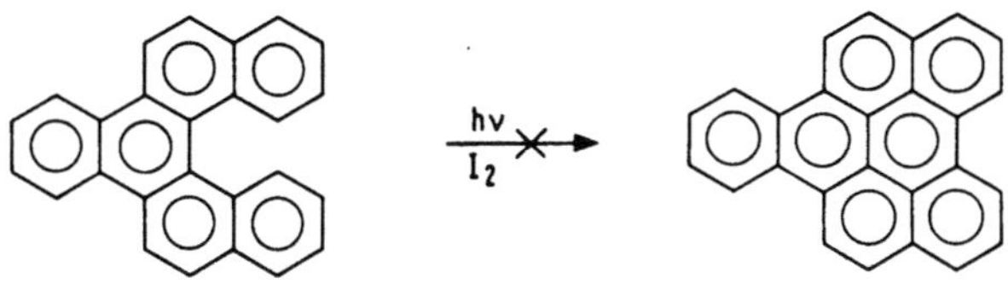

111

Scheme 32

In 1-substituted pentahelicenes the initial ring closure may be followed by a shift of the substituent. Irradiation of 1-fluoropentahelicene (112 R=F) gives 1-fluorobenzoperylene (114) via a 1,3-fluorine shift in the partly oxidized primary cyclization product (113). The 1,3-shift was secured by placing a substituent at C(4) [172]. Irradiation of 1-phenylpentahelicene resulted in the phenylbenzoperylene (17) via a 1,2-phenylradical shift. In this case the 1,2-shift was ascertained by introducing a second phenyl group in the same ring [19]. By another photodehydrocyclization (17) forms benzo[a]coronene as mentioned in 2.1.1. (Scheme 4).

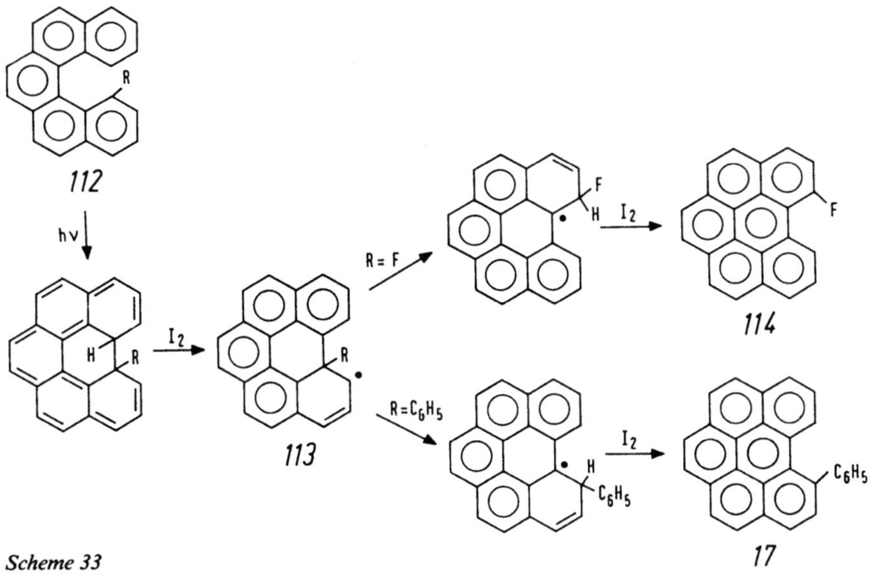

Scheme 33

8.4 Miscellaneous Reactions

The first helical ferrocene 117 has been prepared [177] using diindeno-5.4-c; 4′,5′-g phenanthrene 115, a heptahelicene in which the two terminal rings consist of cyclopentadiene rings (Scheme 34). After treatment with tert. butyllithium in tetrahydrofuran pure $FeCl_2$ was added at −65 °C, 117 was obtained in 60% yield as dark red crystals. The angle between the terminal rings is 19.4°, much smaller than in 115 where this angle is 69.1° [149].

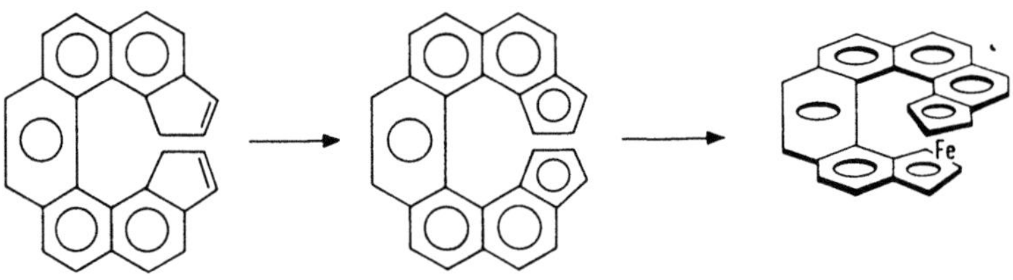

115 116 117

Scheme 34

Nakazaki et al. [178] prepared a crown ether containing the hexahelicene moiety. Starting with 2,15-dibromohexahelicene (118a) the diol 118c was prepared which was condensed with 3,6,9,12-tetraoxatetradecane-1,14-diyl bis toluene p-sulphonate yielding in 50% the 2,15-hexaheliceno-27-crown-6 (118d) (Scheme 35). The enantiomers were separated by HPLC over polytriphenylmetacrylate. A similar product was prepared containing a pentahelicene moiety.

The differential transport of the emantiomers of methyl (±) phenylglycinate hydrochloride and of phenylethylamine hydrochloride through these materials showed that the [5]crown ether has a larger enantiomeric selectivity than 118d. The [5] and [6]-crown ethers having the same sign of optical rotation showed opposite chiral recognition towards both substrates. According to the authors this is due to the fact that the ether part of (M)-(—)-[5] crown has P-helicity whereas (M)-(—)-[6]crown has M-helicity.

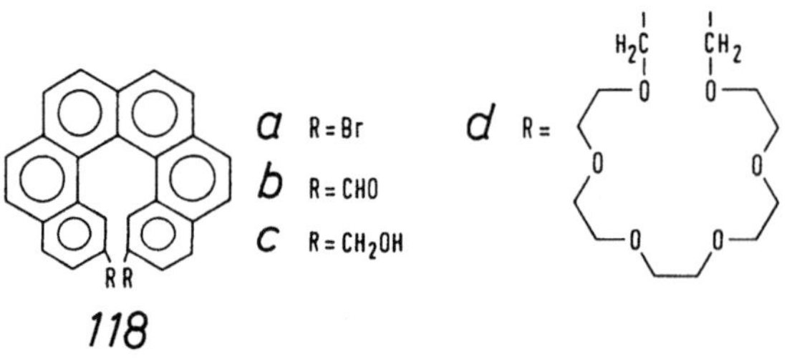

118

Scheme 35

9 Conclusions

The most important conclusion from the extensive investigations on helicenes in the past is, that the benzene ring is much more flexible than previously thought. In helicenes this flexibility comes to light in the strikingly easy racemization of these compounds. In small aromatic compounds such a direct demonstration of the flexibility of aromatic rings is not easy possible [173]. The flexibility of the aromatic rings in helicenes is, however, not a consequence of their special structural characteristics like the helical form or the occurence of overcrowded regions. None of the physical properties of helicenes reflects the presence of any unusual effect.

The relation between UV- and PES-spectra is quite similar as found for planar aromatic compounds. All chemical shifts in NMR-spectra can well be explained by normal ring-current effects and Van der Waals interactions. Polarographic data do not deviate from those for planar compounds.

Indeed, X-ray data show that the distances of carbon-carbon bonds in helicenes deviate from those in planar aromatic compounds, but force-field calculations starting with 'normal' C–C distances lead to exactly the same bond lengths as found in the X-ray analysis.

The main aspects of the chemical reactivity of helicenes (e.g. electrophilic substitution) equally not deviate from those of planar aromatic compounds, and remarkable reactions of helicenes, which are incidentally found (e.g. the transannular bond formation between a C(1)-substituent and a part of the inner helix) can ultimately be reduced to known principles of aromatic reactivity.

Martin wrote at the end of his review in 1974 [6]: 'It is our intimate conviction that further work on these unique molecules . . . should be highly rewarding in many fields of chemistry.' To our opinion this conviction is verified now because the knowledge obtained from these molecules in the field of NMR an UV-spectroscopy, resolution of enantiomers and asymmetric synthesis appears to become more generally useful in several areas of organic chemistry.

Some aspects of the chemistry of helicenes require still more attention. Since the interpretation of the mass spectrum of hexahelicene by Dougherty [159] no further systematic work has been done on the mass spectroscopy of helicenes, to verify the concept of an intramolecular Diels-Alder reaction in the molecular ion. Though the optical rotation of a number of helicenes is known and the regular increase of the optical rotation with increasing number of benzene rings has been shown, the dependence of the rotation on the helicity is still unknown. The asymmetric induction in the synthesis of helicenes by chiral solvents, or in liquid crystals, though small, deserves still more attention because application to other organic compounds will be promoted when the explanation of observed effects is more improved.

The use of an optical pure helicene as an inductor in small scale preparation deserves more attention.

Finally, no report has still appeared about the synthesis of triple helicenes which can be seen as orthoannelated triphenyl derivatives. The first member in the series tribenzo[a,c,e]triphenylene has already been synthesized several years ago [174]. The racemizations and isomerizations of these compounds may reveal interesting new data.

Note Added in Proof

Addition to section 4.1
The proton chemical shifts of [7], [8], [9], [10], [11], [12], [13] and [14] helicenes dissolved in CS_2 at 90 MHz have been assigned by the study of deuterated or substituted derivatives [180].

Compared to previous work the attribution of the chemical shift of the protons H(5) and H(6) of [8] and of [9] and thôse of H(7) and H(8) of [9] have been reversed.

The evolution of the chemical shifts with the growth of the helix is discussed.

The δ-values (in ppm) of the protons of [14] dissolved in $CDCl_3$, assigned with certainty are

H(1)	6.143	H(6)	6.978
H(2)	5.949	H(13)	7.121
H(3)	6.091	H(14)	6.754
H(4)	7.062	H(15)	6.341
H(5)	7.004	H(16)	6.272

10 Acknowledgments

The authors are very grateful to Professor R. J. F. Nivard for careful reading the manuscript and for a number of helpful comments and suggestions. They wish to thank Professor Martin (Brussels) for his valuable remarks after reading the manuscript.

11 References

1. Newman, M. S., Lednicer, D.: J. Am. Chem. Soc. *78*, 4765 (1956)
2. Hugelshofer, P., Kalvoda, J., Schaffner, K.: Helv. Chim. Acta *43*, 1322 (1960)
3. Flammang-Barbieux, M., Nasielski, J., Martin, R. H.: Tetrahedron Lett. 743 (1967)
4. Wijnberg, H., Groen, M. B.: J. Am. Chem. Soc. *90*, 5339 (1968)
5. Wijnberg, H.: Accounts Chem. Res. *4*, 65 (1971)
6. Martin, R. H.: Angew. Chem. Int. Ed. *13*, 649 (1974)
7. Laarhoven, W. H.: Recl. Trav. Chim. (Pays-Bas) *102*, 185 (1983)
8. Mallory, F. B., Mallory, C. W. in: Org. Reactions, J. Wiley & Sons, New York 1984, Vol. 30, p. 1
9. Maercken, A., in: Org. Reactions, J. Wiley & Sons, New York 1965, Vol. 14, p. 270
10. Siegrist, A. E., Meijer, H. R.: Helv. Chim. Acta *52*, 1282, 2521 (1969)
11. Jutz, C., Loebering, H. G.: Angew. Chem. *87*, 415 (1975)
12. Muszkat, K. A.: Topics Curr. Chem. *88*, 89 (1980)
13. Scholz, M., Mühlstadt, M., Dietz, F.: Tetrahedron Lett. 665 (1967)
14. a) Laarhoven, W. H., Cuppen, J. H. M., Nivard, R. J. F.: Recl. Trav. Chim. (Pays-Bas) *87*, 687 (1968)
    b) Laarhoven, W. H., Cuppen, Th. J. H. M., Nivard, R. J. F.: Tetrahedron *26*, 4865 (1970)
15. a) Wismonski-Knittel, T., Fischer, E.: Mol. Photochem. *9*, 67 (1978)
    b) Wismonski-Knittel, T., Fischer, E.: J. Chem. Soc. Perkin II, *1979*, 449
16. Laarhoven, W. H., Cuppen, Th. J. H. M., Brinkhof, H. H. K.: Tetrahedron *38*, 3179 (1982)
17. a) Grellmann, K. H., Hentzschel, P., Wismonski-Knittel, T., Fischer, E.: J. Photochem. *11*, 197 (1979)

b) Tinnemans, A. H., Laarhoven, W. H., Sharafi-Ozeri, S., Muszkat, K. A.: Recl. Trav. Chim. (Pays-Bas) *94*, 239 (1975)
18. Tinnemans, A. H. A., Laarhoven, W. H.: J. Am. Chem. Soc. *96*, 4611 (1974)
19. Laarhoven, W. H., Boumans, P. G. F.: Recl. Trav. Chim. (Pays-Bas) *94*, 114 (1975)
20. Joly, M., Defay, N., Martin, R. H., Declerq, J. P., Germain, G., Soubrier-Payen, B., van Meerssche, M.: Helv. Chim. Acta *60*, 537 (1977)
21. Laarhoven, W. H., Cuppen, Th. J. H. M.: Tetrahedron *30*, 1101 (1974)
22. Somers, J. B. M.: Thesis, Nijmegen 1984
23. Laarhoven, W. H., Veldhuis, R. G. M.: Tetrahedron *28*, 1811 (1972)
24. Gorsane, M.: Thesis, Brussels 1977
25. Laarhoven, W. H., Veldhuis, R. G. M.: Tetrahedron *28*, 1823 (1972)
26. Vanest, J. M., Gorsane, M., Libert, V., Pecher, J., Martin, R. H.: Chimia *29*, 343 (1975)
27. Laarhoven, W. H., Cuppen, Th. J. H. M., Nivard, R. J. F.: Tetrahedron *26*, 4865 (1970)
28. Martin, R. H., Marchant, M. J., Baes, M.: Helv. Chim. Acta *54*, 358 (1971)
29. Martin, R. H., Flammang-Barbieux, M., Cosyn, J. P., Gelbcker, M.: Tetrahedron Lett. 3507 (1968)
30. Martin, R. H., Cosyn, J. P.: Synth. Comm. *1*, 257 (1971)
31. R. H. Martin, Libert, V.: J. Chem. Res. (s) 130 (1980)
32. Martin, R. H., Baes, M.: Tetrahedron *31*, 2135 (1975)
33. Martin, R. H., Morren, G., Schurter, J. J.: Tetrahedron Lett. 3863 (1969)
34. Laarhoven, W. H., Nivard, R. J. F.: Tetrahedron *32*, 2445 (1976)
35. Laarhoven, W. H., Cuppen, Th. J. H. M.: Tetrahedron Lett. 163 (1971)
36. Martin, R. H., Eijndels, Ch., Defay, N.: ibid. 2731 (1972)
37. Laarhoven, W. H., Cuppen, Th. J. H. M.: Recl. Trav. Chim. (Pays-Bas) *92*, 553 (1973)
38. Laarhoven, W. H., de Jong, M. H.; ibid. *92*, 651 (1973)
39. Martin, R. H., Eijndels, Ch., Defay, N.: Tetrahedron *30*, 3339 (1974)
40. Laarhoven, W. H., Cuppen, Th. J. H. M., Nivard, R. J. F.: ibid. *30*, 3343 (1974)
41. Borkent, J. H., Diesveld, J. W., Laarhoven, W. H.: Recl. Trav. Chim. (Pays-Bas) *100*, 114 (1981)
42. Imajo, S., Shingu, K., Kuritani, H., Kato, A.: Tetrahedron Lett. *22*, 107 (1981)
43. Borkent, J. H., Diesveld, J. W., Laarhoven, W. H.: Tetrahedron *38*, 1809 (1982)
44. a) Tinnemans, A. H. A., Laarhoven, W. H.: J. Chem. Soc. Perkin II 1111, 1115 (1976)
b) Tinnemans, A. H. A., Laarhoven, W. H.: Org. Photochem. Synth. Vol. 2, p. 93 (1976)
45. Laarhoven, W. H., Peters, W. H. M., Tinnemans, A. H. A.: Tetrahedron *34*, 769 (1978)
46. Tinnemans, A. H. A., Laarhoven, W. H.: ibid. *35*, 1537 (1979)
47. Voegtle, F., Atzmueller, M., Wehner, W., Gruetze, J.: Angew. Chem. *89*, 338 (1977)
48. Mislow, K., Gust, D., Finocchiaro, P., Boettcher, R. J.: Top. Curr. Chem. *47*, 1 (1974)
49. Voegtle, F., Höhner, G.: Angew. Chem. *87*, 522 (1975)
50. Thulin, B., Wennerstrom, O.: Acta Chem. Scand. *B30*, 688 (1976)
51. Mulliken, R. S.: J. Chem. Phys. *23*, 1833, 1841 (1955)
52. a) Muszkat, K. A., Sharafi-Ozeri, S.: Chem. Phys. Lett. *20*, 397 (1973)
b) Muszkat, K. A., Seger, G., Sharafi-Ozeri, S.: J. C. S. Faraday II *71*, 1529 (1975)
53. Dopper, J. H., Oudman, D., Wijnberg, H.: J. Am. Chem. Soc. *95*, 3692 (1973)
54. Groen, M. B., Wijnberg, H.: ibid. *92*, 6664 (1970)
55. Yamada, K., Ogashiwa, S., Tanaka, H., Nakagawa, H., Kawazura, H.: Chem. Lett. *1981*, 343
56. Fuchs, W., Niszel, I.: Chem. Ber. *60*, 209 (1927)
57. Zander, M., Franke, W. H.: ibid. *102*, 2728 (1969)
58. Rau, H., Schuster, O.: Angew. Chem. *88*, 80 (1976)
59. Chatterjea, J. N., Banerji, K. D.: J. Ind. Chem. Soc. *47*, 576 (1970)
60. Hogberg, H. E.: Acta Chem. Scand. *27*, 2591 (1973)
61. Kagan, H., Moradpour, A., Nicoud, J. F., Balavoine, G., Tsoucaris, G.: J. Am. Chem. Soc. *93*, 2353 (1971)
62. Bernstein, W. J., Calvin, M., Buchardt, O.: ibid. *95*, 527 (1973)
63. Moradpour, A., Kagan, H., Baes, M., Morren, G., Martin, R. H.: Tetrahedron *31*, 2139 (1975)
64. Kagan, H., Moradpour, A., Nicoud, J. F., Balavoine, G., Martin, R. H., Cosijn, J. P.: Tetrahedron Lett. *1971*, 2479
65. Bernstein, W. J., Calvin, M., Buchardt, O.: J. Am. Chem. Soc. *94*, 494 (1972)

66. Muszkat, K. A., Fischer, E.: J. Chem. Soc. *B 1967*, 662
67. Buchardt, O.: Angew. Chem. *86*, 222 (1974)
68. a) Martin, R. H., Defay, N., Figeys, H. P., Le Van, K., Ruelle, J. J., Schurter, J. J.: Helv. Chim. Acta *55*, 2241 (1972)
    b) de Jong, M. H., Laarhoven, W. H.: Recl. Trav. Chim. (Pays-Bas) *92*, 673 (1973)
    c) Somers, J. B. M., Laarhoven, W. H.: ibid. *99*, 160 (1980)
69. Kagan, H. B., Fiaud, J. C., in: Topics in Stereochemistry (Ed. Eliel, E. L., Allinger, N. L., Wiley 1978), Vol. 10, p. 254
70. Bernstein, W. J., Calvin, M., Buchardt, O.: Tetrahedron Lett. *1972*, 2195
71. a) Laarhoven, W. H., Cuppen, Th. J. H. M.: J. Chem. Soc. Chem. Comm. *1977*, 47
    b) Laarhoven, W. H., Cuppen, Th. J. H. M.: J. Chem. Soc. Perkin 2, *1978*, 315
72. Nakazaki, M., Yamamoto, K., Fujiwara, K.: Chem. Lett. *1978*, 863
73. Hibert, M., Solladie, G.: J. Org. Chem. *45*, 5393 (1980)
74. Nakazaki, M., Yamamoto, K., Fujiwara, K., Maeda, M.: J. Chem. Soc. Chem. Comm. *1979*, 1086
75. Cochez, Y., Martin, R. H., Jespers, J.: Isr. J. Chem. *15*, 29 (1977)
76. Vanest, J., Martin, R. H.: Recl. Trav. Chim. (Pays-Bas) *98*, 113 (1979)
77. Cochez, Y., Jespers, J., Libert, V., Mislow, K., Martin, R. H.: Bull. Soc. Chim. Belg. *84*, 1033 (1975)
78. Newman, M. S., Lutz, W. B., Lednicer, D.: J. Am. Chem. Soc. *77*, 3420 (1955)
79. Wijnberg, H., Groen, M. B., Schadenberg, H.: J. Org. Chem. *36*, 2797 (1971)
80. Martin, R. H., Marchant, M. J.: Tetrahedron *30*, 343 (1974)
81. Klemm, L. H., Reed, D.: J. Chrom. *3*, 364 (1960)
82. Teuber, H. J., Vogel, L.: Chem. Ber. *103*, 3319 (1970)
83. a) Nakazaki, M., Yamamoto, K., Maeda, M.: J. Org. Chem. *46*, 1985 (1981)
    b) Nakazaki, M., Yamamoto, K., Maeda, M.: Chem. Lett. *1980*, 1553
84. Okamoto, Y., Honda, S., Okamoto, I., Yuki, H., Murata, S., Noyori, R.: J. Am. Chem. Soc. *103*, 6971 (1981)
85. a) Mikes, F., Boshart, G., Gil-Av, E.: J. Chem. Soc. Chem. Comm. *1976*, 99
    b) Mikes, F., Boshart, G., Gil-Av, E.: J. Chrom. *122*, 205 (1976)
86. Numan, H., Helder, R., Wijnberg, H.: Recl. Trav. Chim. (Pays-Bas) *95*, 211 (1976)
87. Nakagawa, H., Ogashiwa, S., Tanaka, H., Yamada, K., Kawazura, H.: Bull. Chem. Soc. Japan *54*, 1903 (1981)
88. a) Mikes, F., Boshart, G.: J. Chem. Soc. Chem. Comm. *1978*, 173
    b) Mikes, F., Boshart, G.: J. Chrom. *149*, 455 (1978)
89. Wijnberg, H., Lammertsma, K.: J. Am. Chem. Soc. *95*, 7913 (1973)
90. Nakagawa, H., Tanaka, H., Yamada, K., Kawazura, H.: J. Phys. Chem. *86*, 2311 (1982)
91. Tanaka, H., Nakagawa, H., Yamada, K., Kawazura, H.: Bull. Chem. Soc. Japan *54*, 3665 (1981)
92. a) Kim, Y. H., Tishbee, A., Gil-Av, E.: J. Am. Chem. Soc. *102*, 5915 (1980)
    b) Kim, Y. H., Tishbee, A., Gil-Av, E.: Science *213*, 1883 (1981)
93. Lochmuller, C. H., Jensen, E. C.: J. Chrom. *216*, 333 (1981)
94. Kim, Y. H., Balan, A., Tishbee, A., Gil-Av, E.: J. Chem. Soc. Chem. Comm. *1982*, 1336
95. a) Barron, L. D.: J. Chem. Soc. Farad. Trans 2 *71*, 293 (1975)
    b) Barron, L. D.: J. Am. Chem. Soc. *96*, 6761 (1974)
96. Groen, M. B., Wijnberg, H.: J. Am. Chem. Soc. *93*, 2968 (1971)
97. Mason, S. F., Brickell, W. S., Browns, A., Kemp, C. M.: J. Chem. Soc. (A) *1971*, 756
98. Hug, W., Wagniere, G.: Tetrahedron *28*, 1241 (1972)
99. a) Lightner, D. A., Hefelfinger, D. T., Frank, G. W., Powers, T. W., Trueblood, K. N.: Nature *232*, 124 (1971)
    b) Lightner, D. A., Hefelfinger, D. T., Frank, G. W., Powers, T. W., Trueblood, K. N.: J. Am. Chem. Soc. *94*, 3492 (1972)
100. Tribout, J., Martin, R. H., Doyle, M., Wijnberg, H.: Tetrahedron Lett. *1972*, 2839
101. Newman, M. S., Darlak, R. S., Tsai, L. L.: J. Am. Chem. Soc. *89*, 6191 (1967)
102. Yamada, K., Tanaka, H., Nakagawa, H., Ogashiwa, S., Kawazura, H.: Bull. Chem. Soc. Japan *55*, 500 (1982)
103. Martin, R. H., Marchant, M. J.: Tetrahedron *30*, 347 (1974)

104. Lindner, H. J.: ibid. *31*, 281 (1975)
105. Borkent, J. H., Laarhoven, W. H.: ibid. *34*, 2565 (1978)
106. Martin, R. H., Marchant, M. J.: Tetrahedron Lett. *1972*, 3707
107. Martin, R. H., Defay, N., Figeys, H. P., Flammang-Barbieux, M., Cosijn, J. P., Gelbcke, M., Schurter, J. J.: Tetrahedron *25*, 4985 (1969)
108. Haigh, C. W., Mallion, R. B.: Mol. Phys. *22*, 945 (1971)
109. Laarhoven, W. H., Nivard, R. J. F.: Tetrahedron *28*, 1803 (1972)
110. Defay, N.: Org. Magn. Resonance *6*, 221 (1974)
111. Haigh, C. W., Mallion, R. B.: Mol. Phys. *22*, 955 (1971)
112. Somers, J. B. M., Borkent, J. H., Laarhoven, W. H.: Recl. Trav. Chim. (Pays-Bas) *100*, 110 (1981)
113. Diesveld, J. W., Borkent, J. H., Laarhoven, W. H.: Tetrahedron *38*, 1803 (1982)
114. van den Hark, Th. E. M., Noordik, J. H., Beurskens, P. T.: Cryst. Struct. Comm. *3*, 443 (1974)
115. Sato, T., Akabori, S., Muto, S., Hata, K.: Tetrahedron *24*, 5557 (1968)
116. Martin, R. H., Defay, N., Zimmermann, D.: Tetrahedron Lett. *1971*, 1871
117. Martin, R. H., Schurter, J. J.: Tetrahedron *28*, 1749 (1972)
118. Boekelheide, V., Schmidt, W.: Chem. Phys. Lett. *17*, 410 (1972)
119. Heilbronner, E., Maier, J. P.: Helv. Chim. Acta *57*, 151 (1974)
120. Obenland, S., Schmidt, W.: J. Am. Chem. Soc. *97*, 6633 (1975)
121. Deb, B. M., Kavu, G.: Can. J. Chem. *58*, 258 (1980)
122. Clar, E.: Polycyclic Hydrocarbons, Acad. Press New York 1964, Vol. 1
123. Birks, J. B., Birch, D. J. S., Cordemans, E., van der Donckt, E.: Chem. Phys. Lett. *43*, 33 (1976)
124. Boschi, R., Clar, E., Schmidt, W.: J. Chem. Phys. *60*, 4406 (1974)
125. van der Donckt, E., Nasielski, E., Greenleaf, J., Birks, J. B.: Chem. Phys. Lett. *2*, 409 (1968)
126. Brown, J. M., Field, I. P., Sidebottom, P. J.: Tetrahedron Lett. *22*, 4867 (1981)
127. Balan, A., Gottlieb, H. E.: J. Chem. Soc. Perkin 2 *1981*, 350
128. Wijnberg, H., Groen, M. B.: J. Chem. Soc. *D 1969*, 964
129. Martin, R. H., Morren, G., Schurter, J. J.: Tetrahedron Lett. *1969*, 3683
130. Sapir, M., van der Donckt, E.: Chem. Phys. Lett. *36*, 108 (1975)
131. Kim, S. S., Weissman, S. I.: J. Am. Chem. Soc. *101*, 5863 (1979)
132. Weissman, S. I., Chang, R.: ibid. *94*, 8683 (1972)
133. Fey, H. J., Kurreck, H., Lubitz, W.: Tetrahedron *35*, 905 (1979)
134. Tanaka, H., Ogashiwa, S., Kawazura, H.: Chem. Lett. *1981*, 585
135. Allendoerfer, R. D., Chang, R.: J. Magn. Resonance *5*, 273 (1971)
136. Laarhoven, W. H., Brus, G. J. M.: J. Chem. Soc. *B 1971*, 1433
137. Streitwieser, A., in: Molecular Orbital Theory for Organic Chemists, John Wiley, New York 1961, p. 175
138. Diesveld, J. W., Borkent, J. H., Laarhoven, W. H.: Recl. Trav. Chim. (Pays-Bas) *99*, 391 (1980)
140. Stulen, G., Visser, G. J.: J. Chem. Soc. Chem. Comm. *1969*, 965
141. De Rango, C., Tsoucaris, G., Declerq, J. P., Germain, G., Putzeys, J. P.: Cryst. Struct. Comm. *2*, 189 (1973)
142. Frank, G. W., Hefelfinger, D. T., Lightner, D. A.: Acta Crystallogr. Sect. *B 29*, 223 (1973)
143. van der Hark, Th. E. M., Noordik, J. H.: Cryst. Struct. Comm. *2*, 643 (1973)
144. Marsh, W., Dunitz, J. D.: Bull. Soc. Chim. Belg. *88*, 847 (1979)
145. Navaza, J., Tsoucaris, G., Le Bas, G., Navaza, A., De Rango, C.: ibid. *88*, 863 (1979)
146. Kuroda, R.: J. Chem. Soc. Perkin 2 *1982*, 789
147. Doesburg, H. M.: Cryst. Struct. Comm. *9*, 137 (1980)
148. Lindner, H. J., Kitschke, B.: Bull. Soc. Chim. Belg. *88*, 831 (1979)
149. Dewan, J. C.: Acta Crystallogr. Sct. *B 37*, 1421 (1981)
150. Beurskens, P. T., Beurskens, G., van den Hark, T. E. M.: Cryst. Struct. Comm. *5*, 241 (1976)
151. van den Hark, T. E. M., Beurskens, P. T.: ibid. *5*, 247 (1976)
152. Le Bas, G., Navaza, A., Mauguen, Y., De Rango, C.: ibid. *5*, 357 (1976)
153. Le Bas, G., Navaza, A., Knossow, M., De Rango, C.: ibid. *5*, 713 (1976)
154. Konno, M., Saito, Y., Yamada, K., Kawazura, H.: Acta Cryst. Section B *36*, 1680 (1980)
155. Green, B. S., Knossow, M.: Science *214*, 795 (1981)
156. Ramdas, S., Thomas, J. M., Jordan, M. E., Eckhardt, C. J.: J. Phys. Chem. *85*, 2421 (1981)
157. Herraez, M. A., Sanchez, F.: An. Soc. Esp. Fis. Quim. *61 B*, 953 (1965)

158. Kitaigorodskii, A. I., Dashevskii, V. G.: Tetrahedron *24*, 5917 (1968)
159. Dougherty, R. C.: J. Am. Chem. Soc. *90*, 5788 (1968)
160. Martin, R. H., Jespers, J., Defay, N.: Helv. Chim. Acta *58*, 776 (1975)
161. Declerq, J. P., Germain, G., van Meerssche, M., Debaerdemacker, T.: Actacrystallogr. B 32, 335 (1976)
162. Numan, H., Wijnberg, H.: Tetrahedron Lett. *1975*, 1097
163. Martin, R. H., Jespers, J., Defay, N.: ibid. *1975*, 1093
164. Borkent, J. H., Rouwette, P. H. F. M., Laarhoven, W. H.: Tetrahedron *34*, 2569 (1978)
165. For a review see: Balaban and Menitzesen in: Friedel Crafts and Related Reactions, Olah Ed. Interscience Publishers Inc. New York 1963, Vol. II, p. 979
166. Dopper, J. H., Oudman, D., Wijnberg, H.: J. Org. Chem. *40*, 3398 (1975)
167. Dopper, J. H., Wijnberg, H.: ibid. *40*, 1957 (1975)
168. Jespers, J., Defay, N., Martin, R. H.: Tetrahedron *33*, 2141 (1977)
169. op den Brouw, P. M., Laarhoven, W. H.: Recl. Trav. Chim. (Pays-Bas) *97*, 265 (1978)
170. Archer, W. J., El Din Shafig, Y., Taylor, R.: J. Chem. Soc. Perkin 2 *1981*, 675
171. Lehman, P. G., Wijnberg, H.: Aust. J. Chem. *27*, 315 (1974)
172. Mallory, F. B., Mallory, C. W.: J. Org. Chem. *48*, 526 (1983)
173. Wijnberg, H., Nieuwpoort, W. C., Jonkman, H. T.: Tetrahedron Lett. *1973*, 4623
174. Laarhoven, W. H., van Broekhoven, J. A. M.: ibid. *1970*, 73
175. Soubrier-Payen, B.: Thesis, Louvain la Neuve, 1978
176. van Meerssche, M., Germain, G., Declerq, J. P., Soubrier-Payen, B., Figeys, H. P., Vanommeslaeghe, P.: Acta Crystallogr. *B 37*, 1218 (1981)
177. Katz, J., Pesti, J.: J. Am. Chem. Soc. *104*, 346 (1982)
178. Nakazaki, M., Yamamoto, K., Ikeda, T., Kisuki, T., Okamoto, Y.: J. Chem. Soc. Chem. Comm. 787 (1983)
179. Martin, R. H., Morian, J., Defay, N.: Tetrahedron *30*, 179 (1974)
180. Defay, N., Martin, R. H.: Bull. Soc. Chim. Belg. *93*, 313 (1984)

Stereochemistry of the Complexes of Neutral Guests with Neutral Crown Host Molecules

Fritz Vögtle[1], Walter M. Müller[1], and William H. Watson[2]

1 Institut für Organische Chemie und Biochemie der Universität, Gerhard-Domagk-Str. 1, 5300 Bonn 1, FRG
2 Texas Christian University, Department of Chemistry, Forth Worth, Texas 76129, U.S.A.

Table of Contents

1 Introduction

Molecular cavities are of topical research interest because of their ability to enclose and bind guest molecules. They may serve as models for the study of binding sites between, e.g. drugs, odorant/taste substances, antigens, etc. and receptors. Cyclodextrins, as prime examples of host cavities, have found many useful applications. This is due to the guest molecules being bound within the cavity which changes properties such as solubility, volatility and reactivity.

This report is a survey of progress within the last few years in the field of complexation of *uncharged* guest molecules by *uncharged* host molecules of the crown ether type [1]. When the guest or the host bear charges permitting stronger host/guest binding, the conditions are much simpler. Not only have complexes of simple crown ethers with cationic organic guests (ammonium salts, amino acids) [2] been prepared but complexes with more complex hosts [3] and cations [4] have been reported. A characteristic example, confirmed by X-ray analysis, is the inclusion of a guest dication in the cavity of a [3]cryptand, shown in Fig. 1 [5]:

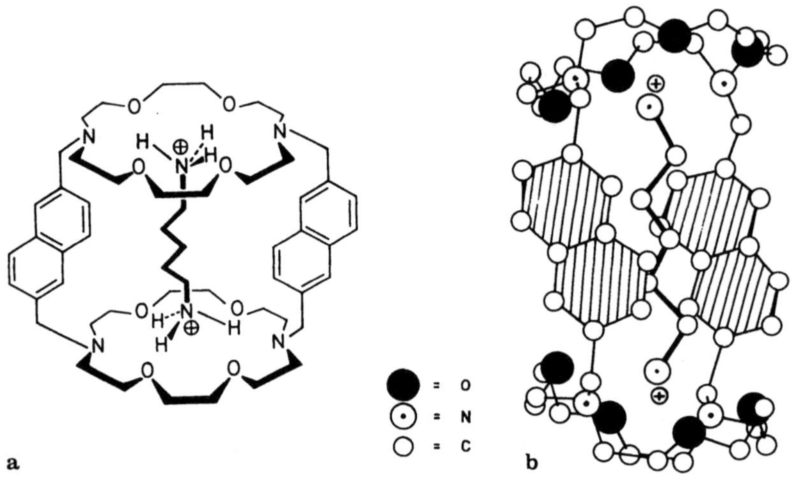

Fig. 1. The inclusion of a *pentamethylene-diammonium* guest cation in an uncharged cryptand [5]: **a** formula, **b** crystal structure

Conversely, inclusion complexes of neutral guest molecules within charged hosts have been synthesized successfully and the structures confirmed conclusively: As an example, Fig. 2 shows the results of the X-ray analysis of *durene* encapsulated in the host cavity of a multimembered tetra(ammonium) cation:

Although complexes between either charged or uncharged host and charged guest molecules can be prepared systematically, the preparation of *uncharged* guest/uncharged host complexes still remains a problem. This is particularly true of molecular encapsulation complexes. In the last few years, however, there has been an increase in the number of approaches utilized to solve this problem. This is of particular

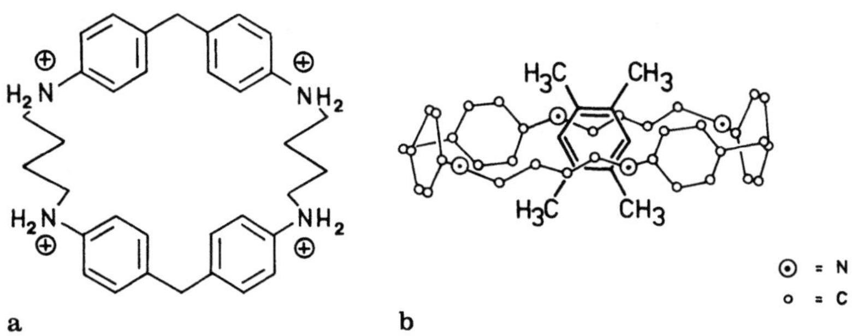

Fig. 2. *Durene* as neutral guest in the cavity of a 30-membered ring host cation [6]: **a** host cation, **b** structure of the complex

importance since weak multiple interactions between host and guest can be studied without perturbation by stronger forces.

For clarity and because of space limitations, we will consider only examples of complexes in which the guest interacts with the host from above (or/and below) the cavity, and the structure has been confirmed by X-ray analysis. These complexes serve as models for the interaction of neutral molecules which are nearly concave/convex complementary in their spatial interaction. Although complementary in shape, some guest molecules simply fill spatial voids in the host lattice. These clathrate-like structures will not be considered.

There are several factors which will be important in describing the stability and conformation of the host/guest complexes. For the [18]crown-6 host, molecular mechanics calculations indicate the biangular conformation *A* found in the solid [7] is 32.8 $kJmol^{-1}$ more stable than the D_{3d} conformation *B* [8] (see Fig. 3). This is attributed primarily to unfavorable dipolar interactions between oxygen atoms (separation 2.8 Å) in the D_{3d} conformation. The conformations can be described in terms of the torsion angles around the crown ether ring. A common notation involves a consideration of the R—O—CH_2—CH_2—O—R unit and whether the torsion angles are primarily synclinal (gauche, g) or antiperiplanar (a). In the D_{3d} conformation *B* the (aga ag^-a aga) pattern is found while in conformation *A* the (aga ag^-g aaa ag^-a) pattern is adopted. Conformation *B* represents a minimum torsional strain structure,

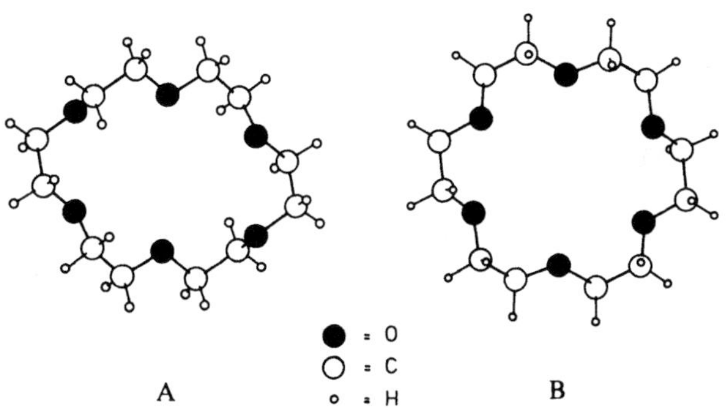

Fig. 3. Conformation of [18]-crown-6(*1*): **A** uncomplexed form, **B** the D_{3d} conformation exhibited by the majority of complexes for which structural data are available

and if it were not for the dipolar interactions, it would be of lowest energy. The insertion of electron accepting groups into the crown cavity reduces these dipolar interactions, and acidic $-XH_3$ and $-XH_2$ groups, which are capable of participating in multiple XH··O interactions, stabilize the D_{3d} conformation. Molecules with electron acceptor sites on several adjacent atoms or with only one electron acceptor group may result in conformations somewhere between that of *A* and *B*. The molecular mechanics calculations indicate at least three conformations lie energetically between conformation *A* and *B*.

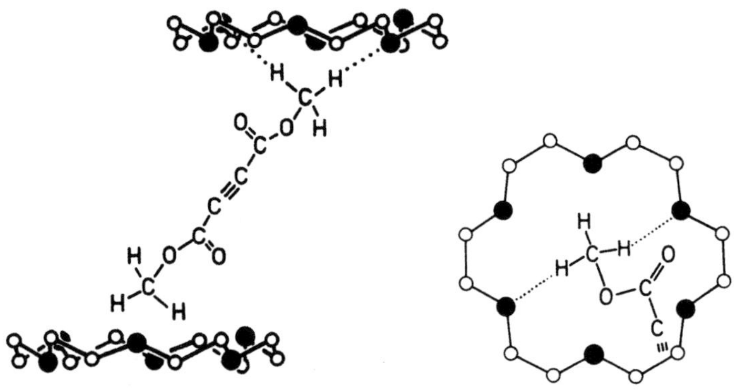

a b

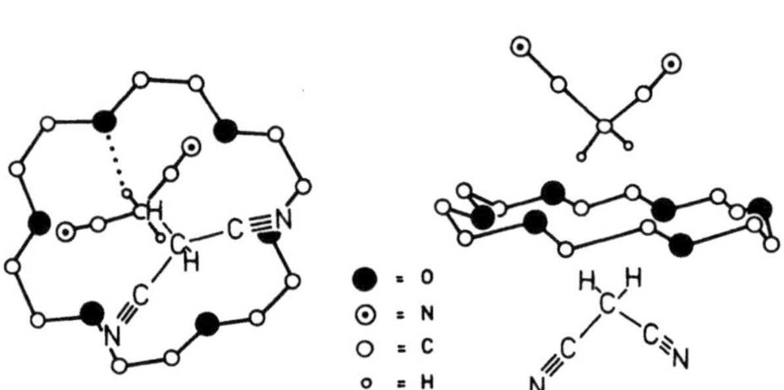

a b

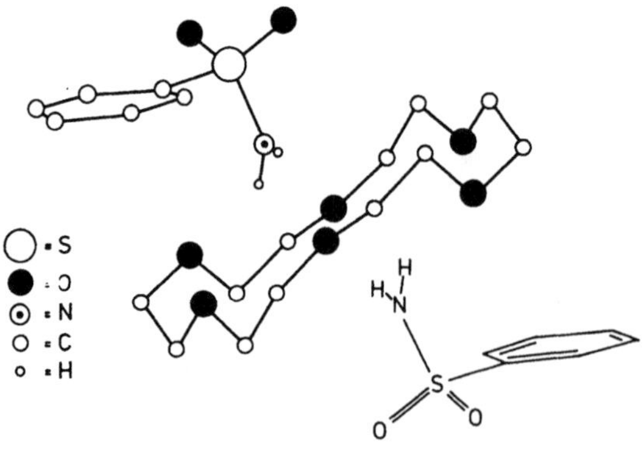

Fig. 4–6. X-ray analyses of [18]crown-6 complexes with *dimethyl acetylene dicarboxylate* [11], *malonodinitrile* [12], and *benzenesulfonamide* [13]. In the latter case the crown is in a conformation intermediate between *A* and *B* (Fig. 3e in [8])

It has been suggested that in the crown-ether complexes, crystal stability is correlated with an increase in the X—H acidity of the guest [9]. Bifurcated hydrogen bonds are also common in these host/guest complexes, and their occurrence has been correlated with an increase in the acidity of the X—H groups [10].

2 Complexes of [18]Crown-6 with CH-, NH-, and OH-Acidic Compounds

The [18]crown-6 (*1*) complexes with *dimethyl acetylene dicarboxylate* (1:1) [11], *malonodinitrile* (1:2) [12], and *benzenesulfonamide* (1:2) [13] were the first to be revealed by X-ray structural analyses. Figures 4–6 show the stereochemistry of these neutral complexes within the crystal[1]. In the first two complexes the CH_3 and CH_2 groups

Table 1. Stoichiometries, melting points, symmetry, and R-values of the hitherto described complexes of [18]crown-6 with uncharged guest molecules

Guest	Stoichiometry Host:Guest:H_2O	melting point [°C]	Structure of [18]crown-6 (cf. Fig. 3)	R-value [%]	Fig. No.	Ref.
dimethylacetylene dicarboxylate	1:1	101	D_3d (B)	3.6	4	11)
malonodinitrile	1:2	137–138	D_3d (B)	4.9	5	12)
nitromethane	1:2	70–85	D_3d (B)	13.0	7	14)
dimethylsulfone	1:2	94–95	D_3d (B)	4.9	8	15)
dimethylsulfate	1:1	88–94	D_3d (B)	8.0	9	16, 18)
N,N′diformylhydrazine	1:2	125–130	D_3d (B)	4.4	11	20, 19)
2,4-dinitrophenylhydrazine	1:2	164–168	D_3d (B)	11.0	10	9, 19)
3-nitroaniline	1:2	55–63	— (A)	6.6	15	27, 21)
4-nitroaniline	1:2	89–97	D_3d (B)	6.4	14	25, 21)
2,4-dinitroaniline	1:2	127–135	— (A)	7.4	13	23, 21)
4-nitro-1,2-phenylene-diamine	1:2	97–110, 180–193	— (A)	4.7	12	22, 21)
dithiooxamide	1:2	149–154	— (A)	4.2	20	29, 21)
formamide	1:2	92–97	D_3d (B)	7.1	21	29, 32)
urea	1:5	148–150	— (A)	3.9	18	31)
thiourea	1:4	168–174	—	—	19	32, 29)
N-methylthiourea	1:1	85–95	D_3d (B) + (A)	6.8	17	29)
N,N′-dimethylthiourea	1:2	86–88	D_3d (B)	4.4	16	28)
benzenesulfonamide	1:2	92	— (A)	5.9	6	13)
phenylcarbamate	1:2	108–115	D_3d (B)	4.0	22	34, 21)
3-nitrophenol	1:2:2	49–56	D_3d (B)	5.0	24	29, 21)
2,4-dinitrophenol	1:2:2	59–67	D_3d (B)	4.9	25	36, 21)
4,4′-biphenyldiol	1:1:2	255–261	D_3d (B)	7.3	27	37)
4-nitrobenzaldoxime	1:2:2	45–65, 68–70	D_3d (B)	7.2	26	29, 21)

1 In the following figures, when there are two guest molecules arranged above and below the crown, one of them will be shown with circles or the ellipses drawn by the computer and the other will be represented by the characters of the formula. This is done to clarify the host/guest arrangements and to save space.

are fitted into the crown ether cavity. This interactions is favored also by the complementary spatial arrangement of the host and guest. The short CH··O and NH··O contacts between host and guest in the three complexes indicate that hydrogen-bond-like interactions play a significant role in stabilizing the complexes through multiple XH··O interactions. These interactions lead to the host adopting the D_{3d} conformation.

2.1 CH-Acidic Guest Molecules

Figures 7–9 show structures determined by X-ray analysis of the complexes of [18]crown-6 with *nitromethane* (1:2) [14], *dimethyl sulfone* (1:2) [15], and *dimethyl sulfate* (1:1) [16]. The crystalline nitromethane complex, first prepared by McLachlane [17], consists of crown rings in a nearly regular D_{3d} conformation (cf. Fig. 3B) [14]. There are close contacts between the methyl carbon atom of the guest and the oxygen atoms of the ring (average C··O 3.8 Å). Although the hydrogen atoms were not located, it was inferred that multiple CH··O interactions are important contributors to the stability of the host/guest complex. The spatial arrangement of the methyl hydrogens and the oxygen atoms of the crown would be particularly favorable for these interactions. Dipole-dipole interactions *also* may contribute to the host/guest binding. In addition the fitting of the methyl group of the guest into the cavity of the crown ring may lead to efficient crystal packing.

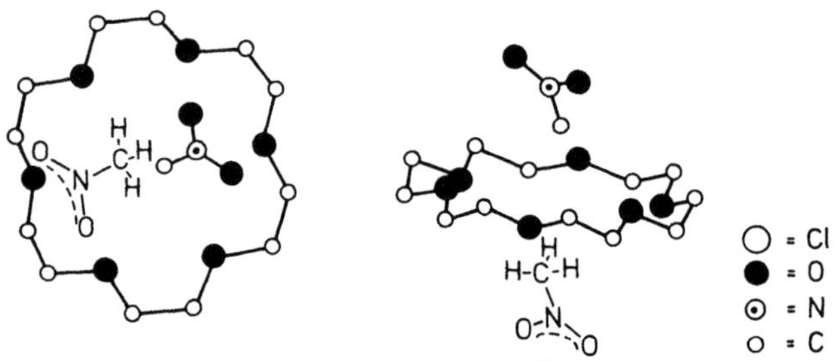

Fig. 7. Crystal structure of the [18]crown-6 *nitromethane* 1:1 complex [14a]

Table 2. Thermodynamic constants of complexes of crown ethers with some CH-acidic guest molecules in deuteriobenzene [11]

Guest	Host	ΔH^0 [kcal mol^{-1}]	ΔS^0 [cal mol^{-1} K^{-1}]
CH_3CN	*1*	− 6.0	− 2.2
CH_3NO_2	*1*	− 7.6	−25
$CH_2(CN)_2$	*1*	−14.2	−37
CH_3NO_2	*2*	− 3.6	−13

In solution, the dynamic ^{1}H-NMR spectrum indicated the presence of 1:1 and 1:2 complexes of [18]crown-6 with nitromethane, acetonitrile and malonodinitrile. The thermodynamic complexation constants for the 1:1 complexes are listed in Table 2.

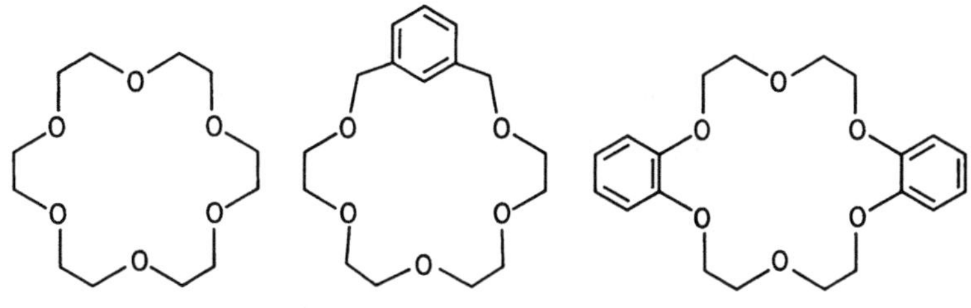

1 *2* *3*

In contrast to [18]crown-6 (*1*) "1,3-xylene[18]crown-6" (*2*) forms only a 1:1 complex with nitromethane in solution. The ΔH° and ΔS° values are listed in Table 2.

Through selective complexation with nitromethane, dimethyl carbonate and dimethyl oxalate, it was possible to isolate [18]crown-6 in 98% purity from a crude reaction mixture [14a]. Although the acetonitrile complex with [18]crown-6 has been known for some time, the structure was determined only recently because of rapid crystal deterioration [14].

As illustrated in Figure 8a, the [18]crown-6 molecule in the 1:2 complex with *dimethyl sulfone* also adopts an approximate D_{3d} conformation. A methyl group of each dimethyl sulphone guest interacts with alternating oxygen atoms giving CH··O contacts of 2.47, 2.48, and 2.62 Å.

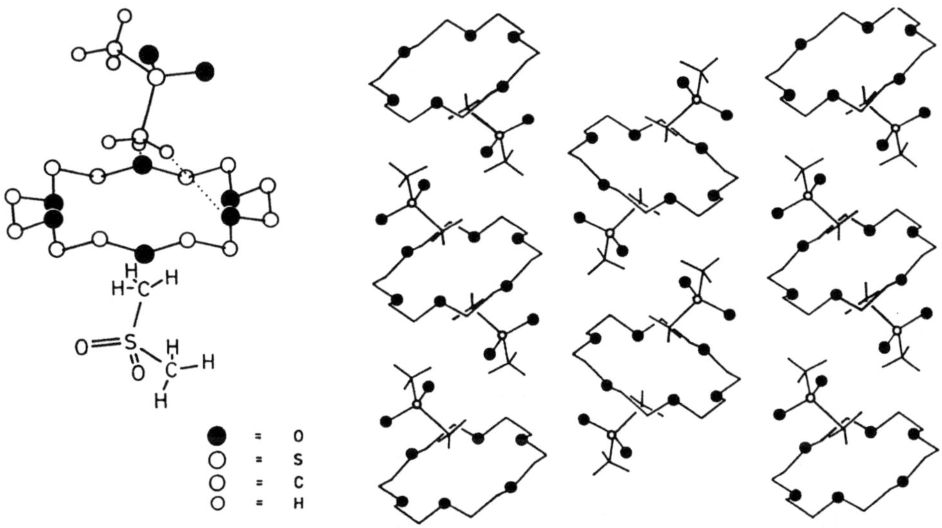

Fig. 8. The [18]crown-6 *dimethyl sulfone* complex [15] (schematically): **a** CH···O distances of less than 2.50 Å are shown with a dotted line; **b** section of the crystal lattice

In the 1:1 complex between [18]crown-6 and dimethyl sulfate [16,18], the three hydrogen atoms of each methyl group lie between 2.35 Å and 2.58 Å of three facial oxygen atoms. While each CH··O interaction is much weaker than a normal hydrogen bond, the three interactions per methyl group lead to stabilization. A combination of increased C—H bond polarity (acidity) and spatial arrangements which permit multiple interactions is necessary in these complexation reactions. Even under these optimum conditions the binding between host/guest is weak as indicated by large thermal anisotropy and easy recovery of host and guest.

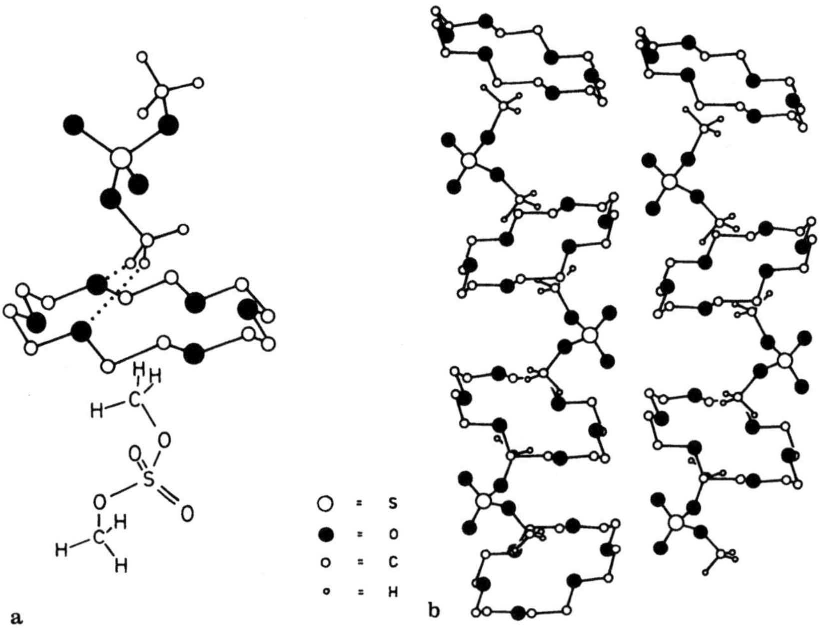

a b

Fig. 9. a Structure of the [18]crown-6 *dimethyl sulfate* complex [16]; **b** details of the infinite chain of alternating host and guest molecules [16]

2.2 NH-Acidic Guest Molecules

In the [18]crown-6 *2,4-dinitrophenylhydrazine* 1:2-complex [19] the two guest molecules are centrosymmetrically arranged above and below the plane of the hexaether. The hydrogen bonding is extensive and unique in the complex. Each hydrazine moiety is involved in four NH··O interactions which range from 2.09 to 2.53 Å and a fifth of 2.66 Å. Two of the hydrogens form bifurcated bonds with ether oxygen atoms while the third forms a bifurcated bond with an ether oxygen and intramolecularly to a nitro oxygen. The five interactions on each face involving two nitrogen centers lead to an elongated conformation of the [18]crown-6 molecule. There may also be an accommodation by the guest in order to maximize the interaction. Four of the crown ether oxygen atoms are bonded from above and below [9].

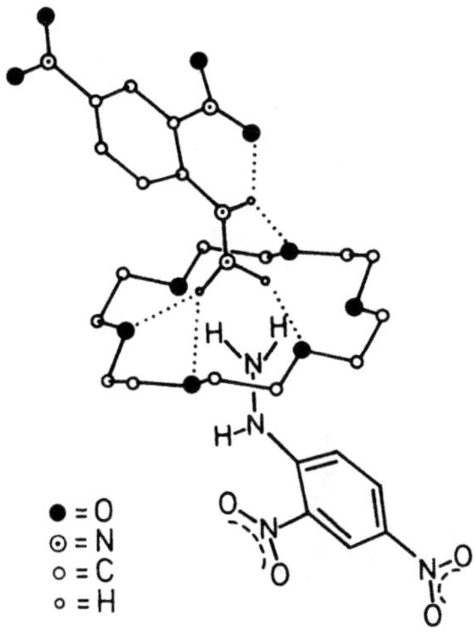

Fig. 10. Structure of the [18]crown-6 *2,4-dinitrophenyl hydrazine* 1:2 complex [9]

In the complex with *1,2-diformylhydrazine* the [18]crown-6 [19] molecule lies on an inversion center, and the conformation deviates only slightly from D_{3d} symmetry (Fig. 3B). The conformation of the guest molecule approaches C_2 symmetry with a torsion angle about the N—N bond of 105.4°. The 1,2-diformylhydrazine guest molecules are linked by two identical NH··O=C hydrogen bonds, which are related through an inversion center (Fig. 11). Each guest is bound to the host by only one NH··O bond. This is unusual since most stable complexes require multiple interactions between host and guest; however, stabilization is achieved through guest-guest hydrogen bonding. In solution the interaction scheme may be modified.

In the 1:2 complex of [18]crown-6 with *4-nitro-1,2-benzenediamine* [21] the crown ether also lies on an inversion center [22]. Both hydrogen atoms of the 1-amino group are bonded to alternate host oxygen atoms. One hydrogen of the 2-amino group also

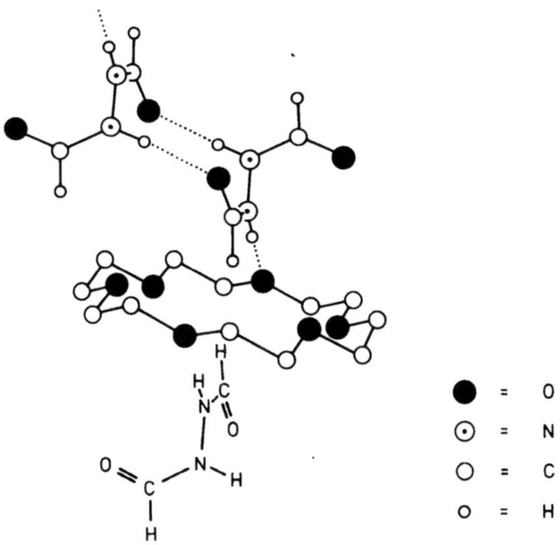

Fig. 11. Structure of the [18]crown-6 1:2 complex with *1,2-diformylhydrazine* [20]

participates in hydrogen bonding and binds to the same oxygen as one of the hydrogens of the 1-amino group. Thus, two oxygen atoms on each face are hydrogen bonded to three different hydrogen atoms. The conformation of the crown ether deviates from D_{3d} symmetry to accommodate the extended hydrogen bonding. The two amine molecules are tilted by 120° with respect to the mean plane of the ether oxygen atoms in order to facilitate the interaction of the 2-amino hydrogen atom.

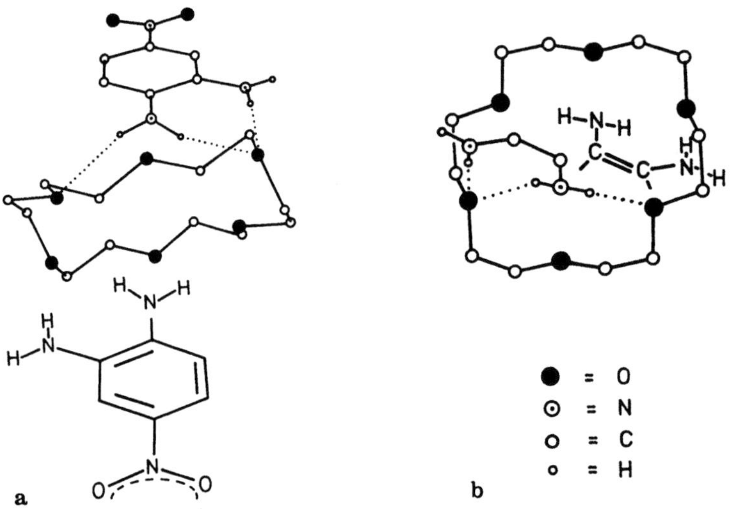

Fig. 12. Structure of *4-nitro-1,2-benzenediamine* [18]crown-6 complex [22]: **a** a perspective view, **b** as viewed almost perpendicular to the plane of the crown, hydrogen bonds marked with dotted lines

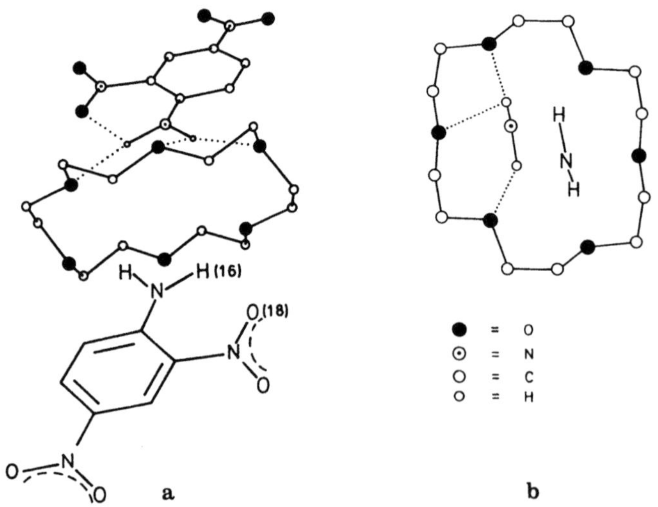

Fig. 13. **a** Structure of the [18]crown-6 *2,4-dinitroaniline* 1:2 complex [23]; **b** as seen nearly perpendicular to the plane of the crown, and the intermolecular hydrogen bonds [23]

The guest molecules in the 1:2 [18]crown-6 *2,4-dinitroaniline* complex [21] are related by an inversion center (Fig. 13) [23]. Both hydrogen atoms of the amine are involved in bifurcated hydrogen bonds.

One hydrogen binds to two adjacent ether oxygen atoms while the other binds to one ether oxygen and intramolecularly to one oxygen of the ortho nitro group. The hydrogen bonding is unusual in that it involves three adjacent oxygen atoms rather than alternate atoms. The acidity of N—H hydrogen atoms leads to bifurcated hydrogen bonding [10] which is most efficiently accommodated by bridging adjacent oxygen atoms. This leads to a crown conformation (aga* aga gga) which differs significantly from either *A* or *B* (Fig. 3). The C(2)—C(3)—O(4)—C(5) torsion angle of 140° (a*) is highly distorted away from the ideal value of 180° and might be better classified as anticlinal. The centrosymmetrically related guest molecules are almost perpendicular to the hexagon consisting of the six oxygen atoms. Atoms of the intramolecularly hydrogen-bonded 6-membered ring of the guest deviate from planarity by +0.12 Å while the same rings in crystalline 2-nitroaniline [24] and in the [18]crown-6 2,4-dinitrophenyl hydrazine [9] are planar. This deviation is attributed to the formation of a bifurcated hydrogen bond by the ring hydrogen. The data indicate these interactions are sufficiently strong that host and guest are able to make conformational accommodations.

[18]crown-6 4-nitroaniline 1:2 complex

The complex is obtained by refluxing the crown ether with *4-nitroaniline* in toluene [21]. The crown ether adopts a D_{3d} conformation although there is a 14.3% contribution in the solid from another conformer in which one centrosymmetrically related pair of oxygen atoms does not adopt the normal "aga" pattern. In the D_{3d} conformation the two amino hydrogen atoms of the guest hydrogen bond to two alternate oxygen atoms [26]. The two centrosymmetric guest molecules therefore coordinate with four of the ether oxygen atoms which usually is sufficient to insure a D_{3d} conformation. In the minor [18]crown-6 conformation, the same four oxygen atoms are coordinated but the remaining two are now coordinated via bifurcated hydrogen bonds. The

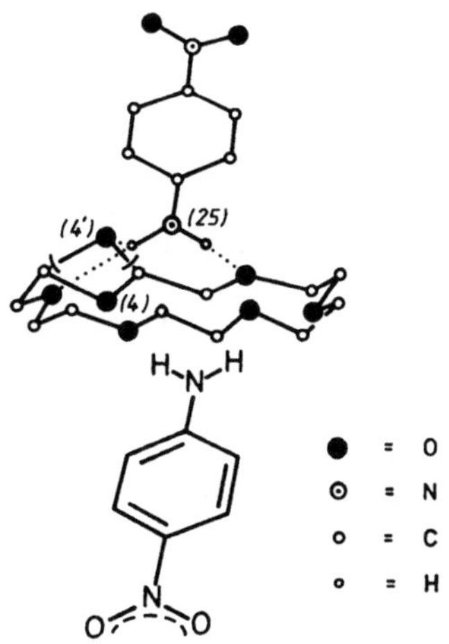

Fig. 14. X-ray structure of the [18]crown-6 *4-nitroaniline* 1:2 complex [25]

new H··O distance is 2.42 Å. The amino group in nitroaniline is less acidic than those in 2,4-dinitrophenyl hydrazine and 2,4-dinitroaniline where bifurcated bonds lead to only one conformation of the host.

[18]crown-6 3-nitroaniline 1:2 Complex but 1:1 Crystal Stoichiometry

Although the stoichiometry of the *3-nitroaniline* complex is 1:1 in the solid, the complexed host/guest are in a 1:2 ratio. Guest molecules with two widely separated functional groups can form 1:1 complexes due to their ability to bridge multiple oxygen atoms in a conformationally related crown ether. This would not lead to the usual centrosymmetric relationship of the guest to each host. Because 3-nitroaniline contains only one electron acceptor group a 1:1 complex indicates some unusual structural accommodation. In the solid the [18]crown-6 molecules are layered in such a manner that a 180° rotation about the C—NO_2 bond followed by a small rotation about an axis through the phenyl ring and a 0.23 Å translation will move the complexing —NH_2 group from complexation with one ether into a complexing orientation with another. The 3-nitroaniline molecules are disordered in the solid state by such a shift. The structure is such that 38.3% of the time molecule I is complexed in a 1:2 fashion while II is uncomplexed, and 61.7% of the time I is free and II is complexed in a 1:2 fashion. All crown ether molecules adopt a D_{3d} conformation.

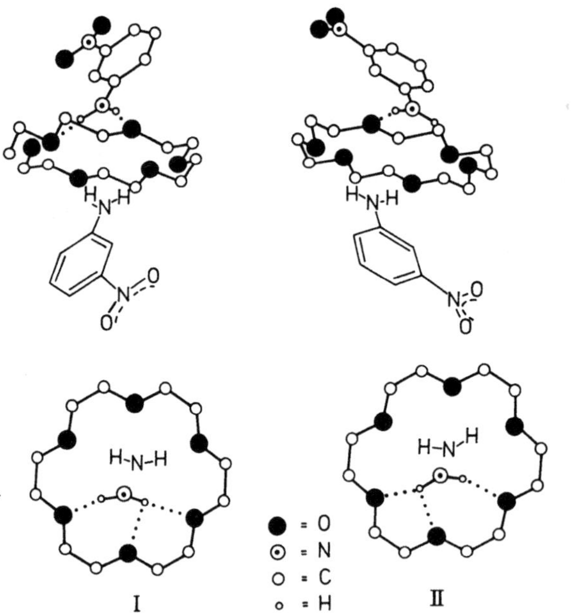

Fig. 15. Structure of the [18]crown-6 *3-nitroaniline* complex [27]: Upper picture: host/guest units I and II; lower picture: as viewed perpendicular to the plane of the crown (hydrogen bonds drawn dotted)

The free crown ethers might be expected to adopt the free ligand conformation (Fig. 3A); however, there is no evidence of disorder or unusual thermal motion associated with the crown ethers. It was suggested that the free crown ethers maintain their D_{3d} conformation because of crystal packing forces. The complexes are formed by utilizing both hydrogens of the amino to interact with alternate oxygen atoms. The centrosymmetrically related guests bind to four of the six host oxygen atoms.

[18]Crown-6 N,N'-dimethyl Thiourea 1:2 Complex

The two centrosymmetrically related guest molecules are coordinated to the host through single strong NH··O hydrogen bonds (2.08 Å). The guest molecules of adjacent complexes are connected via NH··S (2.54 Å) interactions involving the second N—H group. This leads to infinite chains along the Z axis.

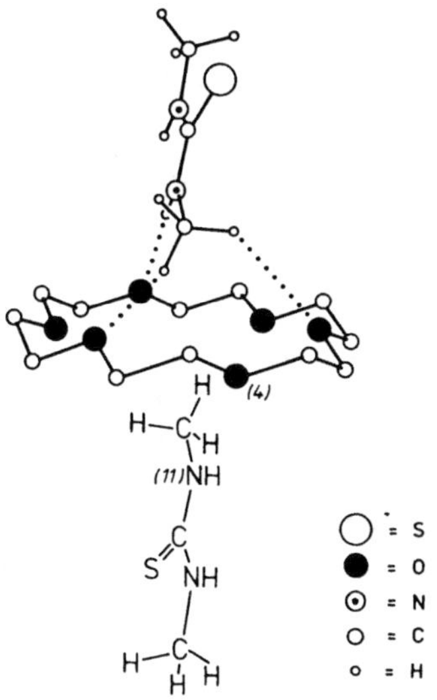

Fig. 16. [18]Crown-6 *N,N'-dimethyl thiourea* 1:2 complex: the guest molecules are related by a center of symmetry [28]

Normally, singly coordinated complexes do not adopt the D_{3d} conformation, and in complexes with urea and thiourea the [18]crown-6 is found in the biangular conformation (Fig. 3A). In the dimethyl thiourea complex two methyl hydrogen atoms of each guest lie 2.60 Å from alternate oxygen atoms of the crown. These hydrogen atoms may reduce dipolar interactions between the oxygen atoms and provide a weak attractive interaction. The hydrogen bonding between guest molecules may also provide sufficient lattice energy to stabilize the conformation.

[18]Crown-6 N-methyl Thiourea 1:1 Complex

A 1:1 complex is possible because both ends of the N-methyl thiourea guest bind centrosymmetrically to separate host molecules. This results in long chains of alternating host/guest molecules.

The electron acceptor geometry at the two ends of the guest have differing spatial demands, and the two independent host molecules adopt different conformations. Each host molecule sits on a center of symmetry. The NH_2 group forms two hydrogen bonds to a host in a distorted biangular conformation (Fig. 3A). The single amine hydrogen is bonded to an oxygen of a host exhibiting the D_{3d} conformation. The CH_3 group lies within the cavity of this host molecule, and one hydrogen atom is

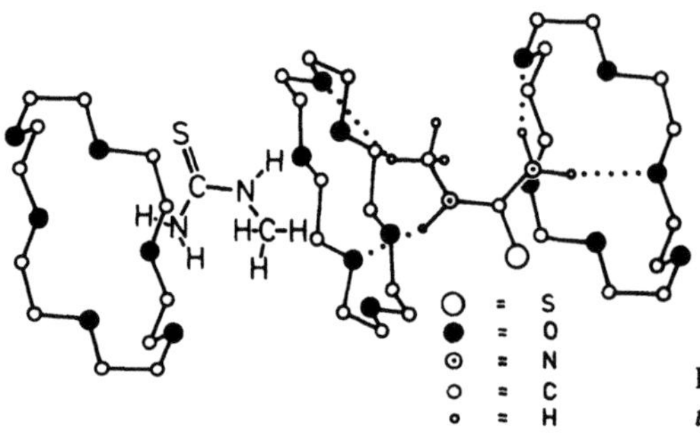

Fig. 17. Structure of the [18]crown-6 *N-methyl thiourea* complex [29)]

oriented toward an ether oxygen. This probably represents an energetically favorable interaction which stabilizes the host conformation. This structure shows that the [18]crown-6 host, to a limited extent, can modify its conformation to accommodate the spatial requirements of the guest.

[18]Crown-6 urea 1:5 Complex

In a second historical paper [30a)] on cyclic oligoethers, Pedersen mentioned several crystalline dibenzo[18]crown-6 adducts with thiourea in which the thiourea to crown ratio varied from 12:3 to 6:1 [30b)]. The urea complex was first prepared in 1981 by the addition of ether to a homogeneous solution of [18]crown-6 and *urea* in a 1:2 methanol/chloroform mixture [31)]. Recrystallization from methanol/ethylacetate yielded the analytically pure 5:1 complex with a melting point of 148–150 °C. An X-ray analysis at —125 °C [31)] revealed a crystalline solid with a 5:1 urea/host ratio containing 2:1 complexes and uncomplexed urea.

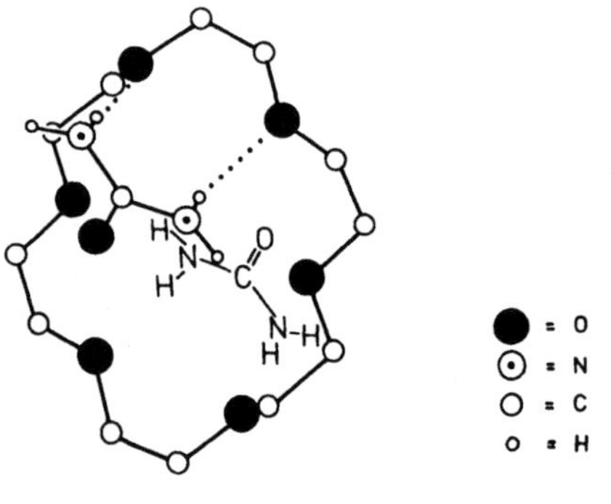

Fig. 18. [18]Crown-6 *urea* 1:5 stoichiometry with 1:2 host/guest complexes. One of the two independent [18]crown-6 molecules is shown with the two complexed guest urea molecules [31)]

The solid contains two independent crown ether molecules lying on centers of symmetry and exhibiting the biangular conformation (Fig. 3A). One hydrogen from each of the amino groups in the guest molecule hydrogen bonds to adjacent oxygen atoms. This ··HNCNH·· bridge across adjacent oxygen atoms is unusual. The N··O distances in the two independent molecules range from 2.84(8) to 3.28(2) Å which is normal for NH··O hydrogen bonds. The oxygen atom of the urea then fits into a two-dimensional layer of hydrogen bonded urea molecules which results in alternating layers of urea molecules and host/guest complexes.

[18]Crown-6 Thiourea 1:4 Complex

The crystalline solid contains a 4:1 stoichiometric guest/host ratio and is composed of 2:1 guest/host complexes with two additional *thiourea* molecules hydrogen bonded to the guest molecules of the complex. There are three independent crown ether molecules in the solid. Only one of these is shown in Fig. 19, and it lies on a center of symmetry.

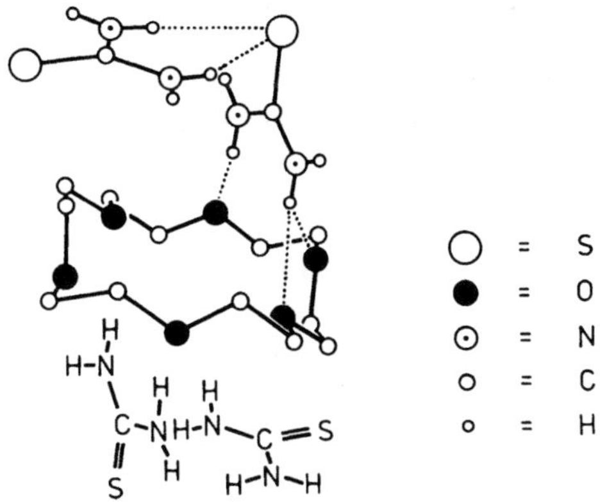

Fig. 19. One of the three host/guest complexes in the [18]crown-6 *thiourea* 1:4 structure (hydrogen bonds shown as dotted lines) [32]

The three crown ethers show minor variations in conformation. As with the previous urea complex both amide groups of a thiourea are hydrogen bonded to adjacent oxygen atoms of the crown ether. In the complex shown in Fig. 19, one of the hydrogen atoms forms a bifurcated hydrogen bond involving an adjacent oxygen of the crown ether. This additional H··O interaction is not observed in the other two complexes. Additional thiourea molecules now bind to the sulfurs of the coordinated guests via hydrogen bonds from the two amide groups. The structure might be described as a hydrogen bonded thiourea adduct of a 2:1 guest/host complex.

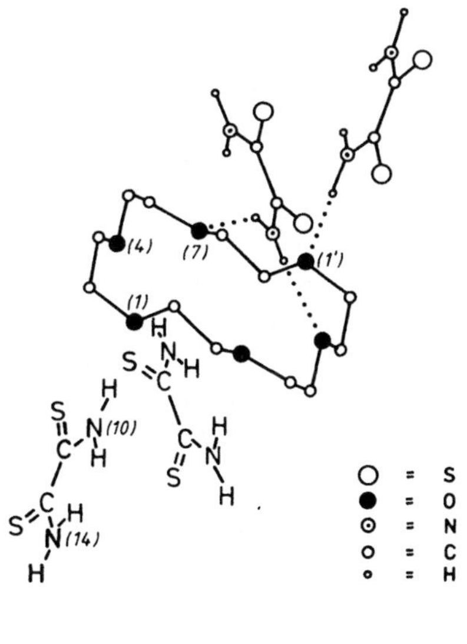

Fig. 20. Structure of the [18]crown-6 *dithiooxamide* 1:2 complex [29]

[18]Crown-6 Dithiooxamide 1:2 Complex

This is an unusual complex because each crown ether is coordinated to four *dithiooxamide* molecules. The crown ether sits on a center of symmetry and is in a distorted biangular conformation (Fig. 3A). One centrosymmetrically related pair of guests interacts with alternate oxygen atoms via both hydrogen atoms of an amide group (H··O, 2.07 and 2.61 Å). The second amide group of these guests binds via one hydrogen to an adjacent crown ether. A second centrosymmetrically related pair of guests binds to the crown ether by a single hydrogen bond (H··O = 1.85 Å). The other amide of these guests binds via both hydrogens to a different crown ether, Fig. 20.

Thus each crown is bound to four guest molecules and through these guests to four other crown ethers. This generates a complex three-dimensional network.

[18]Crown-6 Formamide 2:1 Complex [33]

This is a simple 2:1 complex with the crown sitting on a center of symmetry in the D_{3d} conformation (Fig. 3B). The symmetry related guest molecules interact via one strong hydrogen bond (H··O = 2.09 Å) and a weaker interaction (H··O = 2.84 Å).

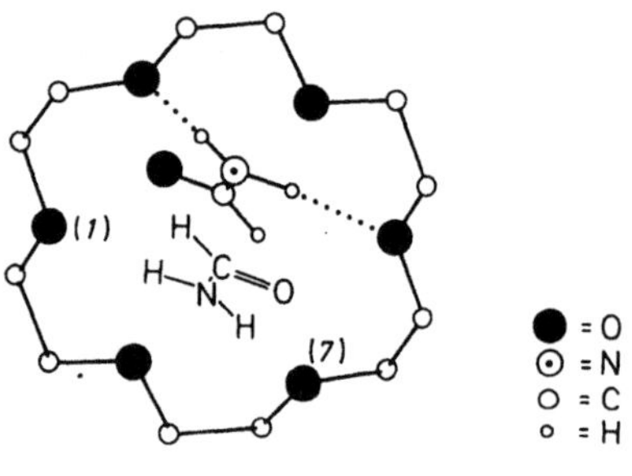

Fig. 21. X-ray structure of the [18]crown-6 *formamide* 1:2 complex [29]

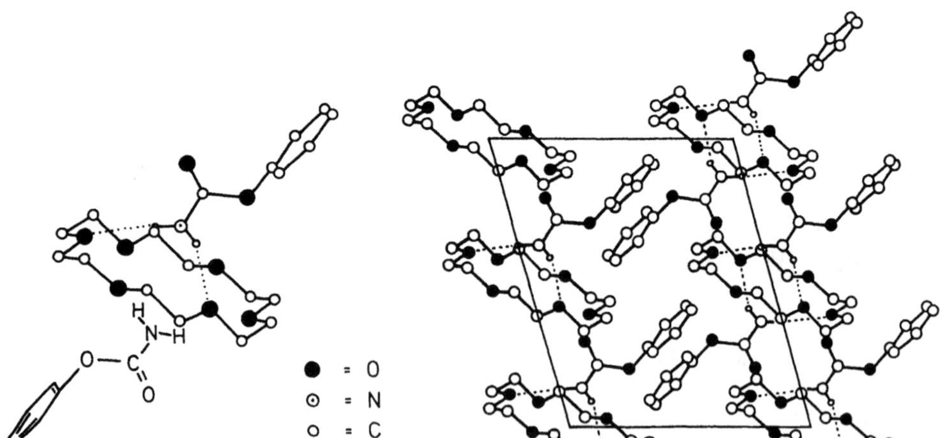

Fig. 22. Structure of the [18]crown-6 *phenyl carbamate* 1:2 complex [34]. **a** Molecular complex; **b** packing showing hydrogen bonds

The second hydrogen is probably more important in reducing the dipolar repulsion between oxygen atoms than in stabilization of the complex through binding.

[18]Crown-6 Phenyl Carbamate 1:2 Complex

The [18]crown-6 sits on a center of symmetry in the D_{3d} conformation. The symmetry related guests interact with the host via both $—NH_2$ hydrogen atoms. The hydrogens bond to alternate oxygen atoms of the crown ether (H··O, 2.12, 2.39 Å).

Dibenzo[18]crown-6 and [18]crown-6 Complexes with [trans-$PtCl_2(PMe_3)NH_3$)]

This is an area of complexation chemistry which should become of increased interest in a number of fields of chemistry. The 1:1 complex between the *trans*-platinum complex and dibenzo[18]crown-6 ether is unusual in that all six ether oxygen atoms are bound from one side by a single $—NH_3$ group forming bifurcated hydrogen bonds, see Fig. 23b. The six N··O distances range from 3.32 to 3.41 Å. The C_{2v} conformation adopted by the crown and the six interactions on one face are probably related to the incorporation of two rigid phenyl rings into the system.

The *trans*-platinum complex forms a 2:1 complex with [18]crown-6 and two centrosymmetrically related $—NH_3$ groups bind 3 alternate oxygen atoms on each face, Fig. 23a. The crown adopts the normal D_{3d} conformation, and the three independent N··O distances range from 3.04 to 3.31 Å.

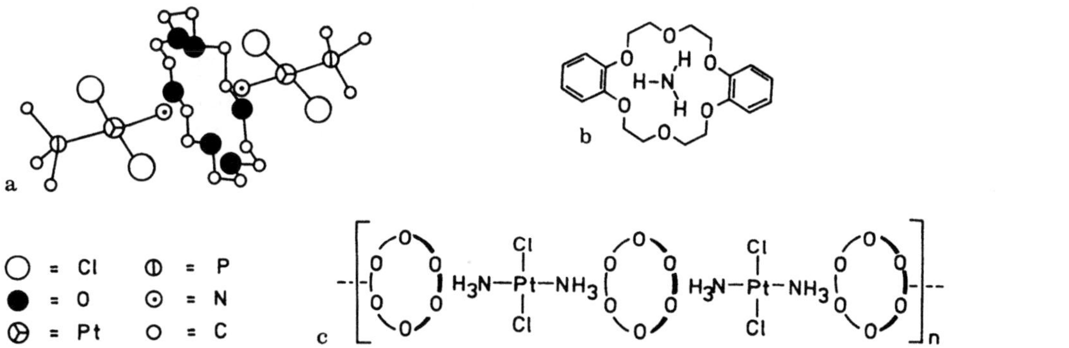

Fig. 23. a the 1:2 complex between [18]crown-6 and *[trans-$PtCl_2(PMe_3)NH_3$]* [35a].
b Schematic representation of the three bifurcated hydrogen bonds in the 1:1 [trans-$PtCl_2(PMe_3)NH_3$] dibenzo[18]crown-6 complex [35a].
c proposed structure of the 1:1 [18]crown-6 [trans-$PtCl_2(NH_3)_2$] complex [35a]

A 1:1 mixture of [18]crown-6 and [trans-$PtCl_2(NH_3)_2$] yielded a solid in which a copolymer structure was proposed, Fig. 23c.

[18]crown-6 [$Cu(NH_3)_4H_2O$] [PF_6]$_2$ 1:1 Complex [35b]

The crown ether adopts a D_{3d} conformation and is coordinated on both faces by NH_3 groups from centrosymmetrically related Cu^{++} complexes. The *trans*: NH_3 groups each bind to a crown ether forming long chain copolymers. In addition to the NH_3

hydrogen atoms in the linear chain each crown face is hydrogen bonded to two atoms from equatorial $:NH_3$ groups. The [18]crown-6 is hydrogen bonded to a total of 10 amine hydrogen atoms. This must be a record for a molecule of the size of [18]crown-6.

[18]crown-6 $[Pt(H_2NCH_2CH_2NH_2)_2]$ $[PF_6]_2$ 1:1 Complex [35c]

The ethylenediamine ligands form the normal chelate structure with Pt^{++}. The acidic $:NH_2$- hydrogen atoms lie above and below the plane of the square-planar Pt^{++} array. A crown ether sits above one chelate ring and is bound to one hydrogen from each of the two nitrogens of the chelate ring. The other chelate ring is hydrogen bonded to a crown ether which is below the plane. Each crown ether is then bound on the other face by hydrogens from other chelate rings. The -[crown ether]-[Cu complex]-[crown ether]- infinite chain forms a stairstep relationship.

2.3 OH-Acidic Guest Molecules

[18]crown-6 3-nitrophenol Diaquo 1:2:2 Complex [21]

This host/guest compound can be regarded as a ternary complex, consisting of *3-nitrophenol* and hydrated [18]crown-6. The crown has a D_{3d} (Fig. 3B) conformation with two centrosymmetrically related hydrogen bonded *water* molecules. Both hydrogen atoms of each water molecule form hydrogen bonds, H··O, 1.86 and 1.92 Å. Each 3-nitrophenol guest molecule is attached to the diaquo-[18]crown-6 by hydrogen bonds between the phenol and a water molecule, H··O = 1.64 Å. The interactions are similar to those in the 4,4'-biphenyldiol complex except the latter is a bifunctional guest and serves as a bridge between host molecules. The complex forms an independent cluster, which interacts with other clusters via van der Waals forces.

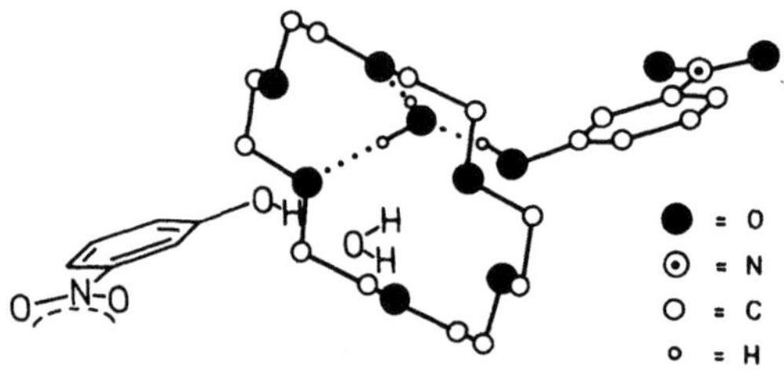

Fig. 24. [18]crown-6 *3-nitrophenol diaquo* 1:2:2 complex [29]

[18]crown-6 2,4-dinitrophenol Diaquo 1:2:2 Complex [21]

In this ternary complex the [18]crown-6 molecule also is situated on an inversion center with approximate D_{3d} symmetry. Two inversion related *water* molecules each act as hydrogen bond donors to two alternate ether oxygen atoms, O···H, 2.05, 2.27 Å. Each water molecule accepts a hydrogen bond from the hydroxyl group of the *2,4-dinitrophenol* guest molecule, H··O 1.67 Å. The binding between host and guest is thus modulated by bridges of water molecules.

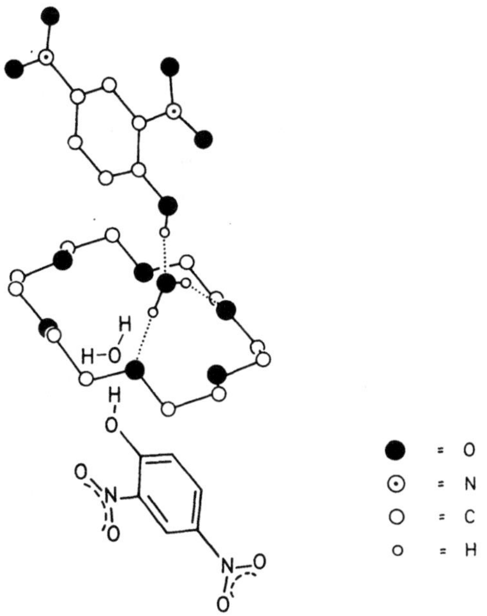

Fig. 25. Structure of the [18]crown-6 *dinitrophenol diaquo* 1:2:2 complex [36)]

[18]crown-6 4-nitrobenzaldoxime Diaquo Complex [21)]

This complex has a structure similar to the 3-nitrophenol diaquo complex described above. The crown is again in the D_{3d} conformation and is centrosymmetrically bound by two water molecules using hydrogen bridges, O··H = 1.94, 1.98 Å. The 4-nitrobenzaldoxime guest molecule is hydrogen bonded to the water molecules through the OH group of the oxime, O··H = 1.69 Å. The complex can be regarded as consisting of clusters, which interact with neighboring clusters by van der Waals forces.

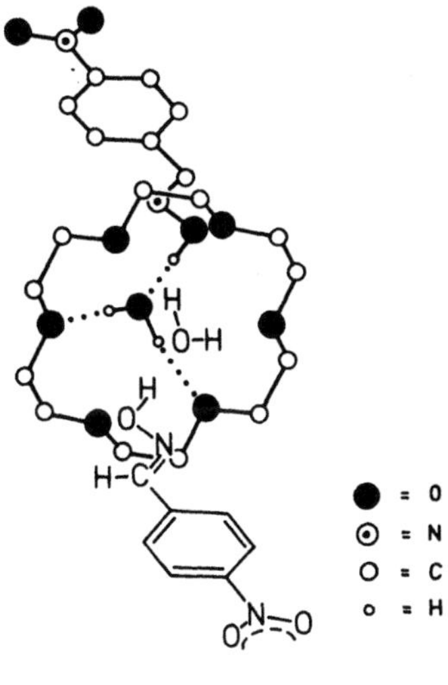

Fig. 26. [18]Crown-6 *4-nitrobenzaldoxime diaquo* 1:2:2 complex [29)]

[18]crown-6 4,4'-biphenyldiol Diaquo 1:1:2 Complex [37]

Here again the [18]crown-6 is situated on a center of symmetry with approximate D_{3d} symmetry. Each of the centrosymmetrically related water molecules is linked to two oxygen atoms of the crown ether O··H = 1.87, 1.92 Å. Each end of the *4,4'-biphenyldiol* units binds to a *water* molecule of a different crown ether. The bridging 4,4'-biphenyldiol leads to the formation of long chains. The spatial arrangement of the three constituents in the crystal are shown in Fig. 27.

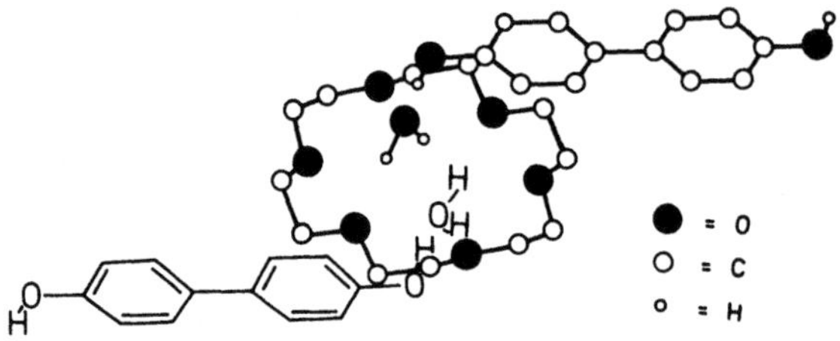

Fig. 27. A section of the crystal lattice of the [18]crown-6 *4,4'-biphenyldiol diaquo* 1:1:2 complex [37]

The last four structures are of interest since it appears that guest molecules containing a single —OH functionality capable of binding to one crown ether, do not form stable complexes in the absence of water. These ternary complexes might be models for secondary binding interactions of enzymes or receptor sites with substrates. The bonds of the three components are strong enough to orient the substrate but are weak enough to be readily broken; e.g., broken by changes of conformation or solvatation.

1:1 Methanol Complex of a (1,2)Cyclophane-Cryptand

The C-isomer of the macrocyclic host molecule *4* (Fig. 28) is bound to a single molecule of methanol via a bifurcated hydrogen bond, O··O = 2.988, 3.072 Å. The shift in the IR frequency is only 89 cm^{-1} indicating a weak interaction. The C-isomer has all four *cis*-hydrogen atoms of the cyclohexane ring junctures lying on the same side

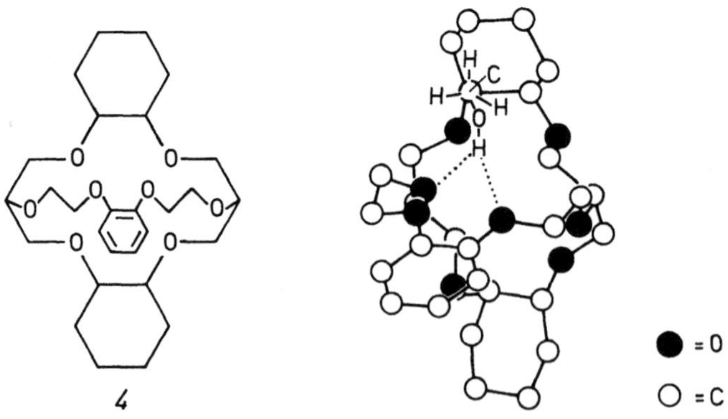

Fig. 28. Structure of the methanol complex of *4* [38]

of the molecule as the bridge containing the phenyl group. The water is not included in the cryptand but binds externally to two oxygen atoms of the bridge. There is no indication of any interaction involving the $—CH_3$ group, and the host exhibits the same conformation as in the uncomplexed state. There is no participation of the oxygen lone pairs in the complexation which is unusual for methanol complexes or adducts.

3 Azacrowns as Host Compounds

Complexation of 1,7,10,16-Tetraoxa-4,13-diazacyclooctadecane ("diaza[18]crown-6", 5a) with Thiourea (1:4)

Although the host/guest stoichiometry is 1:4, the structure can be described as composed of 1:2 host/guest complexes with lattice thioureas hydrogen-bonded to complexed guests. The diazacrown adopts a biangular conformation and is located on a center of symmetry. Centrosymmetrically related thiourea molecules bridge adjacent oxygen-nitrogen sites via a single hydrogen bond from each thiourea amide group (Fig. 29), H··O = 1.91, H··N = 1.99 Å. The conformation is sufficiently distorted to permit the formation of an internal N—H··O bond, H··O = 2.46 Å. Each complexed thiourea is hydrogen bonded to an uncomplexed molecule by two NH··S bonds to form an eight membered ring. The other NH_2 group of the uncomplexed thiourea forms additional hydrogen bonds to create an extensive three-dimensional network. In the [18]crown-6 series guest molecules capable of forming extended hydrogen bonded networks yield a variety of host/guest stoichiometries. The solids are composed of 1:2 complexes in a hydrogen bonded lattice of uncomplexed guests.

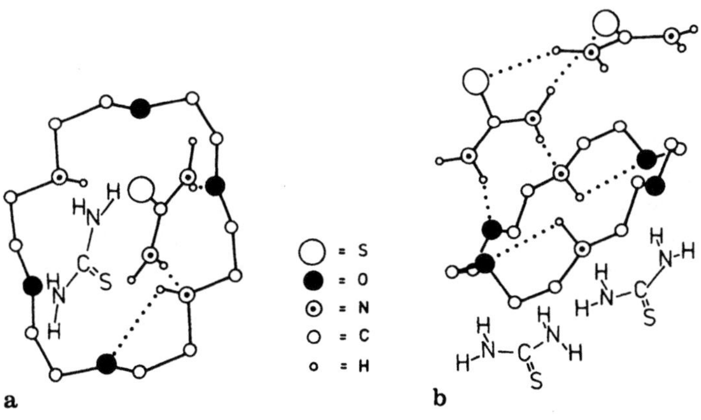

Fig. 29. Structure of the "diaza[18]crown-6" *thiourea* :4 complex [39]

Complexation of 1,7,10,16-tetraoxa-4,13-diazacyclooctadecane (5a) with 2-guanidino-benzimidazole

Unlike the thiourea complex [39], the "diaza[18]crown-6" host adopts the D_{3d} conformation, Fig. 30. The oxygen and nitrogen atoms of the crown lie alternately 0.18(1) Å above and below the mean plane. The crown ring is more flattened than in most complexes of [18]crown-6. The data did not permit the location of the hydrogen atoms; however, the aza nitrogen atoms are oriented such that they might act as donors as well as acceptors. The geometrical parameters for the guanidine moiety indicate the molecule is present in the amino form.

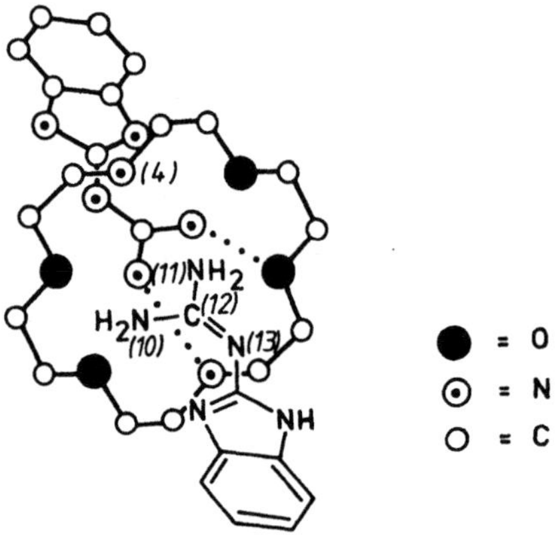

Fig. 30. The *2-guanidinobenzimidazole* complex of the 1,7,10,16-tetraoxa-4,13-diazacyclooctadecane (*5a*) [29]

In order to evaluate the influence of the diaza nitrogens upon the ring conformation and to better describe the guanidine moiety, the structure of the 1:2 [18]crown-6 *2-guanidinobenzimidazole* complex was determined [40]. The two structures are isomorphous and isoelectronic. The [18]crown-6 is located on an inversion center and the oxygen atoms lie alternately 0.18 Å above and below the mean plane which corresponds to the conformation observed for the "diaza[18]crown-6" complex. This implies the finer conformational features of the host are guest-dependent, and that the donor properties of the N—H group are not required to rationalize the crown conformation. The guest molecules are planar, and the guanidine group is in the amino form. The guest binds to the host via one normal (NH··O(1) = 2.27 Å) one weak (NH··O(7) = 2.53 Å) and one bifurcated (NH··O, 2.37, 2.47 Å) hydrogen bond. All six oxygen atoms of the crown are involved in hydrogen bonding and two oxygen atoms are bound to two hydrogens. A crystallographic two-fold axis relates two guests to form base pairs (NH··N = 2.06 Å). The planes of the base pairs make an interplanar angle of 60°. The entire structure is composed of infinite chains of alternate host-guest, guest-guest hydrogen bonds.

Malonodinitrile Complex of N,N'-bis(2-cyanoethyl)-1,10-diaza-4,7,13,16-tetraoxacyclooctadecane (5b)

Colorless crystals of the 1:2 complex are sensitive to moisture. Figure 31 shows the host/guest topology in the crystal as revealed by X-ray analysis. The two *malonodinitrile* guest molecules are related by an inversion center. The donor atoms of the crown are alternately 0.14 Å above and below the mean plane. All C—C bond torsion angles are approximately 65°. The bonding of the malononitrile guest molecule differs from that observed in the analogous [18]crown-6 complex. One of the acidic methylene hydrogen atoms does not participate in bonding (H··O greater than 2.75 Å) while the second forms a bifurcated bond to a ring nitrogen and an adjacent ether oxygen (H··N = 2.41 Å, H··O = 2.39 Å). An additional interaction may occur between a side chain hydrogen and a nitrogen atom of the malononitrile guest. The C(10)··O(5) distance is shorter than the sum of van der Waals radii and may indicate a local dipole-dipole interaction.

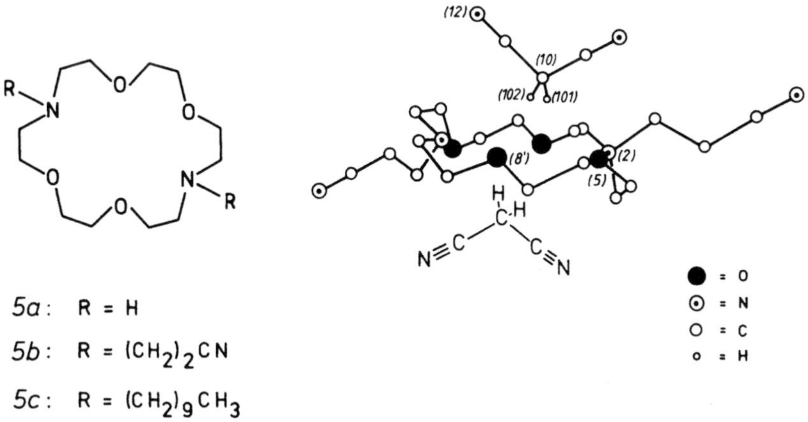

Fig. 31. Structure of the complex between *malonodinitrile* and the diazacoronand *5b* [41]

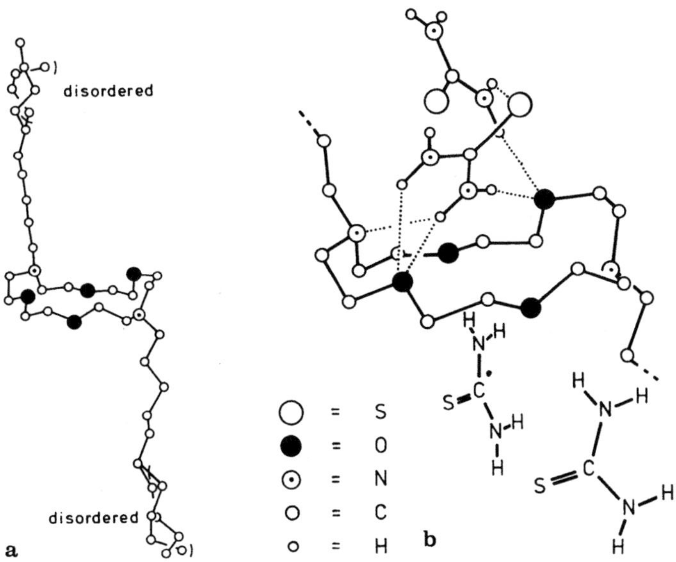

Fig. 32a and b. The conformation of the ligand *5c* in its *thiourea* complex and the disorder of the chain ends. Side view of the *5c*-thiourea complex showing hydrogen bond contacts (parts of the aliphatic chain have been omitted for clarity) [42]

Complex of 4,13-Didecyl-1,7,10,16-tetraoxa-4,13-diazacyclooctadecane (5c) with thiourea (1:14)

The complex of the macrocyclic amino polyether N,N'-didecyl-1.7,10,16-tetraoxa-4,13-diazacyclooctadecane (5c) with *thiourea*[42] (stoichiometry 1:14) has been investigated by X-ray methods[42b]. The unusual 1:14 ratio found in this complex is the highest ever reported in the crown series. Fig. 32a shows the structure of the title ligand together with the twofold disorder of the aliphatic chains. Fig. 32b gives a side view of the 1:14 complex and shows the hydrogen bonds linking host and guest molecules.

4 Water as Guest Molecule(s)

Water Complex of 1,11-dioxo-2,5,8,11,14-pentaoxa[15](2,6)pyridinophane (6)

The dilactone *6* dihydrate isolated from hexane has a melting point of 65–66 °C. At 60° in vacuo the waters of hydration are lost, and the anhydrous compound melts at 83–84 °C. In the dihydrate the donor centers lie within 0.26 Å of the mean plane. One water molecule coordinates directly with the host and lies 0.14(9) Å below the plane, O··O(4) = 2.92 Å (Fig. 33). The pyridino nitrogen, which usually functions as a donor atom, does not participate in the binding of the water. The second water molecule is hydrogen bonded to the first (O··O = 3.01 Å), and does not interact significantly with the crown ether.

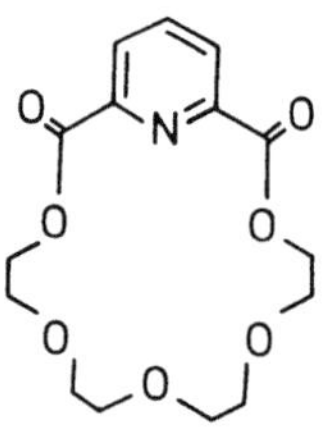

6

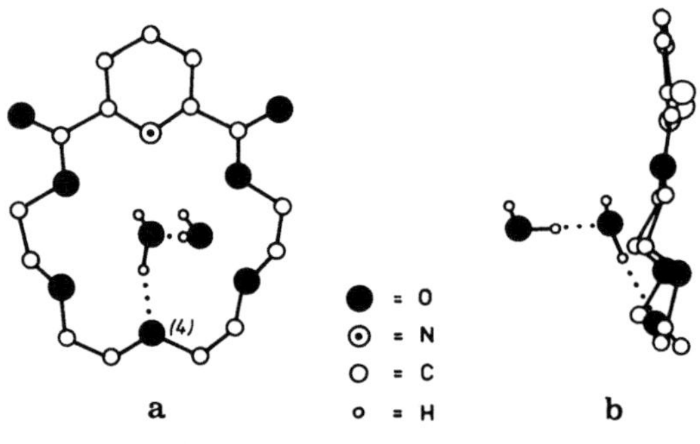

Fig. 33. Structure of pyridino dilactone (*6*) *dihydrate*[43]: **a** as viewed perpendicular to the crown ring plane, **b** as seen in the plane of the crown

These data indicate that one should be careful in interpreting spectroscopic measurements on the crown *6* because of the ready attachment of water and the variable moisture content of nonpolar solvents.

Water Complex of the Dipyridino Ketone Crown 7

The dipyridino ketone crown *7* forms a crystalline monohydrate, in which the crown has almost a C_2 symmetry (Fig. 34).

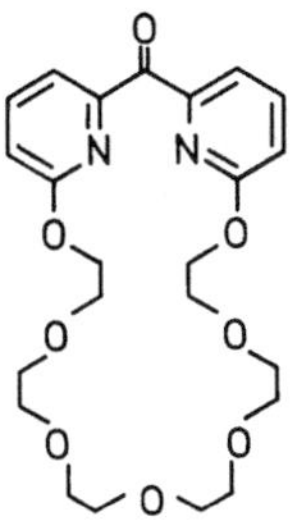

7

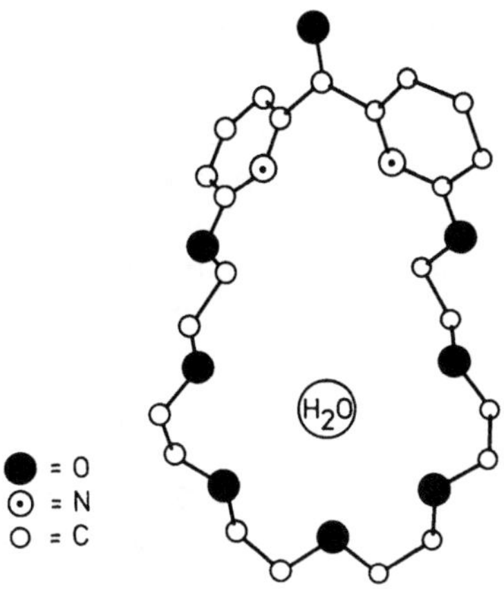

Fig. 34. The crystal structure of the *water* complex of the hexa-ethyleneglycol-bis(2-pyridyl) macrocycle *7* [44]

The pyridino rings of the ligand are turned outward from the carbonyl plane an average of 34.6°. The five oxygen atoms of the crown lie within 0.60 Å of the mean plane. The *water* molecule is out of the plane by 1.34 Å and has a short O··O contact (3.00 Å) with the central oxygen atom O(4). From these data it was inferred that the ability to encapsulate water is dependent on the conformation of the oligoether chain as well as its cavity diameter. The combination of a suitable conformation and an appropriate cavity dimension favors an arrangement in which the water molecule can be linked by several hydrogen bonds.

Water Complex of 1,15-dioxo-5,8,11-trioxa-2,14-diaza[15](6,6')-2,2'-bipyridinophane (8)

Figure 35 shows on ORTEP drawing [45] in which a *water* molecule is encapsulated in the crown cavity of the ligand. Differing from the water complexes described hitherto, a NH··OH hydrogen bond seems to exist; however, the distances suggest the bonding is weak. The water molecule appears to rotate in the cavity.

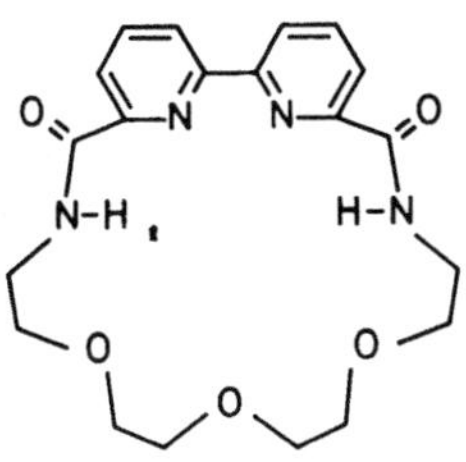

8

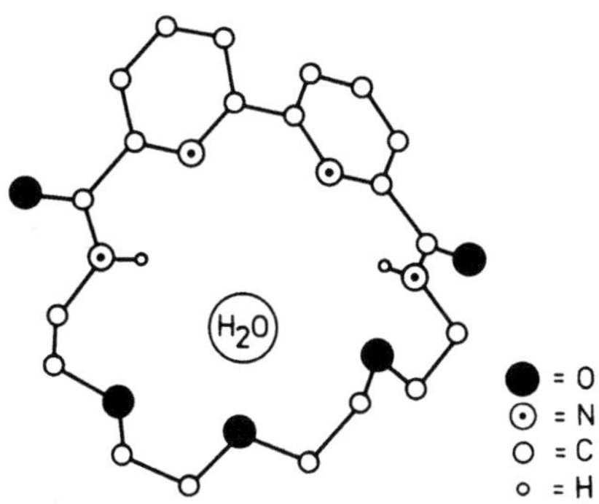

Fig. 35. The structure of the *water* complex of the bipyridino dilactam coronand *8* [45]

Water complex of a chiral crown (1,3:1',3':4,6:4',6'-tetra-O-methylene-2,2':5,5'-bis-O-oxydiethylene-di-D-mannitol) (9)

A 1:1 *aquo* complex was prepared with the chiral crown 1,3:1',3':4,6:4',6'-tetra-O-methylene-2,2':5,5'-bis-O-oxydiethylene-di-D-mannitol (*9*). In the crystal the host molecule has C_2 symmetry, and the hydrogen bonded water guest molecule lies on the twofold axis. The oxygen atom of the water sits above the crown and is hydrogen bonded (H··O = 1.96 Å) to two ether linkages adjacent to the six-membered rings.

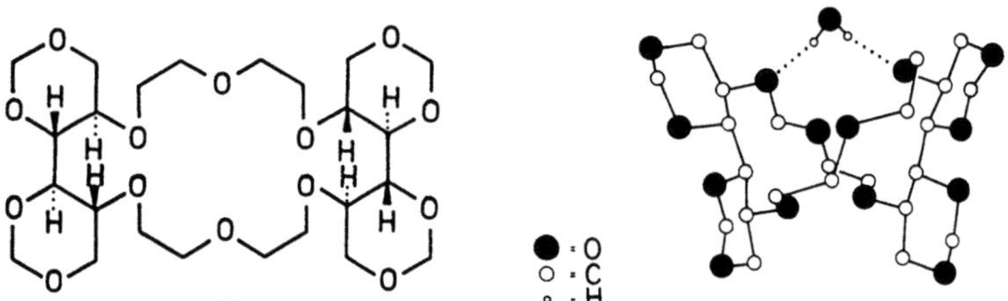

9

Fig. 36. Structure of the *water* complex of D,D-*9* [46]

Water Complex of 1,4,7,10-tetrakis(2-hydroxyethyl)-1,4,7,10-tetraazacyclooctadecane (10)

Figure 37 shows the structure of the *water* complex with the four-armed tetraaza[12]-crown-4 (*10*). The water molecule is located in the center of the cavity and serves as proton donor to two trans nitrogen atoms. The water molecule also accepts hydrogen bonds from OH side groups attached to the other trans nitrogen atoms. Thus a tetrahedral encapsulation of the water molecule occurs. A remarkable aspect is that the conformation of the twelve-membered ring hardly differs from that of the free ligand (*10*).

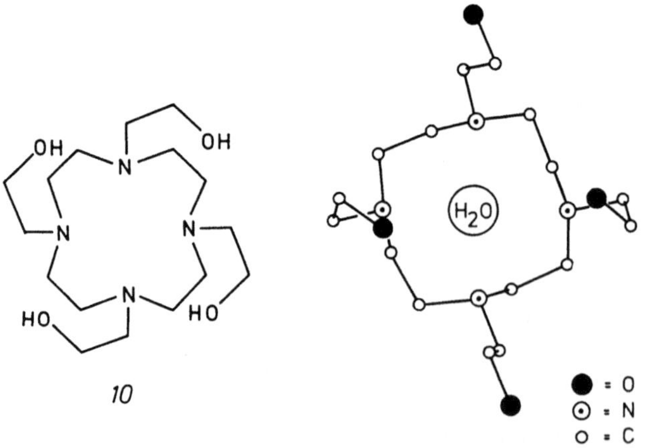

Fig. 37. **a** The four-armed ligand *10*, **b** the structure of its crystalline mono*hydrate* [47]

5 Intramolecular Complexation

The crown ether *11* has a trimethylammonium group reaching into the interior of the cavity and is complexed intramolecularly by short CH···O hydrogen bonds. Additional hydrogen bonds between the acidified CH_3 groups and the iodide anion were confirmed by X-ray analysis.

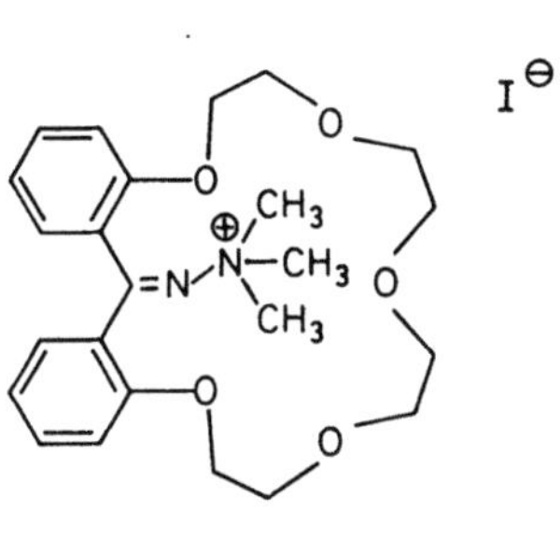

11

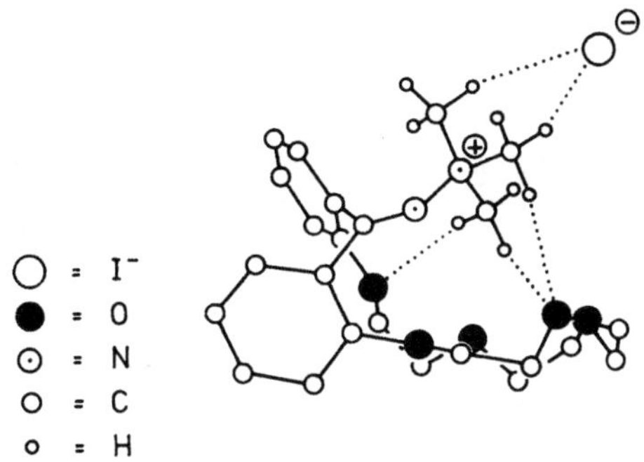

Fig. 38. Structure of the intramolecular *hydrazonium* complex *11* [48]

6 Ternary Crown Complexes

Ternary complexes with water and phenolic compounds as guests were described earlier (section 2.3).

Complex of dibenzo[18]crown-6, Thiourea and KI

The polymeric structure of the 1:1:1 complex of dibenzo[18]crown-6 (*3*), potassium iodide and *thiourea* is shown in figure 39.

Such "supercomplexes" may serve as models for the transition state of crown-assisted reactions, for example nucleophilic substitutions. The observed stereospecifity could be caused by locally ordered structures in which the crown ether cation complex could act as the substrate and as the anion, as well.

In the ternary complex (figure 39), the potassium ion is found in the center of the crown ether cavity, and makes short contacts to the six ether oxygen atoms with $K^+ \cdots O$ distances ranging from 2.71 to 2.80 Å. The iodide ion completes the coordination sphere (3.57 Å). This part of the complex resembles the 1:1 binary complex of dibenzo[18]crown-6 with potassium iodide. The planar thiourea molecules form an extended hydrogen bonded network in the direction of the crystallographic c axis. The hydrogen atom of each amino group is bonded to sulfur of an adjacent thiourea. The remaining amino hydrogens form hydrogen bonds to adjacent iodide ions which interact with the K^+/host complex. The thiourea chain does not participate in binding the potassium ion nor does it interact in any direct way with the crown ether.

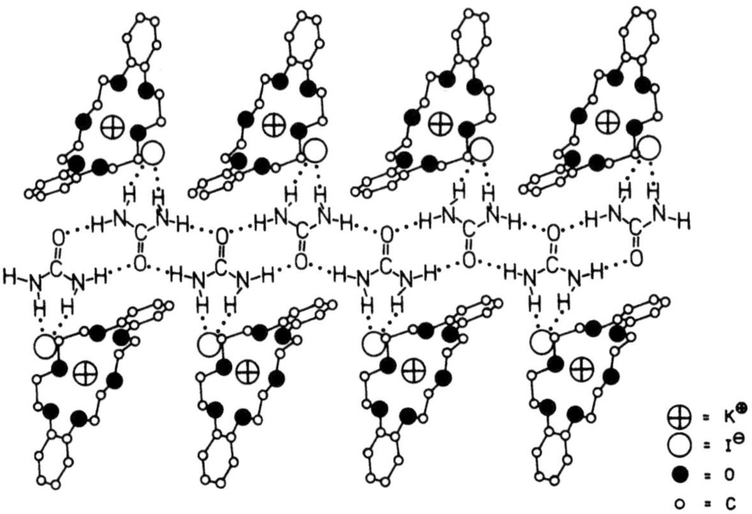

Fig. 39. The structure of the 1:1:1 complex consisting of dibenzo [18]crown-6, potassium iodide, and *thiourea* [49]. (Crown hydrogen atoms are omitted). Hydrogen bonds are drawn as dotted lines while coordinate bonds are indicated by heavy lines

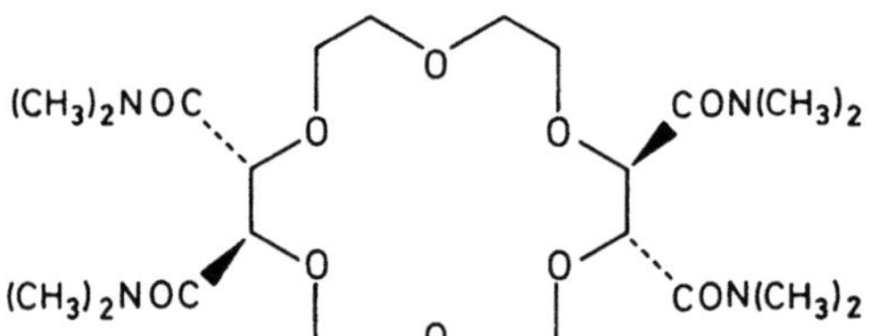

12

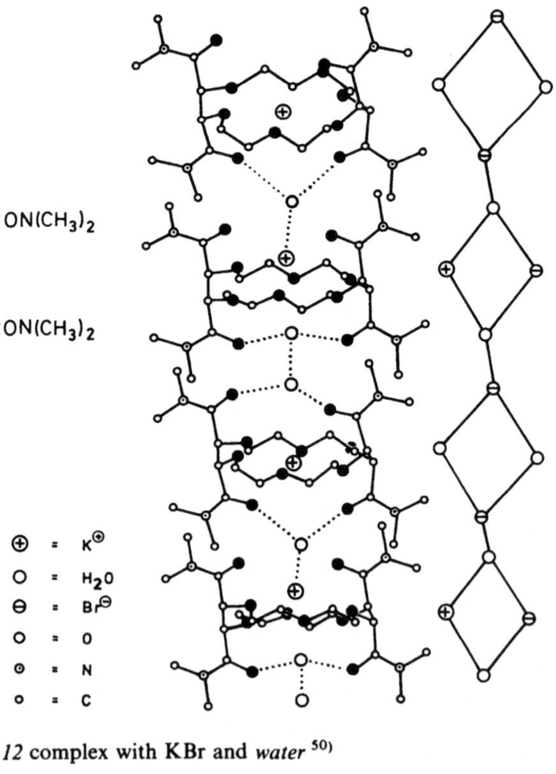

Fig. 40. Molecular packing of the crown *12* complex with KBr and *water* [50]

[18]crown-6 Tetracarboxamide (12) KBr H_2O Complex

The structure of the ternary 1:1.5:3.5 complex with the chiral macrocyclic tetracarboxamide *12* can be regarded as a solid state model for a molecular channel. The macrocyclic units are organized in a polymolecular stack with the potassium ions located alternately inside and on top of successive macrocycles, Fig. 40. The in-plane potassium ions (0.25 Å from mean plane) coordinate with all six ether oxygen atoms while the out-of-plane ions (1.13 Å from mean plane) coordinate to only three oxygen atoms.

The crystal structure was described as a "frozen picture" of potassium ion transport through stacked crown ether units. The structure would represent two steps in a hopping mechanism for cation flow through a pore. There is extensive hydrogen bonding and dipolar interactions within the stacks with O(w)··O(w) = 2.63 Å, O(w)··O=C = 2.50–2.95 Å and K^+··O(w) = 2.83 Å. The counter ion forms a polymeric chain of $[KBr_3(H_2O)]^=$ units and is parallel to the organic chain. It is composed of alternating water-bridged KBr and BrBr fragments (Br··O = 3.3–3.45 Å). The structure is further stabilized by van der Waals interactions between the cationic and anionic chains, (Br··CH_3 (3.68–3.76 Å).

The conformation of the [18]crown-6 ring is similar to that observed in other complexes and side chains are in axial conformations on both sides of the macrocyclic ring. The polar amide carbonyl groups are oriented toward the internal space while the apolar groups composed of methyl and methylenes lie on the outside.

Complex of 2,11-diaza-5,8,15,18-tetraoxa[$8^{2,11}$][8](4,4')-o-terphenylophane (13) with NaSCN and Methanol

The structure of the complex with host *13*, sodium thiocyanate and *methanol* (1:1:1) is shown in figure 41 [51]. The boat conformation of the crown section of the host

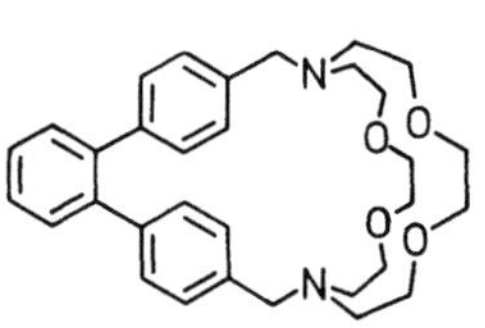

13

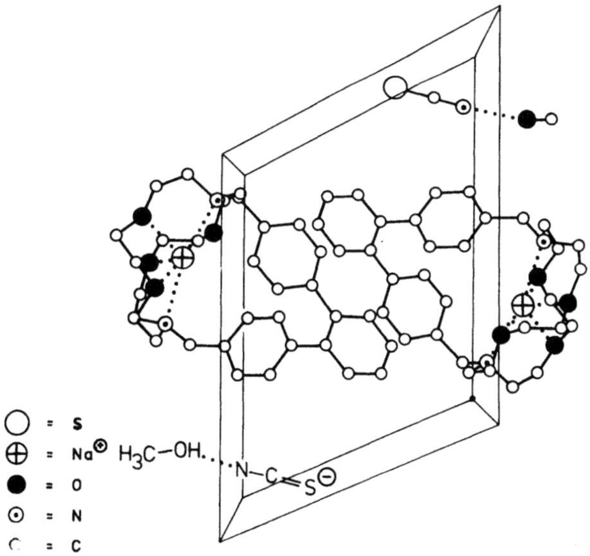

Fig. 41. A perspective view of the ternary complex of *13*, NaSCN and *methanol* [51]

molecule enables the six heteroatoms to coordinate the sodium cation in a bipical square pyramid arrangement. The 1,2-terphenyl group is turned outward and the 1,4-phenylene rings are twisted away. There is no direct contact between the Na^+ and the SCN^-. The anion is connected to the methanol molecule by hydrogen bonds. The compound was recrystallized from a mixture of methanol and ethyl acetate. It is interesting that the complex should contain methanol instead of water since water often replaces methanol in wet solvents. Usually, for oligoether complexes, the inclusion of water is preferred because of the potential for forming an extensive hydrogen bonded system.

Complexation of 1,4-dihydropyridino-crown 14 with Sodium Perchlorate and Acetone

The 1,4-dipyridino crown *14* complexes sodium perchlorate and forms crystals containing a molecule of *acetone* coordinated to a sodium cation. The dihydropyridine ring is in a distinct boat conformation, and is folded relatively to the oligoethylene-glycol chain. The sodium ion lies in the center of the cavity and is additionally coordinated from one side by the carbonyl group of the crown and from the other by an oxygen atom of the perchlorate. The connecting line between the acetone, the oxygen atom, the sodium ion and the oxygen atom of the perchlorate is almost straight. The oxygen atom of the acetone lies approx. 3.6 Å above the dihydropyridine ring. The distance between acetone, oxygen, and sodium is 2.25 Å. The nearest perchlorate oxygen lies 2.31 Å away from a sodium ion.

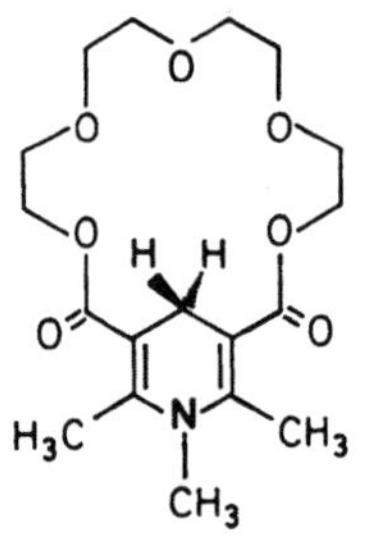

14

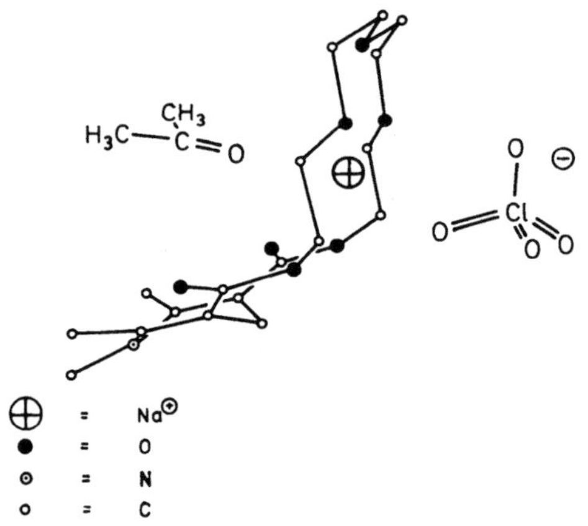

Fig. 42. Projection of the ternary complex *14* $NaClO_4$ *acetone* [51]

Dibenzo[18]Crown-6 $K[Al_2(CH_3)_6O_2]$ 1,5 Benzene Complex

This complex was prepared from KO_2 with aluminiumtrimethyl and dibenzo[18]-crown-6 in benzene. In boiling toluene it is stable for more than 24 h [53]. It depicts a new coordination type of the superoxide ion. The 1.47 Å bond is the longest yet reported for an O_2^- superoxide ion.

1,5,12,16,23,29-Hexaoxa[$7^{3,14}$] [5.5] (1,2)Cyclophane (15) · Naphthalene-2,3-diol · · Monohydrate

The 1:1:1 complex [54] is obtained by reaction of the cryptand *15* with *2,3-naphthalenediol* in the presence of *water*. The water molecule is bound by bifurcated hydrogen bonds to four oxygen atoms of the cryptand, one binds symmetrically (H··O = 2.33 and 2,56 Å) while the other bind assymmetrically (2.03 and 2.56 Å). The 2,3-naphthalenediol forms one intramolecular hydrogen bond while the second diol is coordinated to the water molecule (H··O = 1.79 Å).

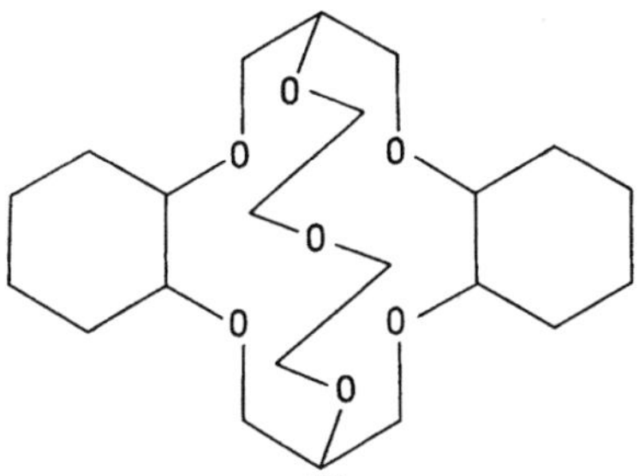

15

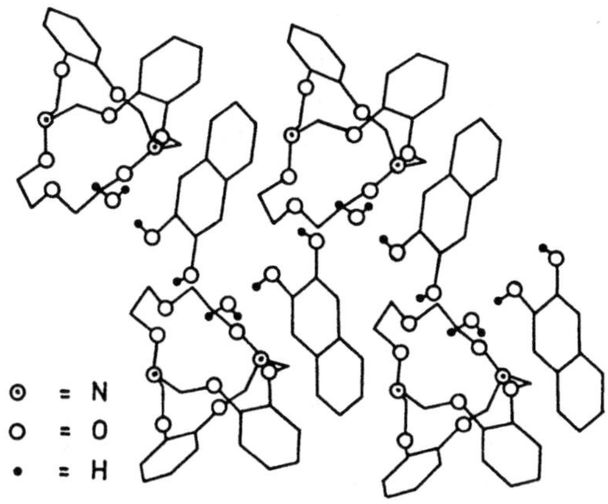

Fig. 43. Monohydrate of the cryptand *15*: projection in direction of the 001 axis [54]

Dibenzo[18]crown-6 · tert-butylammonium perchlorate · DDQ complex

This crystalline ternary complex is obtained by treating the tert-butylammonium complex of dibenzo[18]crown-6 with *DDQ (2,3-dichloro-5,6-dicyano-1,4-benzophenone)* in chloroform. The X-ray analysis of the red crystals, melting point 177

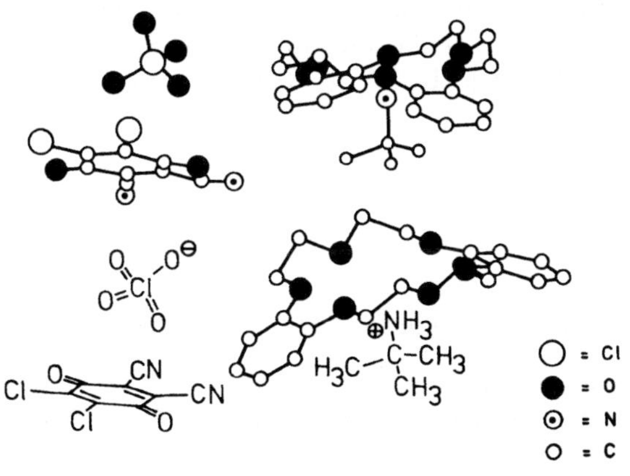

Fig. 44. Structure of the ternary complex of dibenzo[18]crown-6 with tert-butylammonium perchlorate and *DDQ* [55]

to 178 °C, revealed that a close contact between the anion and DDQ exists. The structure of the complex, as seen in figure 44, can be described as a packing of two layers with different charges, one containing DDQ molecules and perchlorate anions, the other being built of alternating layers of tert-butylammonium ions and crown ether molecules. The tert-butylammonium ions are connected to one another by NH··O hydrogen bonds and electrostatic N^{+}··O interactions.

7 Outlook

As this report shows, crown ethers are suitable host substances not only for encapsulating and masking cations, but for binding neutral substrate molecules as well. They are useful as models for the study of recognition processes on the molecular level. Additional applications involving the alteration of solubility, volatility, or reactivity of guest substances and the enhancement of membrane transport of small organic molecules can be envisioned. A specific example which may indicate the importance of these interactions is the binding of the physiologically significant molecule urea by suitable hosts. Tailored cryptands, which can bind urea in water, or with which urea could be extracted from water, seem to be of topical interest. Crown- or cryptand hosts, which are selective for specific neutral molecules could be bound to resins and used in affinity chromatography. The crown ethers may be useful in separating inorganic and organic isomers containing amine functionalities. The separation of optical isomers might be achieved through use of chiral hosts.

In the future many new uncharged complexes between crown ether/cryptand compounds with uncharged organic molecules will be synthesized or their existence in solution will be confirmed. High field NMR studies of solids and liquids should provide additional insight into solution conformations and dynamic processes. This should lead to the design of more elegantly shaped host molecules which are specifically functionalized to yield with more sophisticated receptor models, tailor-made organic catalysts or efficient enzyme model systems.

8 Acknowledgements

We thank the Deutsche Forschungsgemeinschaft and the Fonds der Chemischen Industrie for support of our own work. We are indebted to Dipl.-Chem. P. Mayenfels, M. Palmer, G. Mahr and Mrs. B. Jendrny and U. Werner for their cooperation in preparing the figures and the manuscript.

9 References

1. Vögtle, F., Sieger, H., Müller, W. M.: Top. Curr. Chem. *98*, 107 (1981), and references cited therein
2. Cram, D. J., Trueblood, K. N.: ibid. *98*, 43 (1981)
3. Stoddart, J. F.: in Progress in Macrocyclic Chemistry, Vol. 2, p. 173 (Eds. Izatt, R. M., Christensen, J. J.), Academic Press 1982

4. Lehn, J. M.: Pure Appl. Chem. *52*, 2303, 2441 (1980)
5. Pascard, C., Riche, C., Cesario, M., Kotzyba-Hibert, F., Lehn, J. M.: Chem. Commun. *1982*, 557
6. Odashima, K., Itai, A., Iitaka, Y., Koga, K.: J. Am. Chem. Soc. *102*, 2504 (1980)
7. Dunitz, J. D., Dobler, M., Seiler, P., Phizackerley, R. P.: Acta Crystallogr. *B30*, 2733 (1974). Maverick, E., Seiler, P., Schweizer, W. B., Dunitz, J. D.: ibid. *B36*, 615 (1980)
8. Bovill, M. H., Chadwick, D. J., Sutherland, I. O., Watson, D.: J. Chem. Soc. Perkin II, *1982*, 1529
9. Hilgenfeld, R., Saenger, W.: Z. Naturforsch. *36b*, 242 (1981)
10. Donohue. J.: Structural Chemistry and Molecular Biology (ed. Rich, A., Davidson, N.), pp. 443–465, San Francisco, Freeman 1968
11. Goldberg, I.: Acta Crystallogr. *B31*, 754 (1975)
12. Kaufmann, R., Knöchel, A., Kopf, J., Oehler, J., Rudolph, G.: Chem. Ber. *110*, 2249 (1977). See also Basyony, A., Klimes, J., Knöchel, A., Oehler, J., Rudolph, G.: Z. Naturforsch. *31b*, 1192 (1976)
13. Knöchel, A., Kopf, J., Oehler, J., Rudolph, G.: Chem. Commun. *1978*, 595
14. a) de Boer, J. A. A., Reinhoudt, D. N., Harkema, S., van Hummel, G. J., de Jong, F.: J. Am. Chem. Soc. *104*, 4073 (1982)
 b) See also van Zon, A., de Jong, F., Reinhoudt, D. N., Torny, G. J., Onwezen, Y.: Rec. Trav. Chim. Pays-Bas *100*, 453 (1981)
15. Bandy, J. A., Truter, M. R., Vögtle, F.: Acta Crystallogr. *B37*, 1568 (1981)
16. Weber, G.: J. Mol. Struct. *98*, 333 (1983)
17. McLachlan, R. D.: Spectrochim. Acta *30A*, 2153 (1974)
18. Vögtle, F., Müller, W. M.: Naturwissenschaften *67*, 255 (1980)
19. Vögtle, F., Müller, W. M.: Chem. Ber. *113*, 2081 (1980)
20. Caira, M. R., Watson, W. H., Vögtle, F., Müller, W. M.: Acta Crystallogr. *C40*, 136 (1984)
21. Vögtle, F., Müller, W. M.: Chem. Ber. *114*, 3179 (1981)
22. Weber, G.: Acta Crystallogr. *B38*, 629 (1982)
23. Weber, G., Sheldrick, G. M.: ibid. *B37*, 2108 (1981)
24. Dhaneshwar, N. N., Tavale, S. S., Pant, L. M.: Acta Crystallogr. *B34*, 2507 (1978)
25. Weber, G.: Z. Naturforsch. *36b*, 896 (1981)
26. The number of the atoms corresponds to that in the original literature
27. Weber, G.: Z. Naturforsch. *37b*, 1250 (1982)
28. Weber, G.: Acta Crystallogr. *C39*, 896 (1983)
29. Watson, W. H., Galloy, J., Grossie, D. A., Vögtle, F., Müller, W. M.: J. Org. Chem. *49*, 347 (1984)
30. a) Pedersen, C. J.: J. Am. Chem. Soc. *89*, 7017 (1967)
 b) Pedersen, C. J.: J. Org. Chem. *36*, 1690 (1971)
31. Harkema, S., van Hummel, G. J., Daasvatu, K., Reinhoudt, D. N.: Chem. Commun. *1981*, 368
32. Weber, G.: J. Incl. Phenom. *2* (1984), in press
33. Vögtle, F., Müller, W. M., Weber, E.: Chem. Ber. *113*, 1130 (1980)
34. Galloy, J., Watson, W. H., Vögtle, F., Müller, W. M.: Acta Crystallogr. *B38*, 1245 (1982)
35. Colquhon, H. M., Stoddart, J. F., Williams, D. J.: Chem. Commun. *1981*, (a) 847, (b) 849, (c) 851; J. Am. Chem. Soc. *104*, 1426 (1982)
36. Caira, M. R., Watson, W. H., Vögtle, F., Müller, W. M.: Acta Crystallogr. *C40*, 491 (1984)
37. Grossie, D. A., Watson, W. H., Vögtle, F., Müller, W. M.: ibid. *B38*, 3157 (1982)
38. Bandy, J. A., Hughes, D. L., Truter, M. R.: ibid. *B38*, 2648 (1982)
39. Weber, G.: ibid. *B38*, 2712 (1982)
40. Caira, M., Watson, W. H., Vögtle, F., Müller, W. M., unpublished results
41. Deuten, K., Knöchel, A., Kopf, J., Oehler, J., Rudolph, G.: J. Chem. Res. (S) *1979*, 358; (M) *1979*, 4035
42. a) Vögtle, F., Müller, W. M.: unpublished results
 b) Weber, G.: personal communication
43. Newkome, G. R., Fronczek, F. R., Kohli, D. K.: Acta Crystallogr. *B37*, 2114 (1981)
44. Newkome, G. R., Taylor, H. C. R., Fronczek, F. R., Delord, J., Kohli, D. K., Vögtle, F.: J. Am. Chem. Soc. *103*, 7376 (1981)
 See also Newkome, G. R., Taylor, H. C. R.: J. Org. Chem. *44*, 1362 (1979)

45. Newkome, G. R.: unpublished results
Buhleier, E., Wehner, W., Vögtle, F.: Chem. Ber. *111*, 200 (1979)
46. Fuller, S. E., Stoddart, J. F., Williams, D. J.: Tetrahedron Lett. *23*, 1835 (1982)
47. Buoen, S., Dale, J., Groth, P., Krane, J.: Chem. Commun. *1982*, 1172
48. v. Itter, F. A., Vögtle, F., Weber, G., Sheldrick, G. M.: Z. Naturforsch. *38b*, 262 (1983)
49. Hilgenfeld, R., Saenger, W.: Angew. Chem. *93*, 1082 (1981); Angew. Chem., Int. Ed. Engl. *20*, 1045 (1981)
50. Behr, J. P., Lehn, J. M., Dock, A. C., Moras, D.: Nature *295*, 526 (1982)
51. Weber, G.: Acta Crystallogr. *B37*, 1832 (1981)
See also Rossa, L., Vögtle, F.: Liebigs Ann. Chem. *1981*, 459
52. van der Veen, R. H., Kellogg, R. M., Vos, A.: Chem. Commun. *1978*, 923
53. Hrncir, D. C., Rogers, R. D., Atwood, J. L.: J. Am. Chem. Soc. *103*, 4277 (1981)
54. Herbert, J. A., Truter, M. R.: J. Chem. Soc. Perkin II, *1980*, 1253
55. de Boer, J. A. A., Reinhoudt, D. N., Kiterwijk, J. W. H. M., Harkema, S.: Chem. Commun. *1982*, 194

Asymmetric Syntheses with Amino Acids

Jürgen Martens

Degussa AG, Weisfrauenstrasse 9, D-6000 Frankfurt am Main 11, FRG

Table of Contents

1 Introduction

1.1 Why Asymmetric Syntheses? Definition

The synthesis of enantiomerically pure organic compounds presents a challenge to academic and industrial chemist alike. The art of organic synthesis has reached a point where it is perfectly conceivable to prepare by total synthesis many sophisticated compounds especially useful for their biological properties. For example, captoril (1) is an orally active antihypertensive agent having a unique inhibitory action on angiotensin-converting enzyme. Dipeptide (2) is a sweetening agent (aspartame) which recently appeared on the market. L-dopa (3) is active against Parkinson's disease and amino acid D-penicillamine (4) has shown favorable effects on the clinical aspects of rheumatoid arthritis. Many biologically active compounds have one or more asymmetric centers, the specific activity being related only to one stereoisomer. The drug (1) having (S,S)-configuration, for example, is about 100 times more active than the corresponding diastereomer with (R)-configuration in the side chain [1]. Other examples for molecules exhibiting different biological activities of the stereoisomers include pheromones [2], herbicides [3] etc. However, surprisingly, of the numerous drugs prepared by total synthesis that contain at least one asymmetric center, only about 20% have so far been used in sterically pure form.

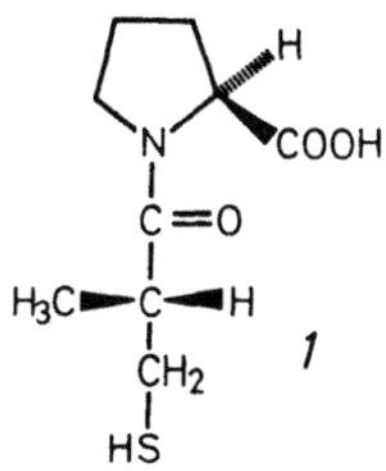

1

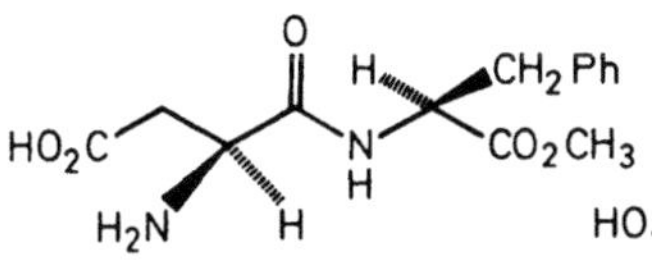

2

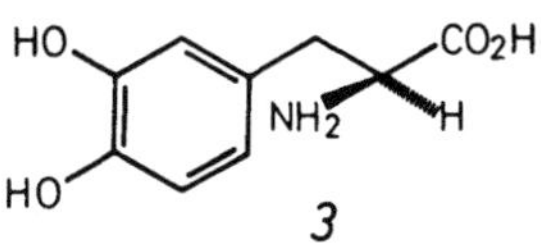

3

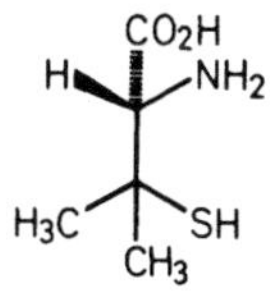

4

It is therefore of prime importance to be able to control the formation of asymmetric centers during the course of the synthesis. Especially delicate is the formation of the first asymmetric center in an achiral molecule. The methods of asymmetric synthesis try to solve this problem.

"Asymmetric synthesis" is a term first used in 1894 by E. Fischer and defined [4] in 1904 by W. Markwald as "a reaction which produces optically active substances from symmetrically constituted compounds with the intermediate use of optically active materials but with the exclusion of all analytical processes". A modern definition was proposed [5] by Morrison and Mosher: "An asymmetric synthesis is a reaction in which an achiral unit in an ensemble of substrate molecules is converted by a reactant into a chiral unit in such a manner that the stereosiomeric products (enantiomeric or diastereomeric) are formed in unequal amounts. This is to say, an asymmetric synthesis is a process which converts a prochiral [6] unit into a chiral unit so that unequal amounts of stereoisomeric products result". When a prochiral molecule

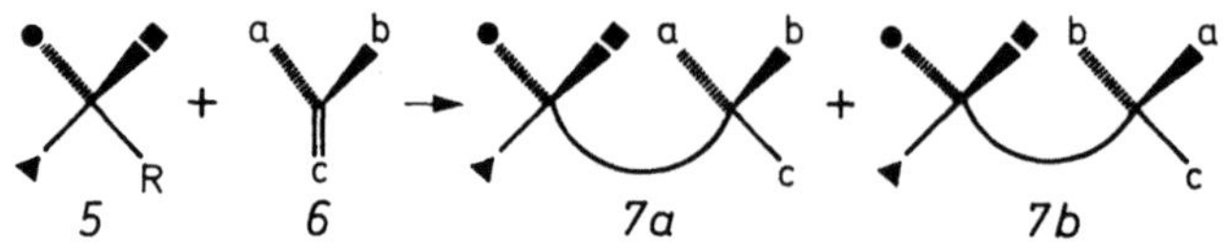

Eq. (a)

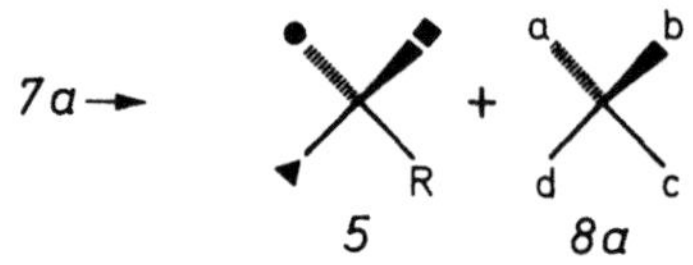

Eq. (b)

Eq. (c)

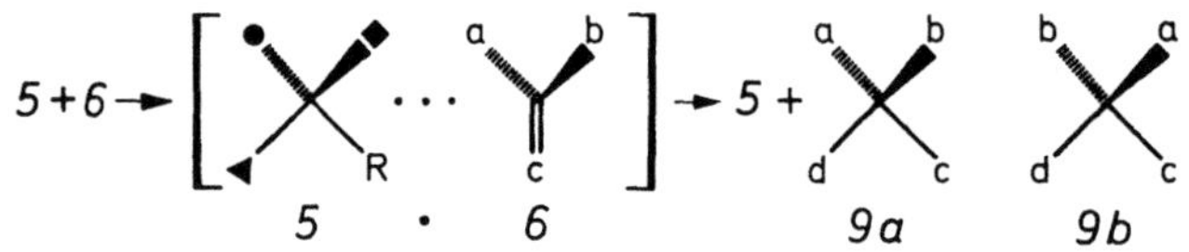

Eq. (d)

(6) reacts with a chiral reagent (5) to form a new center of chirality, two diastereomers, (7a) and (7b) [Eq. (a)], are formed. The ds value (%) is used as a measure of the diastereoselectivity of the reaction:

$$\text{ds} = \frac{[7\text{a}] - [7\text{b}]}{[7\text{a}] + [7\text{b}]} \times 100\%$$

If only one of the two diastereomers (7a) or (7b) is formed, the reaction is considered stereospecific (ds = 100%). In the subsequent course of the synthesis (7a) or (7b) may be cleaved with the formation of a new enantiomer (8a) or (8b), whereby chiral reagent (5) is either recovered [Eq. (b)] or decomposed [Eq. (c)].

If equivalent or catalytic amounts of a chiral molecule (5) react with a prochiral molecule (6) — the complex (5) · (6) is formed as an intermediate — to give the new chiral compounds (9a) and/or (9b), one speaks of an enantioselective reaction [Eq. (d)]. The enantioselectivity of such a reaction is given by the enantiomeric excess (ee) or by the optical yield:

$$\text{ee} = \frac{[9\text{a}] - [9\text{b}]}{[9\text{a}] + [9\text{b}]} \times 100\%$$

$$\text{optical yield} = \frac{\alpha}{\alpha_0} \times 100\%$$

α = measured rotation of the product
α_0 = rotation of the pure enantiomer

By far, the best asymmetric synthesis is done in nature by enzymes [7]. These have also found industrial application [8], e.g. the stereospecific amination of fumaric acid (10) to (S)-aspartic acid (11):

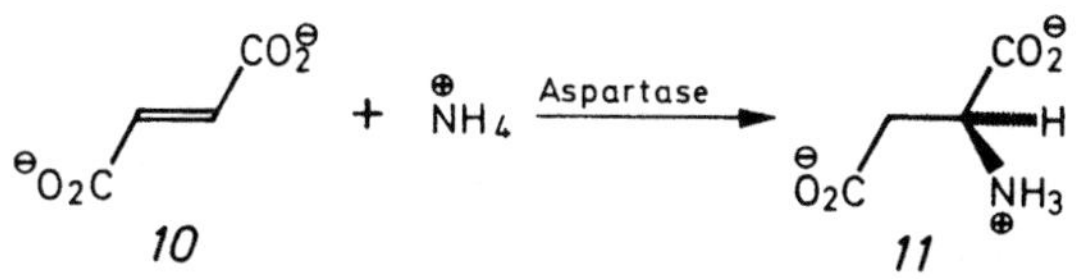

10 11

However, a considerable effort has been put forward by chemists to achieve comparable results. There is the challange to develop chemical systems as efficient as enzymic ones. For many years it had been questioned whether high optical yields could be effectively attained by organic chemists without the aid of enzymes. *However*, an increasing amount of recent results demonstrates that versatile and efficient non-enzymatic asymmetric syntheses are indeed possible.

The ever-increasing knowledge in synthetic methology for asymmetric synthesis is exemplified by the numerous reviews that have appeared in literature, in particular, the recent ones by Morrison and Mosher [5], Scott and Valentine [9], Meyer [10], Kagan and Fiaud [11], Otsuka and Tani [12], Apsimon and Sequin [13], Fischli [14], Wynberg [15], Masamune and Choy [16] and Heathcock [17]. These reviews, however, did not cover in depth the recent advances in asymmetric synthesis by chiral amino acid reagents. The enantiomerically pure educts of asymmetric syntheses, the chiral units (5), can either be prepared in the laboratory or be isolated from natural products. Thus, the chemist has at his disposal a large number of chiral units among the natural products. In particular, carbohydrates [18], terpenes [19], α-hydroxycarboxylic acids [20], alkaloids [21], biogenic amines [22] and amino acids [23] have found application as building blocks for asymmetric syntheses. A disadvantage in the use of carbohydrates, alkaloids and terpenes, which can rarely be obtained in both enantiomeric forms, does not exist in the case of α-hydroxycarboxylic acids such as lactic, mandelic, malic and tartaric acid.

In the present review we concentrate on the induction of asymmetry for the case in which the chiral reagent (5) is represented by an *amino acid* or a derivative thereof. Only those papers are considered in which the formation of a *new* center of asymmetry is induced. This can take place with the simultaneous incorporation of the chiral amino acid (or a derivative thereof) in the target molecule or by the action of catalytic amounts of this amino acid on a prochirale molecule. Reactions in which only the asymmetric center of the amino acid is modified without the stereoselective appearance of a new chiral center, have not been considered. Enzymatically catalyzed transformations [24] of molecules are not treated here.

1.2 Commercial Significance of Amino Acids

During recent years, the industrial manufacture of chiral amino acids has been actively developed throughout the world. The total amino acid market volume is estimated at US $ 1.2 billion in 1982 [25]. Their chief consumers have been the pharmaceutical industry, the food industry, in which sodium (S)-glutamate is used as a flavour intensifier, animal husbandry, and poultry farming, where (S)-lysine and (RS)-methionine are used to increase the food value of protein feeds [26]. The (S)-amino acids used most frequently [27] are obtained by fermentation [28], extraction of protein hydrolyzates, enzymic syntheses [28], or enzymic cleavage of racemates. There are also specific production processes [29] for the nonnatural (R)-amino acids, including homogeneous asymmetric hydrogenation of N-acyl dehydroamino acids [30]. Thus, most amino acids are available in significant quantities in two enantiomeric forms. However, in most cases, the natural enantiomer is available at a lower price than the unnatural one. This is mainly the result of a small demand for these products.

An interesting relationship exists between supply and current market price of natural chiral amino acids [31]. The greater demand induces lower cost, and the lower cost stimulates the greater demand. Therefore, one can also expect lower prices for unnatural amino acids if the demand expands.

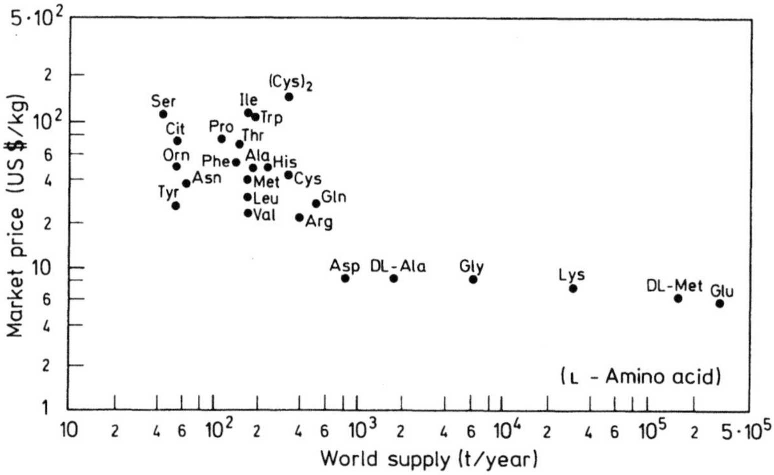

Fig. 1. Relationship between Supply and Current Market Price of chiral Amino Acids (1980)

Sources of Amino Acids

The chiral amino acids which are appropiate for use in synthesis must either be available commercially or be prepared readily from cheap precursors. Apart from the usual chemical suppliers, Ajinomoto (Tokyo, Japan), Degussa AG (Frankfurt, Federal Republic of Germany), Kyowa Hakko and Tanabe Seiyaku (both Tokyo, Japan) can offer the full range of amino acids and a number of derivatives from technical scale production.

2 Induction of Asymmetry by Catalytic Amounts of Amino Acids or their Derivatives

Although the use of enzymes as chiral catalysts will undoubtedly increase as they become more available, nonenzymic catalytic asymmetric synthesis is a very powerful tool in organic chemistry.

Until 1968, not a single nonenzymic catalytic asymmetric synthesis had been achieved with a yield above 50%. Now, barely 15 years later, no fewer than six types of reactions can be carried out with yields of 75–100% using amino acid catalysts, i.e., catalytic hydrogenation, intramolecular aldol cyclizations, cyanhydrin synthesis, alkylation of carbonyl compounds, hydrosilylation, and epoxidations.

In all reactions treated in this section the chiral educts are used only as catalysts. Thus, the reactions closely resemble an enzymatic process, i.e., an optimal process in which a small amount of chiral information is sufficient to direct a large number of molecules through the desired transformations in accordance with Eq. (d).

2.1 Homogeneous Asymmetric Hydrogenation

Asymmetric hydrogenation has been reviewed several times [32], and a very complete review through 1980 has appeared recently [30]. We present only the main results and new developments here.

2.1.1 Preparation of the Catalysts

Besides a considerable number of chiral Wilkinson-catalysts prepared from chiral α-hydroxy acids [33], carbohydrates [34], stereoids [35], and ferrocenylphosphines [36], many functionalized chiral biphosphanes based on amino acids have also been developed. The first optically active tertiary phosphane was synthesized in 1961 by Horner et al. [37].

Knowles et al. [38] prepared the first chiral Wilkinson complexes with an outstanding inductive activity. The vigorous development in this area is covered by an excellent review article [39].

In 1976, Achiwa [40] synthesized the first functionalized chiral biphosphanes starting with an optically active amino acid, namely 4-hydroxy-(S)-proline (12).

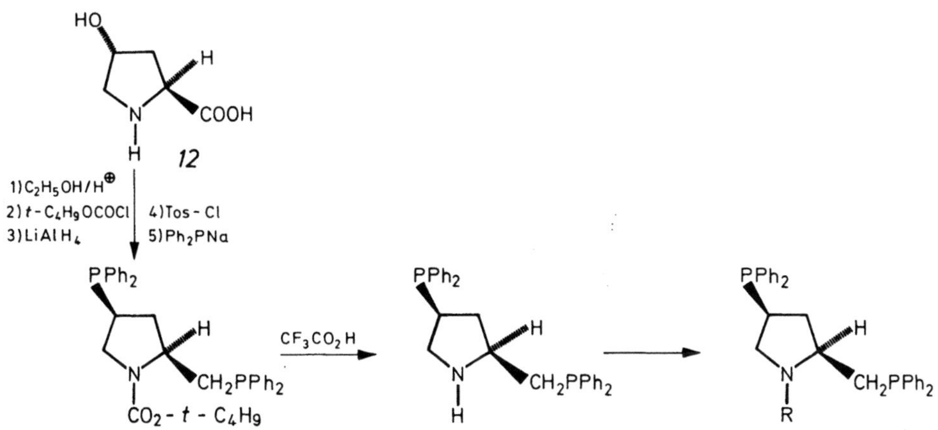

13 *14* *15*

R = $OCOC_{27}H_{45}$, $COCH_3$, CO-*t*-C_4H_9, COC_6H_5, CO—NHC_6H_5, CONH(4-Br—C_6H_4), CONH(4-Cl—C_6H_4), CO—NH(3,4-Cl_2—C_6H_3), CONH(4-NO_2—C_6H_4), CO—$NHCH_3$, CONHCH=$CHCH_3$, CONH-*cyclo*-C_6H_{11}, CHO

(2S,4S)-N-tert-butoxycarbonyl-4-diphenylphosphino-2-diphenylphosphino-methyl-pyrrolidine (BPPM) (13) is an excellent bisphosphane and is also the starting material for the synthesis of PPM (14). The chiral bisphosphanes (15) have been prepared by reaction of (14) with either acid chlorides or isocyanates.

The α-amino acids (S)-alanine and (S)-valine can be converted into N,N-bis-(phosphinomethyl) derivatives (16) in a modified Mannich reaction [42].

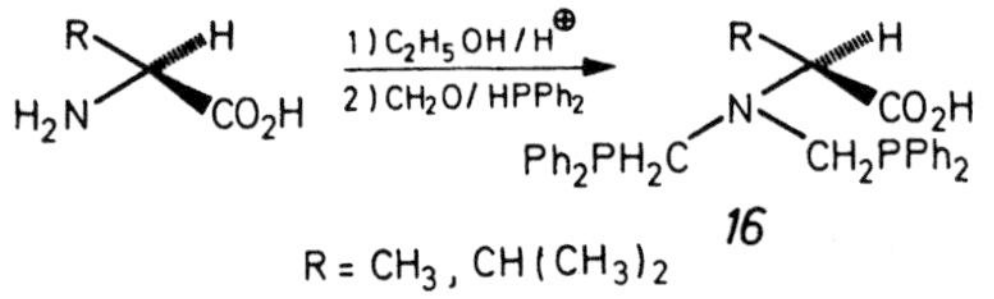

16

The chiral ligands (R)-Val-Phos(17a) and (R)-Phe-Phos (17b) have been synthesized from (S)-valine and (S)-phenylalanine, respectively, in an industrial laboratory [43].

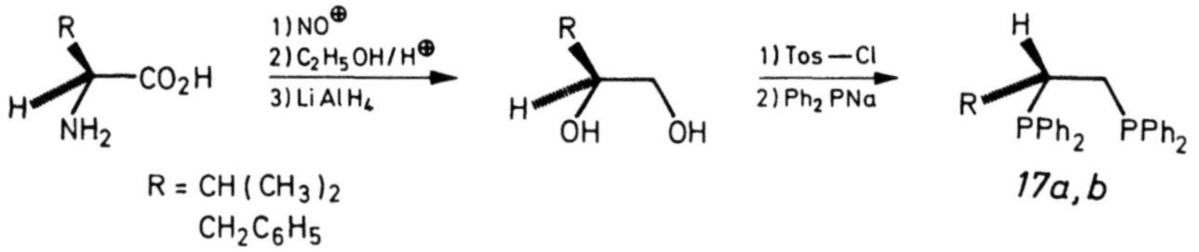

17a, b

Chiral, N-substituted diphenylphosphinoacetamides (18) were prepared by the acylation of (S)-amino acid esters with diphenylphosphinoacetic acid [44].

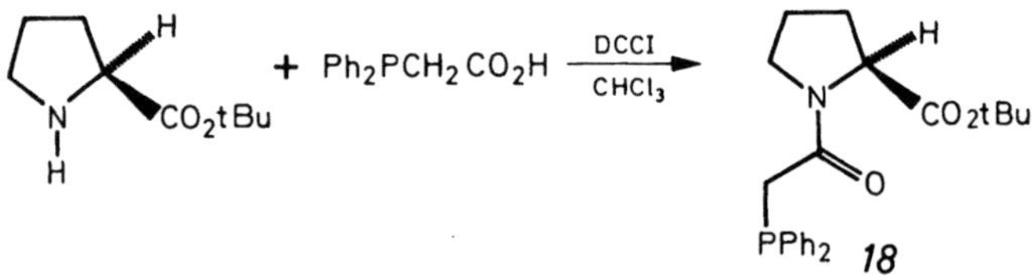

18

New aminophosphines (19) were obtained from (S)-ornithine recently [45].

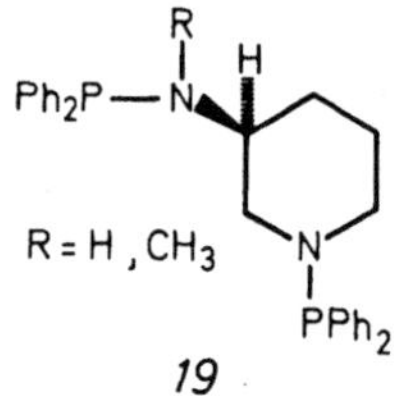

19

Takeuchi and Ohgo have reported a less stereoselective catalyst system based on (S)-N-methylproline [46].

2.1.2 Homogeneous Asymmetric Hydrogenation of C=C Double Bonds

In recent years, asymmetric hydrogenation of prochiral olefins, using rhodium (I) complexes with chiral phosphine ligands as catalysts, has been extensively investigated. Namely the asymmetric hydrogenation of α-acylaminocinnamic acid or α-acylamino-acrylic acids for obtaining optically active N-acyl-α-amino acids (20) has found interest [47]. In this reaction it is especially gratifying to see that e.e. 80—99% could be obtained [40, 41, 43, 45, 48] employing rhodium (I) complexes formed from $[Rh(COD)Cl]_2$ (COD = cyclooctadiene) and the chiral phosphines (13–19). L-Dopa (3) is now manufactured by such a process [38]. This represents the first nonenzymic industrial asymmetric synthesis.

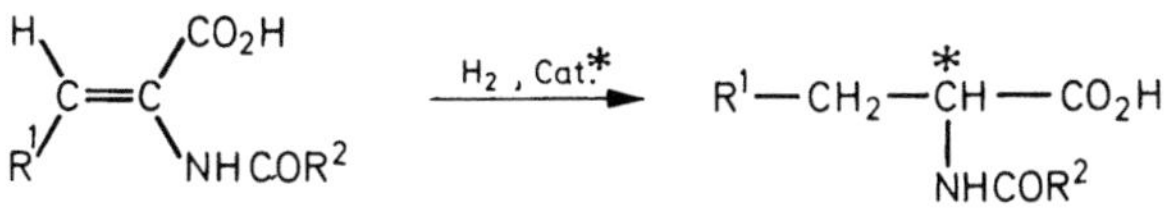

20

R^1 = H, alkyl, aryl R^2 = CH_3, $OCH_2C_6H_5$, C_6H_5

It has recently been demonstrated that a *stereoselective synthesis of dipeptides* by hydrogenation of the corresponding monodehydropeptides (N-protected free acids or methyl esters) is possible. In this reaction, chiral catalysts, for example BPPM (13), in the form of a Wilkinson complex have been used. These are superior to the corresponding DIOP complexes (DIOP = P,P′-[2,2-dimethyl-1,3-dioxolane-4,5-bis(methylene)] bis(diphenylphosphane). A d.s. value of 90—99% was generally obtained [49].

Dehydropeptides (21) were employed for the asymmetric hydrogenation, catalyzed by chiral rhodium complexes of the hydroxyproline derivative (13). It was reported that the stereoselectivity is satisfying (ds = 90–95%) [50].

t-BOC–Gly–Δ–Phe–(S)–Leu–OMe $\xrightarrow[\text{Cat.*}]{H_2}$ t-BOC-Gly-(S)-Phe-(S)-Leu-OMe + t-BOC-Gly-(R)--Phe-(S)-Leu-OMe

21

Stereoselective Hydrogenation of Other Prochiral Olefins

Optical yields are always low in asymmetric reduction of simple olefins where no polar groups are close to the double bond [41a, 51].

2.1.3 Homogeneous Asymmetric Hydrogenation of C=O Double Bonds

Carbonyl groups are not reduced with classical Wilkinson catalysts. However, some cationic rhodium complexes show catalytic activity [52]. There are only a few examples of asymmetric hydrogenation of ketones. Addition of base to a neutral rhodium complex is also a way to produce a catalyst for ketone reduction [44]. Acetophenone was hydrogenated with low enantioselectivity, using rhodium complexes of (S)-proline derivative (18) [44].

Especially high optical inductions have been achieved with catalysts which contain BPPM (13), PPM (14) or (15) as chiral ligands. Pyruvic acid esters were usually hydrogenated to lactic acid esters in quantitative chemical yield with an enantiomeric excess of 65–75% [41a,53]. However, this method is not of technical significance, since chiral lactic acid derivatives may be produced more conveniently by biotechnological processes.

(R)-(—)-Pantolactone (22), a key intermediate for the preparation of the vitamin pantothenic acid, has been obtained with high stereoselectivity (e.e. = 87%) by the asymmetric hydrogenation of the corresponding ketopantotyllactone in the presence of a rhodium complex of BPPM (13) [54].

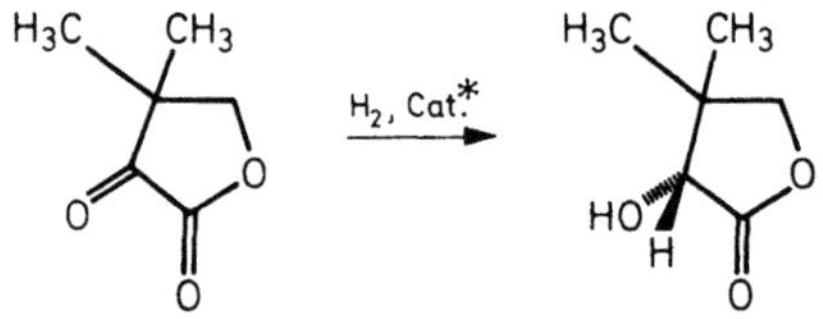

22

A smaller e.e. value is obtained with Rh-(14) and Rh-(15) [54] in this reaction. Moreover, in one case the undesired enantiomer of (22) is obtained.

2.2 Heterogeneous Asymmetric Hydrogenation

Nickel and other transition metal catalysts, when "modified" with a chiral compound such as (R,R)-tartaric acid [55], become enantioselective. All attempts to modify solid surfaces with optically active substances have so far resulted in catalysts of only low stereoselectivity. This is due to the fact that too many active centers of different structures are present on the surface of the catalysts. Consequently, in asymmetric hydrogenations the technique of homogeneous catalysis is superior to heterogeneous catalysis [56]. However, some carbonyl compounds have been hydrogenated in the presence of tartaric-acid-supported nickel catalysts in up to 92% optical purity [55].

An excellent review of the problems of the enantioselective heterocatalytic hydrogenation of prochiral double bonds, covering the literature up to 1970, has been compiled by Izumi [57]. Raney nickel catalysts modified with chiral amino acids or dipeptides gave only very moderate enantiomeric excesses of between 0 and 10% in the hydrogenation of olefins, carbonyl compounds or oximes [57]. Only Raney nickel modified with (S)-tyrosine furnished a higher enantiomeric excess in the products [58].

2.3 Electrochemical Asymmetric Syntheses

Pioneer work in the field of electrochemical asymmetric synthesis was done by Gourley et al. [59] using optically active alkaloids as chiral auxiliaries. Afterward,

Miller and his co-workers [60] reported surprisingly high optical yields, close to 50%, in the reduction of 2-acetylpyridine in the presence of strychnine. They also prepared chemically modified electrodes with optically active amino acids and attempted asymmetric induction in both reduction and oxidation [61]. The best optical yield, only 14.5%, seemed to be obtained in the reduction of 4-acetyl-pyridine on a graphite cathode modified with (S)-phenylalanine methyl ester.

Nonaka et al. [62] examined the asymmetric reduction of open-chain olefins in the presence of optically active amino acids. In the best case (R)-methylsuccinic acid (23) was formed in 2.4% and 53% optical and chemical yields, respectively, from citraconic acid (24) in the presence of (R)-cysteine.

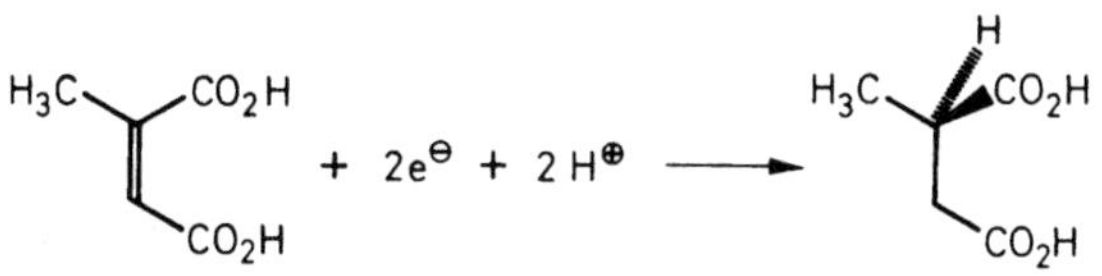

24 23

As reported in 1983, the same group was able to improve the optical yield of the reduction of (24) to 25%, while carrying out the electrochemical reduction at a poly-(S)-valine-coated graphite cathode [63]. In the reduction of 4-methylcoumarin (25) on this cathode an optical yield of 43% was achieved; the (S)-configurational enantiomer (26) was formed in excess [63].

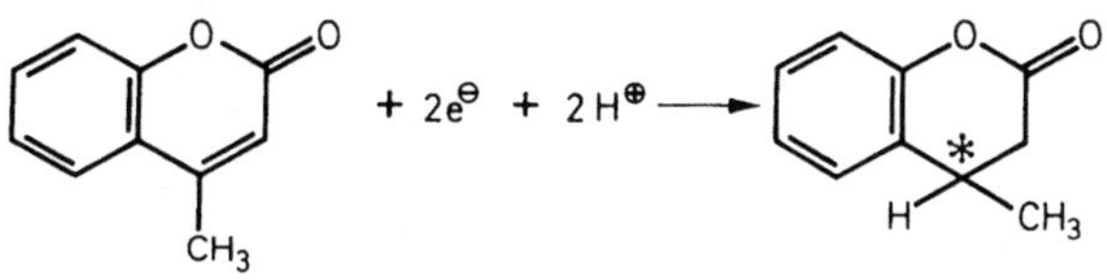

25 26

2.4 Enantioselective Hydrosilylation [64]

Acetophenone was enantioselectively hydrosilylated using a catalyst prepared from $[(COD)Rh\,Cl]_2$ and the (R)-cysteine derivative (27). (27) is the condensation product of (R)-cysteine methyl ester and 2-pyridinecarboxaldehyde.

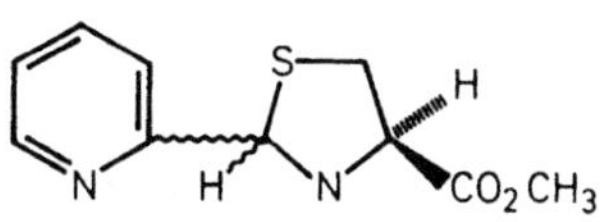

27

The enantioselective hydrosilylation is carried out between −20 and +20 °C; 1-phenylethanol (28) is obtained in 98% chemical yield with an e.e. value of 87%. The inductive power of the catalyst system $[(COD)Rh\,Cl]_2/(27)$ (Rh: (27) ≈ 1:13) is surprising, since (27) was only obtained as a mixture of two diastereomers (58:42), according to ^{1}N-NMR-data.

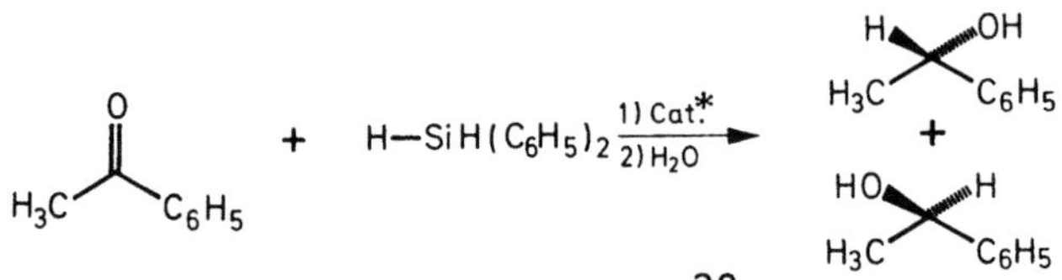

28

The main product is always the (R)-enantiomer of (28) [64]. Employing other chiral catalysts, e.g. Schiff bases prepared from (S)-alaninemethyl ester or (S)-valinemethyl ester and 2-pyridinecarboxaldehyde in form of their rhodium complexes, in the same reaction, no or only very low asymmetric induction was observed.

2.5 Asymmetric Aldol Addition

The efficient asymmetric intramolecular aldolization of certain triketones with a reflective symmetry axis using chiral amino acid catalysts has been reported with a view at obtaining optically active steroids.

Starting with the fundamental work of Wiechert et al. [65] and Hajos et al. [66], more than 15 amino acids have so far been used as chiral auxiliaries. It is remarkable that in most cases catalysts of the (S)-series, to which most natural α-amino acids belong, induce (S)-configuration. Chiral auxiliaries of the (R)-series, on the other hand, lead predominantly to products having an (R)-configuration.

Most workers in the field have investigated the asymmetric cyclization of 2-methyl-2-(3-oxobutyl)-1,3-cyclopentadione (29). The resulting enedione (30) was used as the C,D-unit in the total syntheses of steroids having a natural configuration [67]. With (S)-proline, (30) was obtained with an enantiomeric excess of 95% in almost quantitative chemical yield [67d].

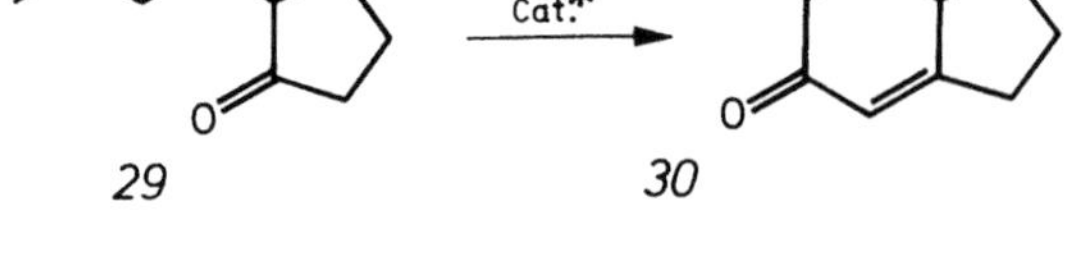

29 *30*

Hajos and Parrish [66] have shown that in the presence of only 3 mol-% of (R)- or (S)-proline in DMF at room temperature (29) is converted into the aldols (31a) and (31b), respectively. These readily eliminate water upon heating with p-toluenesulfonic acid in benzene.

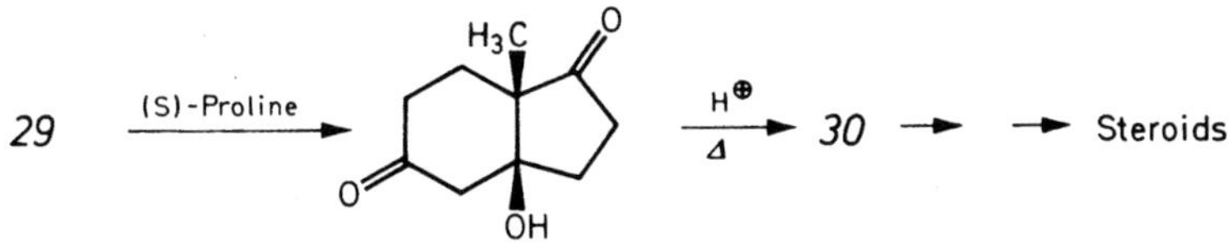

31a

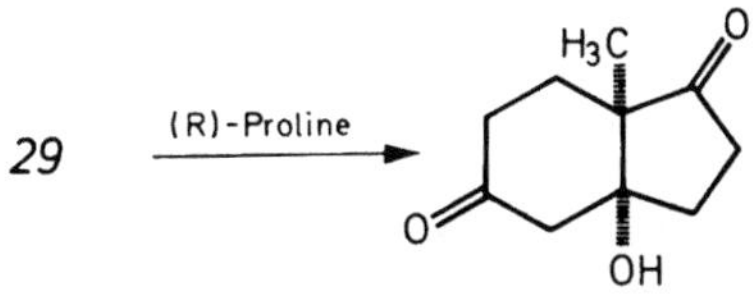

31b

The best chemical and optical yields in the above reactions are obtained by using (S)- or (R)-proline. Some 19-norsteroids are prepared on an industrial scale from products of intramolecular aldol additions catalyzed by (S)-proline [68].

Takano et al. [69] exploited the asymmetric aldolization for the synthesis of more functionalized chiral products which possess units suitable for the construction of certain tetracyclic triterpenes, such as gibberellins and kaurenes. They described the enantioselective synthesis of the tricyclic enone (33) from the symmetric triketone (32) and its conversion into the gibbane framework. Again, (S)-proline was used as the catalyst.

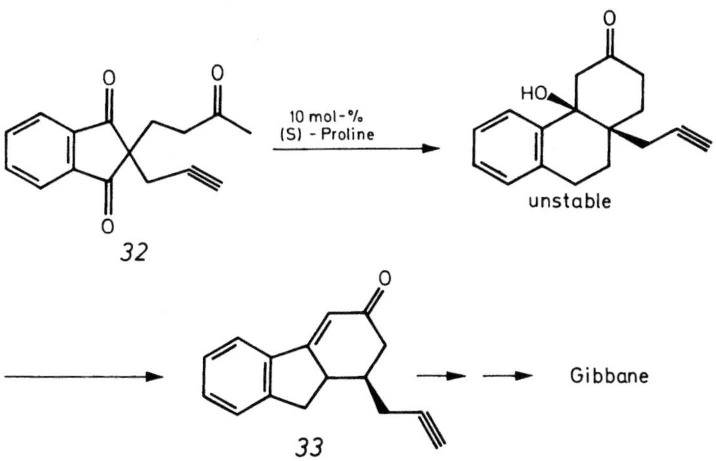

In his 1956 article in "Perspectives in Organic Chemistry", the late Professor R. B. Woodward characterized the macrolide antibiotic erythromycin as a synthetic challenge which is "... quite hopelessly complex, especially in view of its plethora of asymmetric centers" [70]. Twenty-five years after making this dismal prognosis,

R. B. Woodward and forty-eight coworkers have overcome this obstacle by using an ingenious strategy.

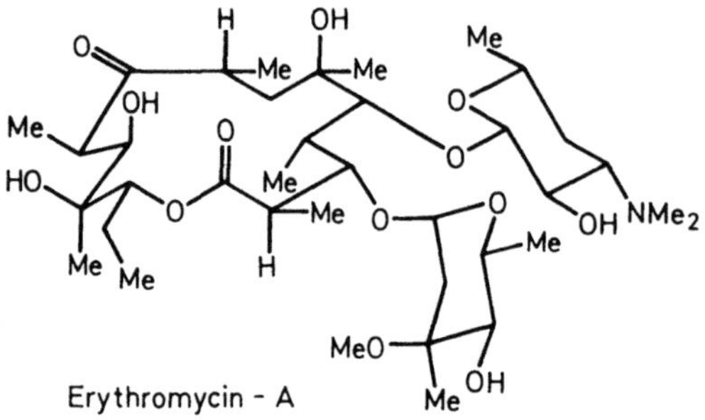

The dithiadecalin (34) was used to provide the carbon backbone for C-3 to C-8 and C-9 to C-13. Compound (34) was obtained in an optically active form by a route involving an enantioselective (36% e.e.) aldol cyclisation catalyzed by (R)-proline [71].

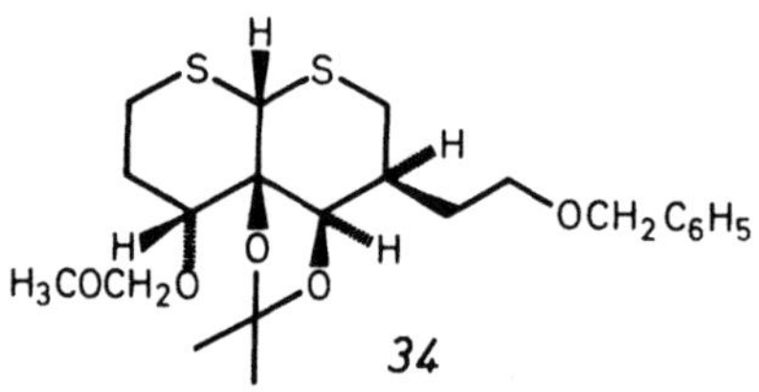
34

Until 1968, not a single nonenzymic catalytic asymmetric synthesis had been achieved with an enantiomeric excess above 50%. Now, the intramolecular aldol cyclisation, catalyzed by chiral amino acids has proven to be a very useful synthetic tool. This reaction was extensively covered by two reviews [23,68]. Two more papers [72], published recently, should also be cited.

2.6 Asymmetric Grignard Cross-Coupling Catalyzed by Chiral Phosphine-Nickel Complexes [73]

Hayashi et al. [74] described a process of kinetic resolution in the coupling of Grignard reagents R*Mgx (having a chiral center at the point of attachment to the metal) with various alkenyl halides under the influence of chiral phosphine-nickel complexes. Chiral amino acid derivatives (35) were used as ligands.

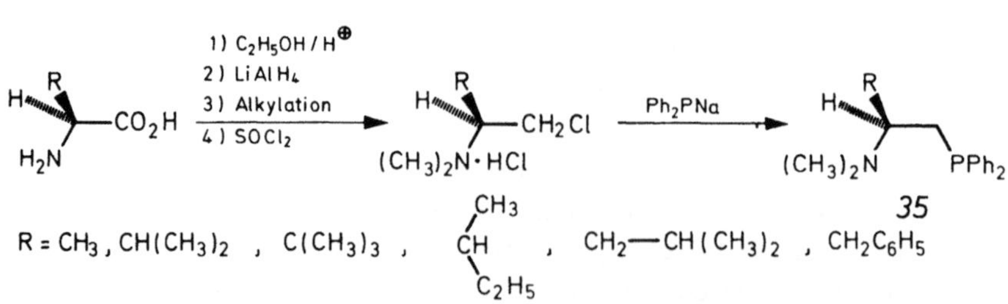

The reaction of 1-phenylethyl-, 2-octyl-, and 2-butyl-magnesium chloride (36a, b, c) with vinyl bromide (37a), (E)-β-bromostyrene (37b), 2-bromopropene (37c), and bromobenzene (37d) was carried out in the presence of 0.5 mol-% of a nickel catalyst prepared in situ from nickel chloride and the chiral ligand (35).

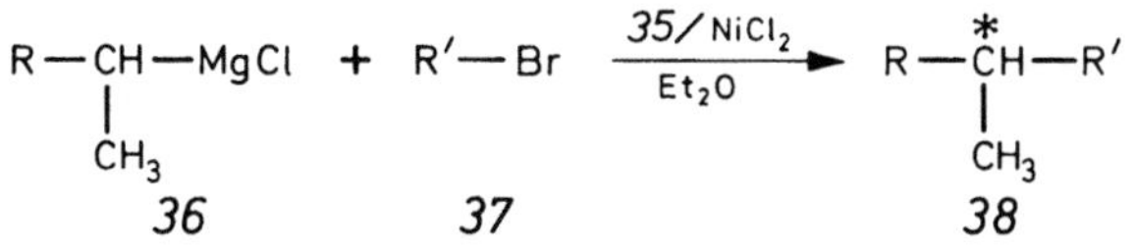

R = C_6H_5, n-C_6H_{13}, C_2H_5

R′ = $CH{=}CH_2$, CH=CH–C_6H_5, $C(CH_3){=}CH_2$, C_6H_5

* Indicates a chiral center

The products (38) where obtained with relatively high stereoselectivity (in some cases e.e. ≧ 80%).

2.7 Asymmetric Cyanohydrin Syntheses

Chemists in Japan have studied an asymmetric cyanohydrin synthesis: addition of hydrogen cyanide to benzaldehyde using synthetic peptides as catalysts [75].

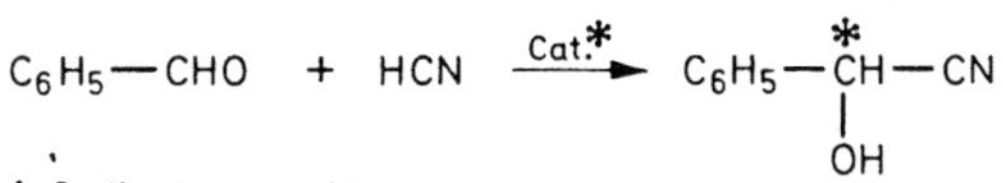

* Indicates a chiral center

This reaction is known to be catalyzed by the enzyme oxynitrilase to produce the optically pure cyanohydrin [76]. Since this reaction proceeds with a base catalyst, Jnoue et al. [75] used cyclic and linear dipeptides containing (S)-histidine. The catalysts employed are as follows: benzyloxycarbonyl-R-(S)-histidine methyl ester with R = (S)-alanyl, (R)-alanyl, (S)-phenylalanyl,[Z-(S)-Ala-(S)-His-OCH_3, Z-(R)-Ala-(S)-His-OCH_3, and Z-(S)-Phe-(S)-His-OCH_3] as linear dipeptides, and cyclic (S)-histidine containing dipeptides Gly–(S)–His,

(S)–Ala–(S)–His, (R)–Ala–(S)–His, (S)–His–(S)–Phe, and

(S)–His–(S)–His.

When (S)–His–(S)–Phe was used in a mole ratio to benzaldehyde of about 1/50, the enantiomeric excess of mandelonitrile was very high (90%, rich in (R)) in the early stage, but decreased with reaction time.

In sharp contrast, the corresponding linear dipeptide Z-(S)-Phe-(S)-His-OCH_3 exhibited only very low stereospecificity (0.6% e.e., rich in (S)) in the addition of hydrogen cyanide to benzaldehyde. Thus, the rigid structure of the 2,5-piperazine-dione ring of the cyclic dipeptide is very important for the asymmetric addition. The flexible linear dipeptide is disadvantageous [75].

2.8 Enantioselective Epoxidations

The asymmetric epoxidation of several chalcones (39) and other electron-poor olefins in a triphase system (water/organic solvent/chiral polyamino acid) afford optically active oxirans with optical yields of up to 96%. The influence of the molecular structure of the catalysts and substrates, the solvent, and the temperature on the stereochemistry was investigated by a group of chemists from Italy and Spain [77].

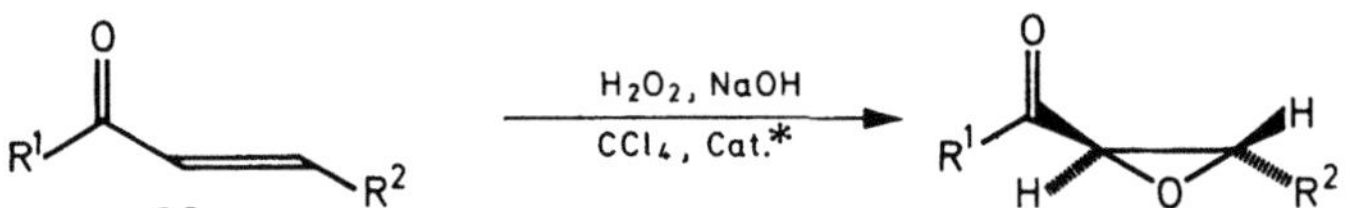

39 40

R^1, R^2 = phenyl or substituded aryl

The epoxidation of (39) generally occurs with good chemical conversions and high optical yields [77]. When poly-(S)-alanine is employed as chiral catalyst, the reaction is practically stereospecific. Similar good results were obtained with poly-(S)-leucine and poly-(S)-isoleucine, whereas poly-(S)-valine not only reduces the chemical yields, but also greatly affects the asymmetric synthesis. Both poly-(S)-phenylalanine and the dipeptide (S)-Ala-(S)-Ala give almost racemic oxiranes (40). The structure of the substrate also plays an import role in determining the amount of asymmetric induction. Poly-(S)-amino acids are much less effective catalysts in the epoxidation of systems other than chalcone (39) and other related electron-poor olefins, such as 2-methyl-1,4-naphthoquinone (41) and 1-phenyl-2-nitro-propene (42).

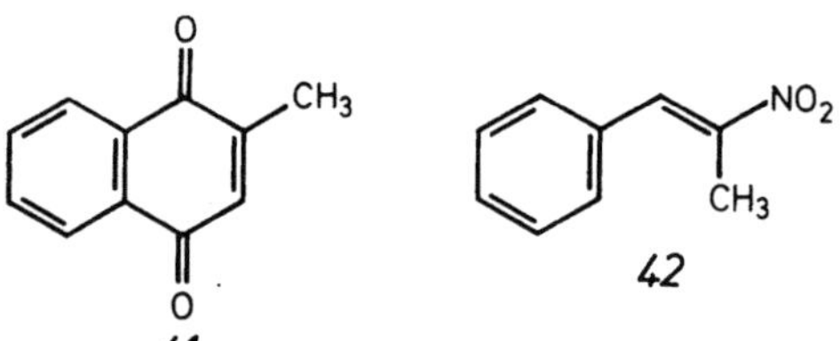

41 42

Enantioselective epoxidation of allylic alcohols using hydrogen peroxide and chiral catalysts was first reported for molybdenum [78] and vanadium [79] complexe. In 1980, Sharpless [80] reported a titanium system. Using a tartaric acid derivative as chiral auxiliary it achieves almost total stereoselection in this reaction.

During the investigation of the molybdenum-catalyzed epoxidation of the allylic alcohol (43) mediated by a chiral (S)-proline derivative (44). S. Coleman-Kammula and E. Th. Duim-Koolstra observed that the stereoselectivity decreased with increasing conversion.

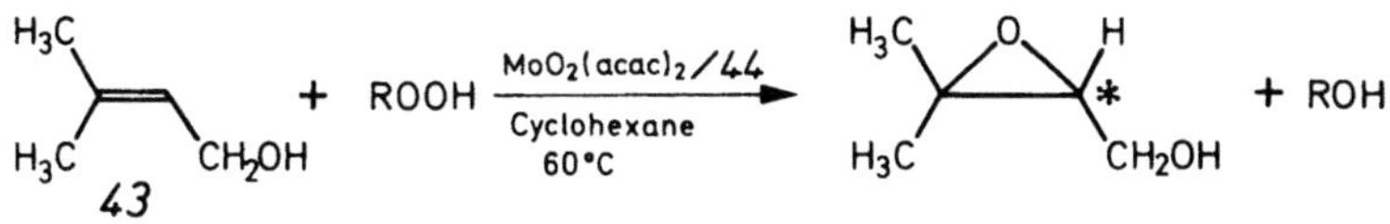

43

The ligand (S)-N-methylprolinol was used in a 2:1 molar ratio to the $MoO_2(acac)_2$ catalyst (1 mol % on the allylic alcohol (43) in the epoxidation of 3-methyl-2-buten-1-ol (43) with cumene hydroperoxide in cyclohexane solvent.

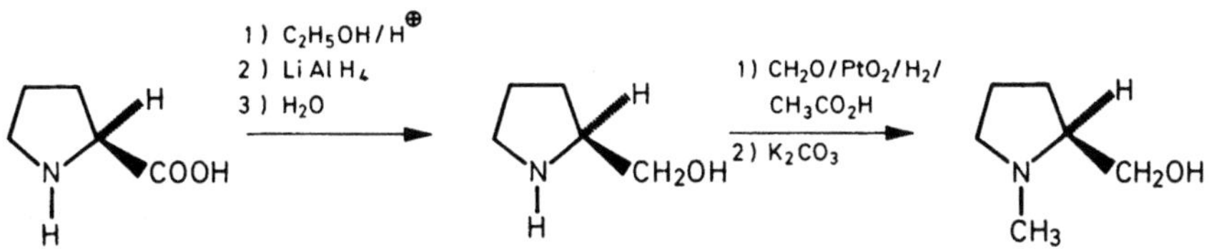

(S)-Proline (S)-Prolinol 44

The chiral ligand (44) was prepared starting from the cyclic α-amino acid (S)-proline [80]. Recently, similar chiral catalysts and related molybdenum complexes involving optically active N-alkyl-β-aminoalcohols as stable chiral ligands and acetylacetone as a replaceable bidentate ligand, were designed for the epoxidation of allylic alcohols with alkyl hydroperoxides which could be catalyzed by such metal complexes [81].

2.9 Enantioselective Michael Addition of Thiols to 2-Cyclohexen-1-one

Chiral aminoalcohols (45), derived from (2S, 4S)-4-hydroxyproline and (S)-proline, respectively, were found to be superior catalysts for the enantioselective 1,4-addition of arylthiols to 2-cyclohexen-1-one to yield 3-arylthiocyclohexanones (46) [82].

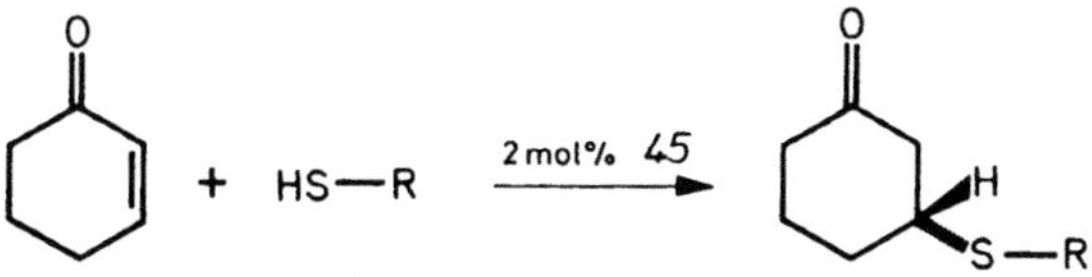

R = aryl 46

In the best case the optically active adduct (46) was obtained in 88% optical yield. The ketone (46) was made optically pure when recrystallized twice from pentane.

The same authors [83] used the chiral ketone (46) as substrate for the preparation of optically active cyclohexanol derivatives (47) which may be useful intermediates in the synthesis of chiral natural products, such as (—)-mesenbranoene, (+)-2-carene etc.

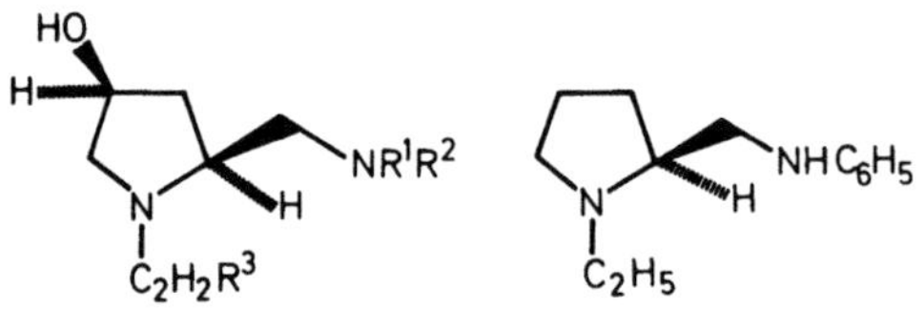

45a 45b

R^1 = H, CH_3
R^2 = Aryl, Cyclohexyl
R^3 = H, C_2H_5

It is noteworthy that (47a) was obtained almost exclusively (e.e. = 96%) by employing $LiALH(O^+Bu)_3$, while (47b) was obtained by reduction with K-selectride in 92% e.e. [83].

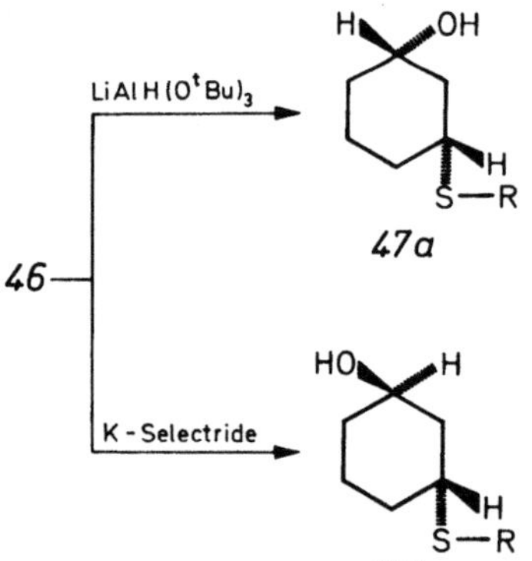

47a 47b

3 Stoichiometric Asymmetric Syntheses

In the preceding Section we considered the catalytic asymmetric synthesis. In this connection the induction of asymmetry by catalytic amounts of "chiral information" (= amino acids or their derivatives) was treated. The chiral information was transferred into a prochiral substrate.

In Section 3 we turn to reactions which require at least equimolar amounts of chiral information for the induction of asymmetry in the products. The newly formed asymmetric center can be induced by either intramolecular or by intermolecular interactions. Having served its stereochemical purpose, the amino acid moiety may be destroyed so that it does not exist as a discrete entity in the product, although sections of it may survive.

In another case the amino acid is used as a chiral reagent in stoichiometric amounts. Thus, the reagent is preferably recovered and recycled.

We do not address the chiral template method in this article.

3.1 Hydrogenations

3.1.1 Asymmetric Hydrogenation of C=C Double Bonds

If an asymmetric hydrogenation of C=C bonds is desired in the presence of achiral catalysts, chiral information is required to be present in the substrate. Peptides and cyclopeptides containing dehydroaminos acid units are very good substrates achieving quite high stereoselectivities upon asymmetric hydogenation on 10% Pd-C or other achiral catalysts [49a, 84].

The optical yields have been as high as 61% for the hydrogenation of open-chain dehydrogeneous or heterogeneous phase.

However, excellent optical yields have been obtained for the hydrogenation of cyclodipeptides. As early as 1944, M. Bergmann and J. E. Tietzmann [84m] obtained the diketopiperazine of (S)-phenylalanyl-(S)-proline (49), while hydrogenating dehydrophenylalanyl-(S)-proline diketopiperazine (48).

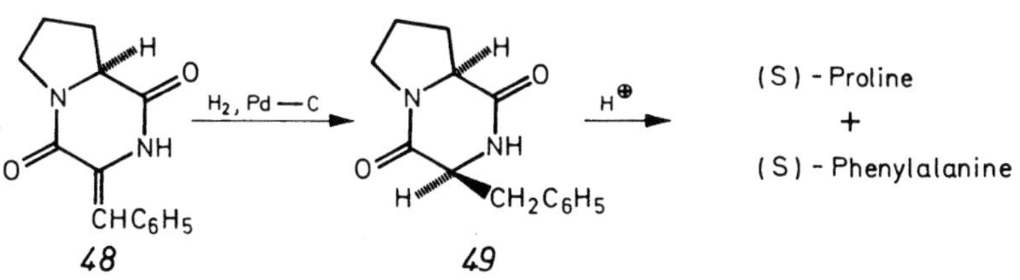

Hydrolysis of the resulting (S,S)-cyclodipeptide (49) (e.e. >90%) affords (S)-phenylalanine and the chiral inductor (S)-proline [84h].

Captopril (1), an approved antihypertensive drug, was prepared from (S)-proline. The key step in an elegant asymmetric synthesis of (1) was a hydrogenation of (50) to give (51) which is hydrolyzed to afford (1) [85].

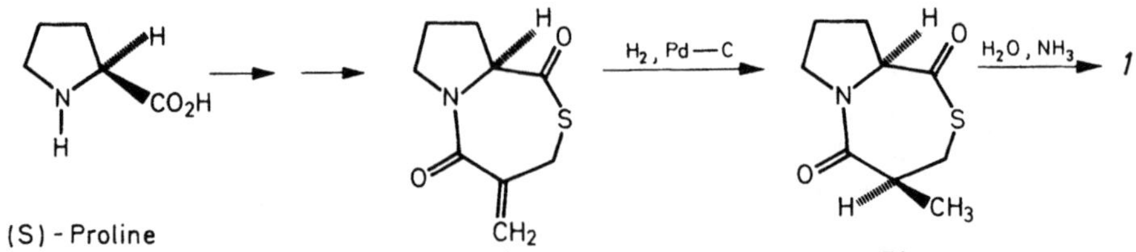

50 51

(S)-2-carbethoxy indolene hydrochloride (52) was catalytically hydrogenated (Pd-C) in ethanol to (2S, 3aS, 7aS)-2-carbethoxy perhydroindole hydrochloride (53) which was purified by crystallization [86]. (53) was used in the synthesis of a potent inhibitor of the angiotensin converting enzyme.

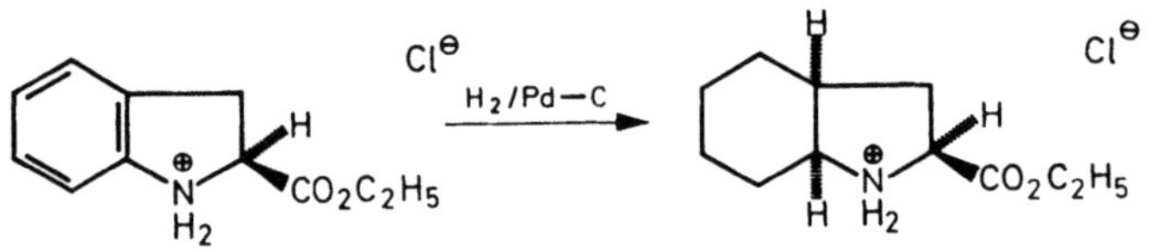

52 53

Stereoselective addition of hydrogen to a C=C double bond of an (S)-proline derivation was applied in the total synthesis of gephyrotoxin, a biologically active alkaloid [87]. Optically active pyrrolizidine bases have been synthesized by Robins and Sakdarat [88] from chiral hydroxyproline derivatives by hydrogenation (ds > 60%).

In the asymmetric hydrogenation of the (R)-phenylglycine derivative (54) in the presence of an achiral catalyst the stereoselectivity was reported [89] to be low. The lactone (55) could subsequently be converted into (S)-aspartic acid [89]. This reaction sequence is an example of the intramolecular transfer of chirality with subsequent disappearance of the original chiral center.

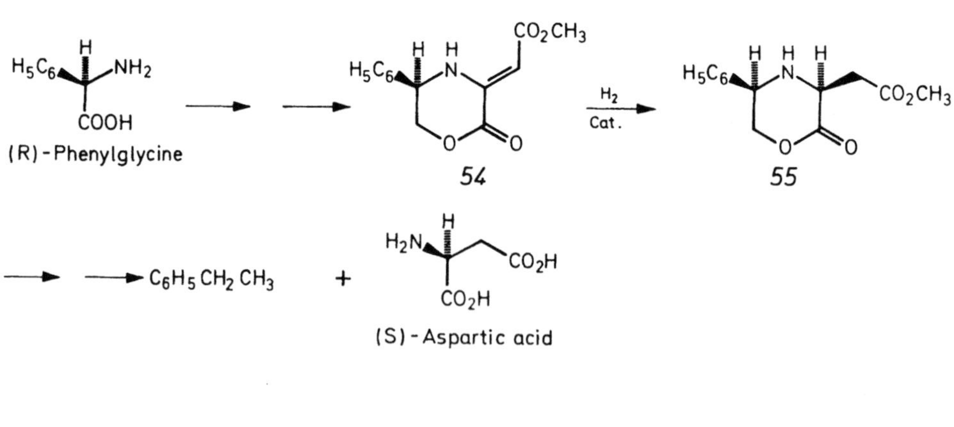

The reducing agents (*56*) to (*62*), derived from (S)-proline, are eminently suitable for the hydrogenation of prochiral CN and CO double bonds.

The prochiral ketones (63) were reduced with (57) at low temperature in high chemical (82–97%) and optical (31–96%) yields [90] to the (S)-alcohols (64). The chiral precursor or (57) can be recovered.

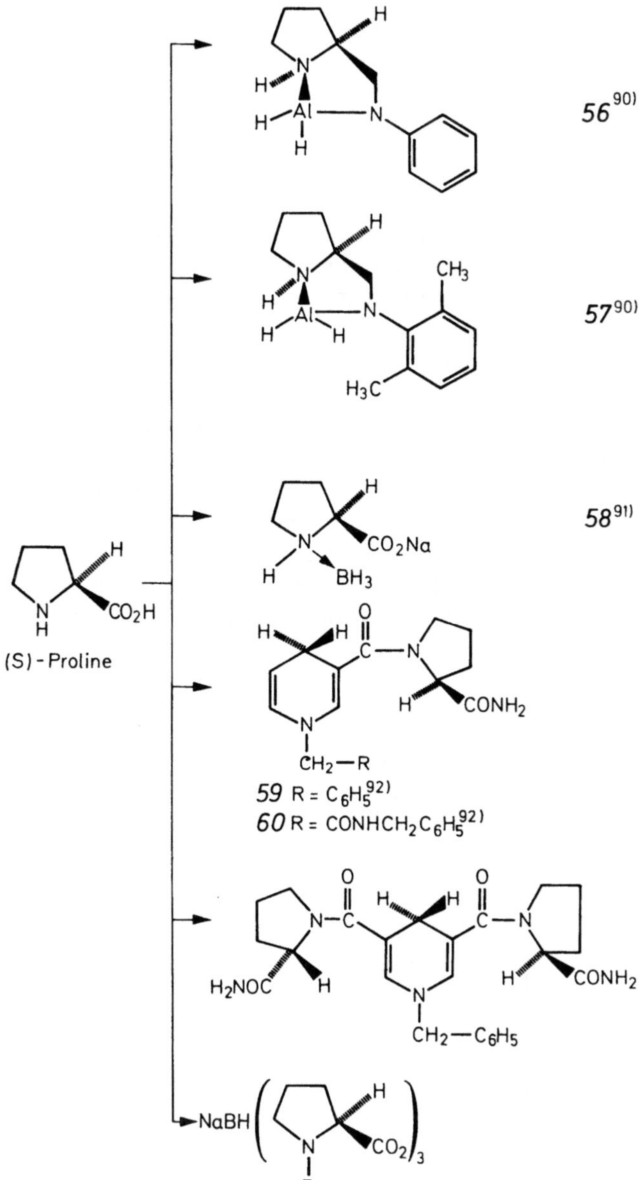

56 [90)]

57 [90)]

58 [91)]

59 R = C_6H_5 [92)]

60 R = $CONHCH_2C_6H_5$ [92)]

61 [92]

62a R = Cbz [93,94)]

62b R = CH_3–CO [94)]

62c R = C_6H_5–CO [94)]

62d R = Boc [94)]

When the ketone (65) was treated with the chiral reagent produced by decomposing $LiAlH_4$ with (S)-2-(anilinomethyl) pyrrolidine, the alcohol (66) obtained in 60% yield. The optical yield could be determined as 62% e.e. [95].

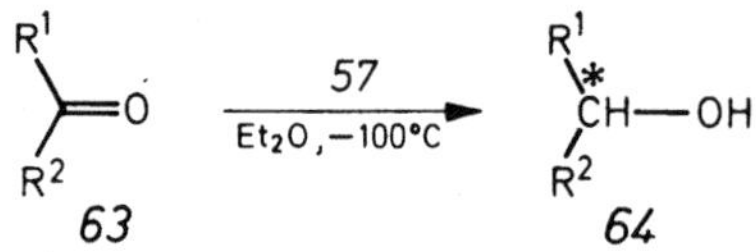

63 64

$R^1 = n\text{-}C_6H_{13}, CH_2C_6H_5, C_6H_5$
$R^2 = CH_3, C_2H_5, i\text{-}C_3H_7$

The chiral alcohol (66) is a valuable intermediate for the asymmetric synthesis of optically active anthracyclinones (67). These aglycones of anthracycline antibioties are currently attracting much attention because of their promising anticancer activities.

(58) reduces aryl alkyl ketones as well as prochiral dialkylketones to the corresponding chiral alcohols in chemical yields of 66 to 92% and optical yields of up to 62% [91].

High stereoselectivities (94–100%) are attained in the reduction of aromatic ketones by use of a new chiral borane complex with (S)-2-amino-3-methyl-1,1-diphenylbutan-1-ol,(S-68) readily prepared in two steps from (S)-valine, in an experimentally convenient procedure [96]. (S)-Valine methyl ester hydrochloride was converted with excess of phenylmagnesium bromide into (S-68). The same treatment of (R)-valine gave (R-68). In a typical asymmetric reduction the reagent, prepared from (S-68) and borane, and the ketone (69) in tetrahydrofuran were kept at 30 °C for some hours. The corresponding alcohols were obtained in high optical purity. (S-68) could be recovered to more than 80% without racemization [96].

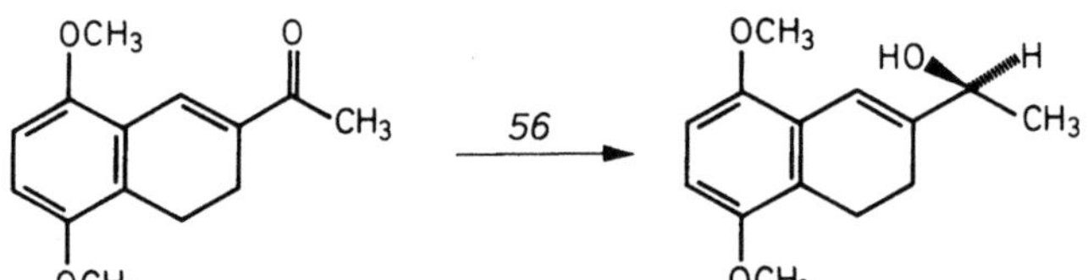

65 66

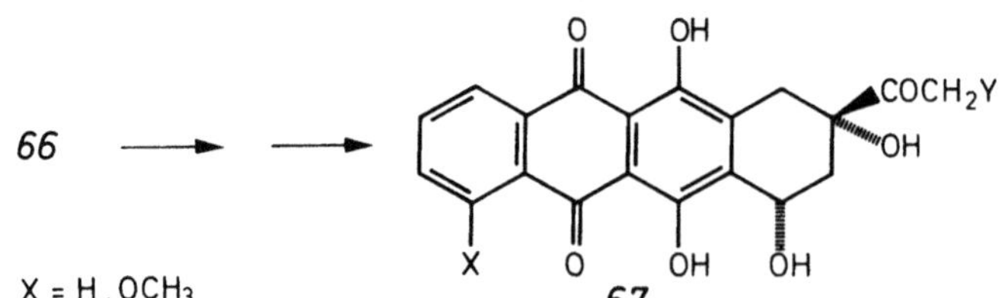

$X = H, OCH_3$
$Y = H, OH$

67

It is of particular interest that the borane complex of (S-68) produces the (R)-alcohol, while the reversed stereoselectivity with the same degree of asymmetric induction was achieved by use of the reagent from (R-68). Since both (S-68) and (R-68) are readily accessible from the corresponding amino acids (S)-valine and (R)-valine, respectively, this method allows both enantiomers of secondary alcohols to be synthesized readily from aromatic ketones.

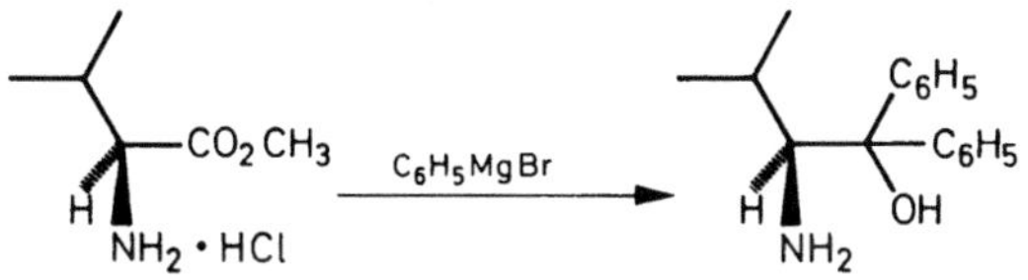

(S)-Valine methyl ester hydrochloride   S-*68*

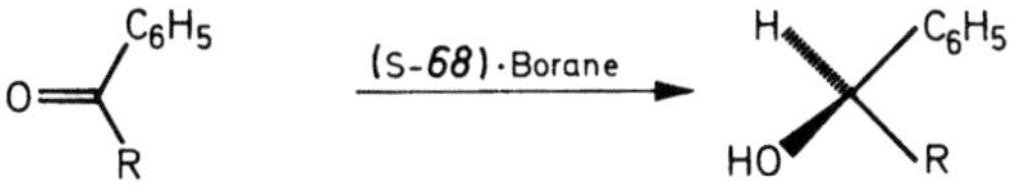

69   *70*

R = CH_3, C_2H_5, n-C_3H_7, n-C_4H_9

A number of chiral macrocyclic crown ligands have been synthesized by using amino acids as the chiral natural source. Thus, macrocyclic compounds which have a 12-, 15-, or 18-membered ring have been prepared from α-amino acids to introduce chirality into their rings or the side chains [97]. Most chiral crowns derived from amino acids have not as yet been evaluated for their ability to effect chiral recognition or act as catalysts. Kellog [97a] used the dihydropyridine crowns (71) to cause asymmetric reduction of a number of aromatic ketones. Optical yields as high as 86% have been achieved using these macrocycles.

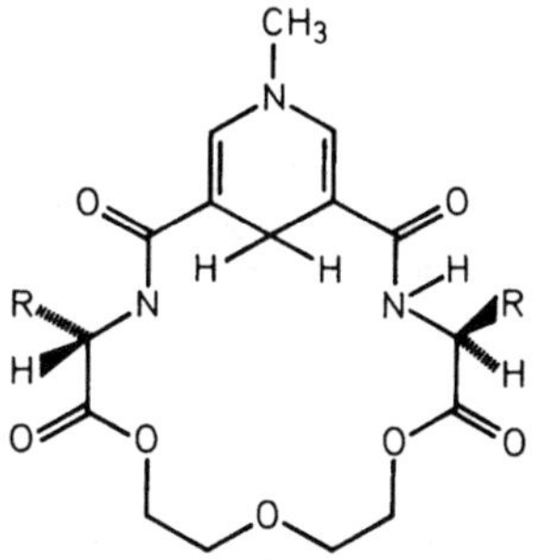

71

Recently, Inouye et al. [92] reported that NADH model compounds (59, 60, 61) carrying one or two (S)-prolinamite moieties as the asymmetric center showed virtually complete stereoselectivity in the asymmetric reduction of several ketones.

The natural product (+)-(S)-gingerol (72) was synthesized using 3,5-disubstituted isoxazole as masked β-ketol. Reductive fission of the labile N—O bond of the isoxazole gave an enamino-ketone which was converted into the vinylogous imide (73) using N-tosyl-(S)-prolyl chloride. Reduction of (73) gave (72) in 30–40% optical yield [98].

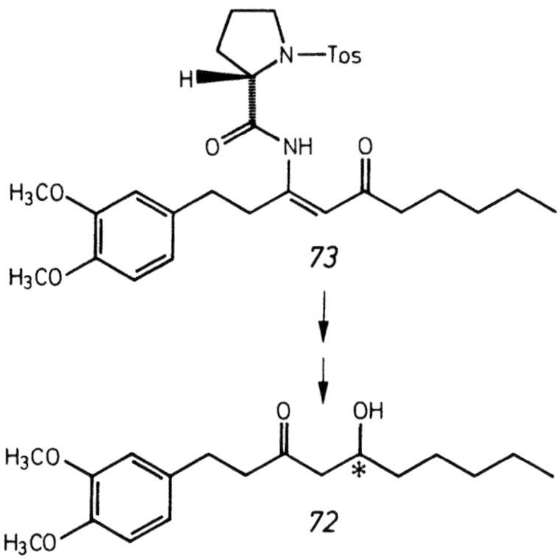

α-N-Acylamino acids have been developed as useful reagents for the preparation of optically pure α-aminoalkyl aryl ketones. Protection of the amino group as the ethoxycarbonyl derivative (74) [99] allows (S)-alanine to serve as an effective educt for the asymmetric synthesis of a variety of structures containing the phenylethylamine backbone. Thus, N-acyl-(S)-alanine chloride undergoes Friedel-Crafts acylation.

Table 1. Reduction of (S)-2-[(Ethoxycarbonyl)amino]propiophenone (74) to Ephedrine (75a) and Pseudoephedrine (76b)

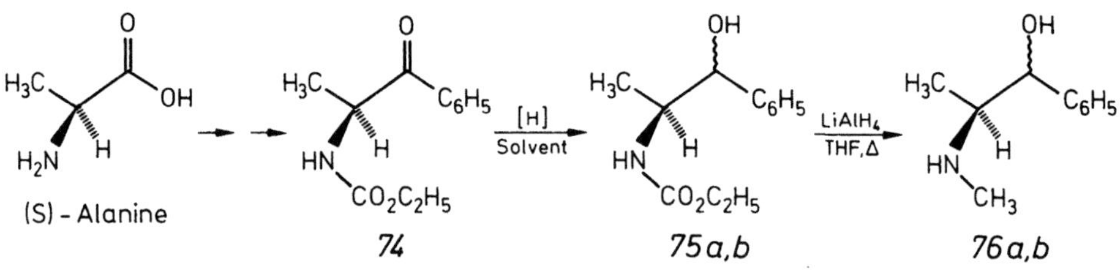

Reducing agent [H]	Solvent	Ephedrine 75a / Pseudoephedrine 76b
$NaBH_4$	CH_3OH	4 : 1
$LiAlH_4$	THF	4 : 1
Li selectride	THF	1 : 1
Pd - C/H_2	C_2H_5OH	2 : 1

These intermediates were used to synthesize optically pure ephedrines and amphetamines without recourse to resolution, since the chirality of the amino acid educt was entirely conserved throughout the process. The reduction of (S)-2-(ethoxycarbonyl)amino-propiophenone (74) first produced a mixture of alcohols (75a, b) [99]. Lithium aluminium hydride reduction then produced the desired secondary amino alcohols (76a, b). Table 1 illustrates the reduction scheme and the diastereomer ratios obtained [99].

Dieckmann cyclization of (S)-N-2-di(carboethoxymethyl)pyrrolidine, made from (S)-proline, produced (8S)-1-carboethoxypyrrolizidin-2-one (77). Upon catalytic hydrogenation, (78) was obtained as the main product in high diastereoselectivity [100].

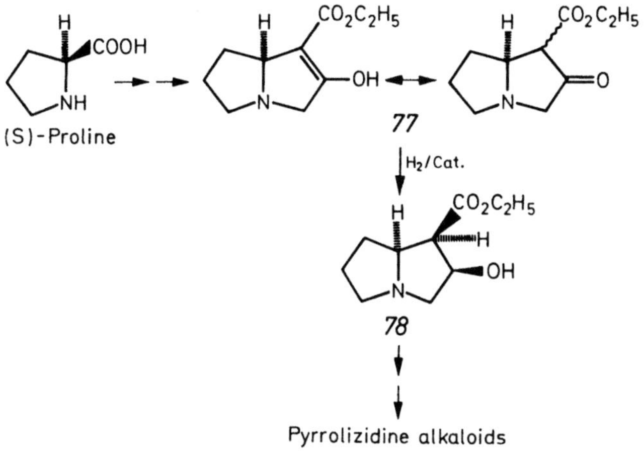

(78) serves as a key intermediate in the synthesis of (—)-isoretronecanol, (—)-trachelanthamidine, (—)-supinidine, and other pyrrolizidine alkaloids [100].

A related asymmetric reduction directed by a chiral center from a (S)-proline moiety was reported by Budzikiewcz et al. earlier [101].

3.1.3 Asymmetric Hydrogenation of C=N Double Bonds

Knoop and Martius [102a)] reported the asymmetric hydrogenation of the Schiff base obtained from (S)-arginine and pyruvic acid as early as 1939.

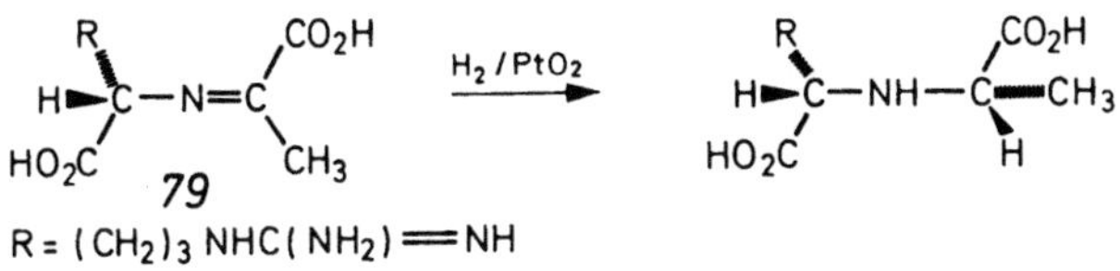

The asymmetric transamination from chiral α-amino acids [102)] and amino acid derivatives (57) (esters [86, 103)], amino alcohols [104)]) to carbonyl functions in prochiral substrates (58) (α-keto acids [102)], α-keto esters [86, 103)], ketones [103b, d)] was described in many papers. In each case, a Schiff base (80) is an intermediate. The optical yield in the reduction was as high as 87% and was measured either from compounds (81) or (82).

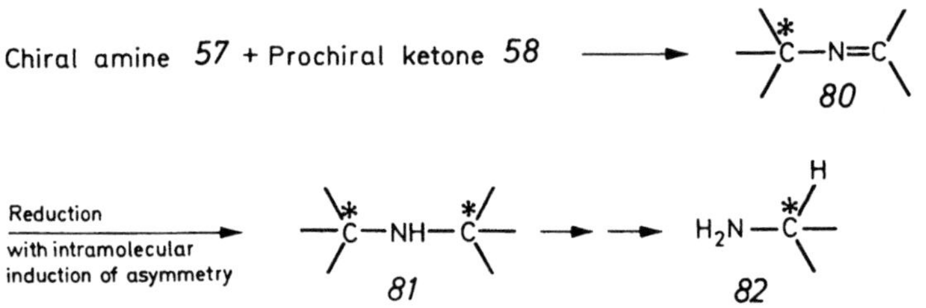

The asymmetric hydrogenation of the phenylhydrazone of methyl-N-(3,3-dimethyl-2-oxobutanoyl)-(S)-valinate (83) was investigated using palladium catalysts [105]. The absolute configuration of the newly formed t-leucine moiety in (84) was found to be (S). The diastereoselectivity in the hydrogenation step was shown to be as high as 56%.

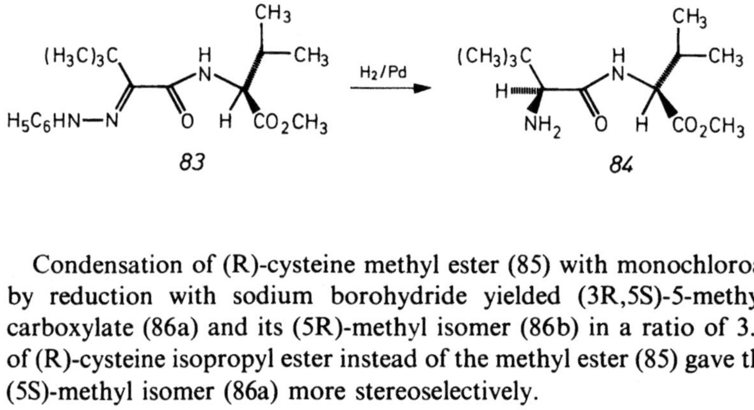

Condensation of (R)-cysteine methyl ester (85) with monochloroacetone followed by reduction with sodium borohydride yielded (3R,5S)-5-methyl-1,4-thiazane-3-carboxylate (86a) and its (5R)-methyl isomer (86b) in a ratio of 3.1:1 [106]. The use of (R)-cysteine isopropyl ester instead of the methyl ester (85) gave the corresponding (5S)-methyl isomer (86a) more stereoselectively.

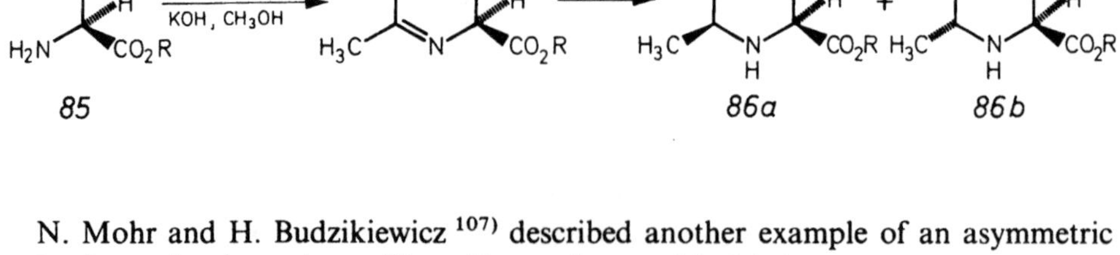

N. Mohr and H. Budzikiewicz [107] described another example of an asymmetric reductive amination using a (S)-proline moiety as chiral inductor.

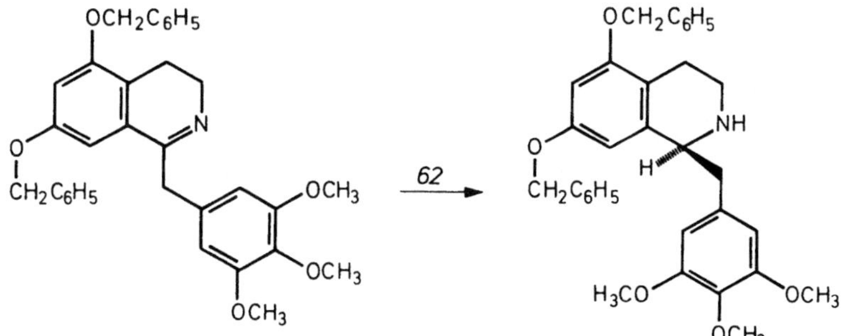

87 88

The 1-substituted 3,4-dihydroisoquinoline (87) was reduced with chiral reducing agent (62) to the corresponding alkaloids (88) in excellent optical yields (e.e. = 60 to 87%) [94].

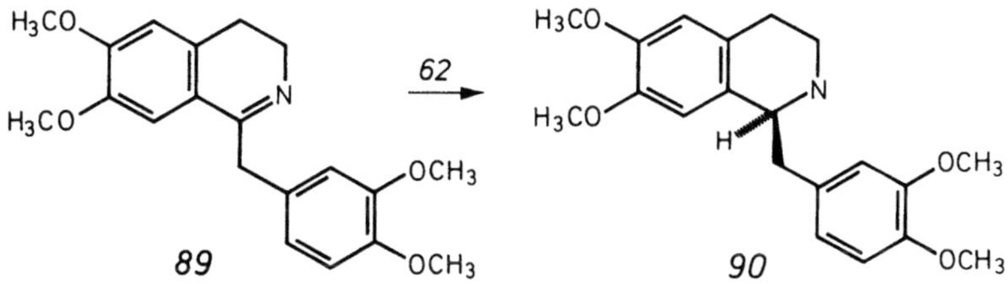

89 90

(88) is an intermediate in the synthesis of a positional isomer of trimetoquinol with respect to the hydroxy groups and is currently under clinical trials as an orally effective bronchodilator [94].

Triacycloxyborohydrides (62), derived from $NaBH_4$ (1 equiv.) and (S)-N-acyl-proline (3 equiv.), were found to reduce 3,4-dihydropapaverine (89) in tetrahydrofuran to (S)-norlaudanosine (90) hydrochloride in 60% optical yield [93]. In some cases (90) was obtained in even higher optical purity (e.e. = 70–86%).

3.2 CC Bond-Forming Reactions

The development of efficient and highly selective methods for carbon-carbon bond formation has been and continues to be a challenging and exciting endeavor in organic chemistry. Especially in the field of asymmetric synthesis, the chemical community has seen a real breakthrough in the past 10 years. Several processes routinely allowing asymmetric inductions of greater than 90% e.e. are now at our disposal. In this Section, asymmetric C–C bond-forming reactions are considered in which amino acids or amino acid derivatives are used as chiral auxiliary reagents.

3.2.1 Reactions at the Carbonyl Group

1,2-Additions of Organometallic Compounds:

The formation of chiral alcohols from carbonyl compounds has been fairly widely studied by reactions of aldehydes or ketones with organometallic reagents in the presence of chiral ligands. Mukaiyama et al. [108] obtained excellent results (up to 94% e.e.) in at least stoichiometric addition of the chiral auxiliary to the carbonyl substrate and the organometallic reagent.

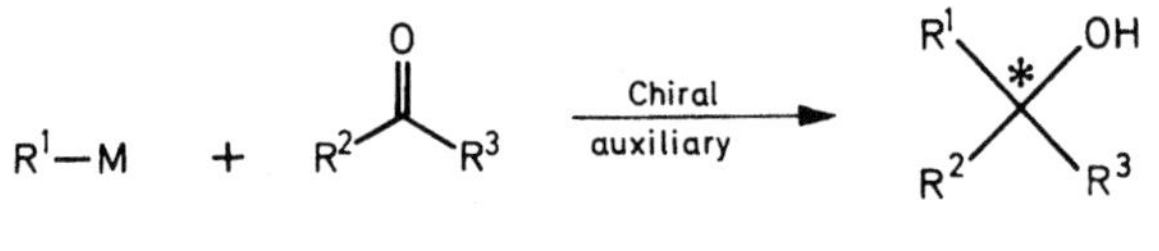

Various chiral diaminoalcohols (91, 92, 93, 94) were synthesized starting from commercially acailable (S)-proline. The enantioselective addition of organometallic compounds to aldehydes in the presence of the aminoalcohols was investigated.

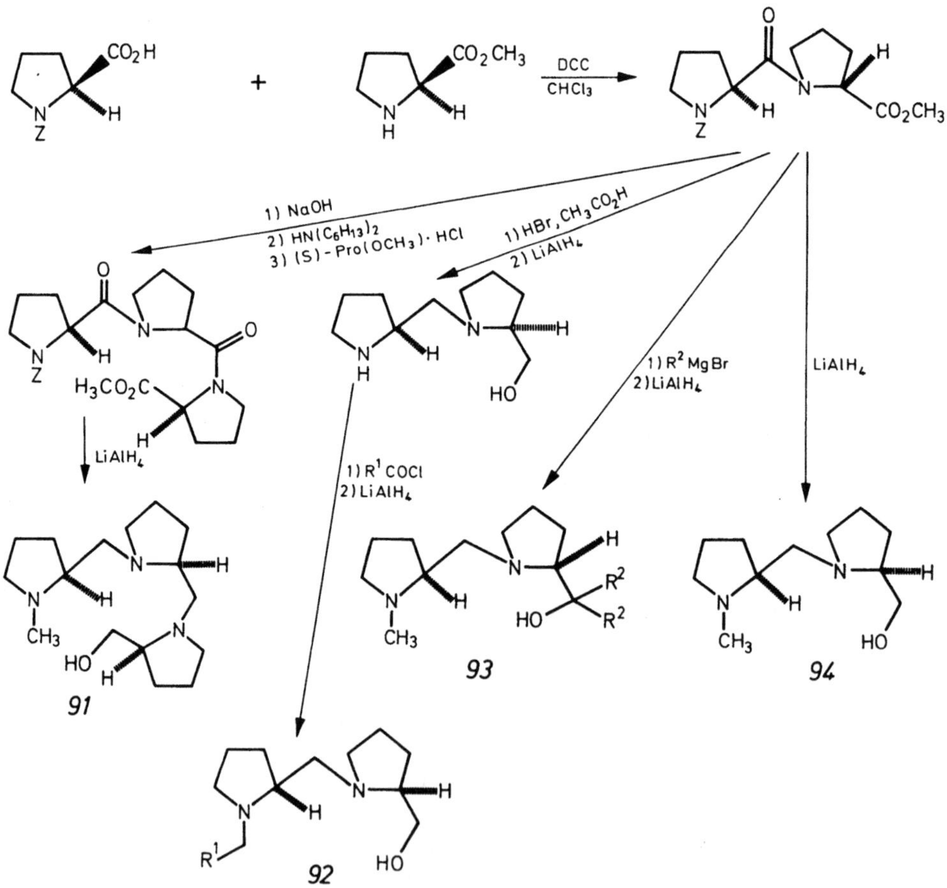

The enantioselective addition of alkyllithium to aldehyde in the presence of the lithium salt of diaminoalcohol (94) yielded optically active secondary alcohols as shown in Table 2.

Table 2. Enantioface-differentiating Addition of Alkyl Lithiums to Aldehydes [109]

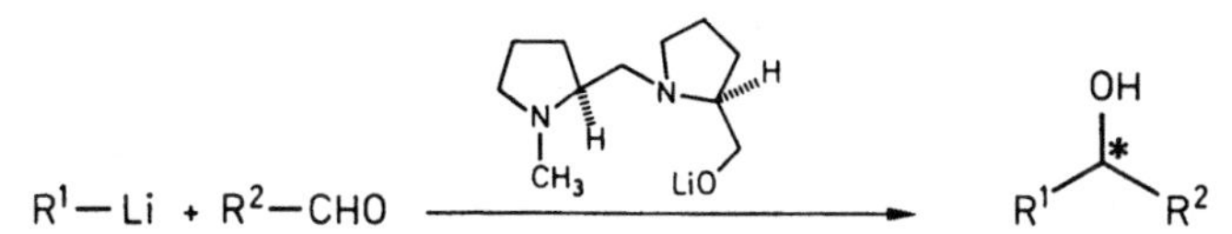

R^1	R^2	Yield (%)	ee (%)	Config.
CH_3	C_6H_5	81	40	R
C_2H_5	C_6H_5	59	54	S
n-C_3H_7	C_6H_5	64	60	S
n-C_4H_9	C_6H_5	77	95	S
n-C_4H_9	i-C_3H_7	57	80	S

L. Colombo et al. [110] synthesized two related (S)-proline derivatives and used them as chiral ligands for lithium in reactions of n-butyllithium with benzaldehyde. 1-Phenyl-1-pentanol was obtained with moderate optical purity (4–33% e.e.). Both nitrogen atoms as well as the free hydroxy group in ligands (91) to (94) appear to be essential centers for coordination with the alkali metal.

By the extension of the above-mentioned stereoselective asymmetric addition of alkylithiums to other organolithium reagents such as lithium salts of methyl phenyl sulfide, 2-methylthiazoline, trialkylsilylacetylene, N-nitroso-dimethylamine, and acetonitrile, chiral oxiranes (95) [111], thiiranes (96) [111], acetylenic alcohols (98) [112], and amino alcohols (97) [111] were readily obtained.

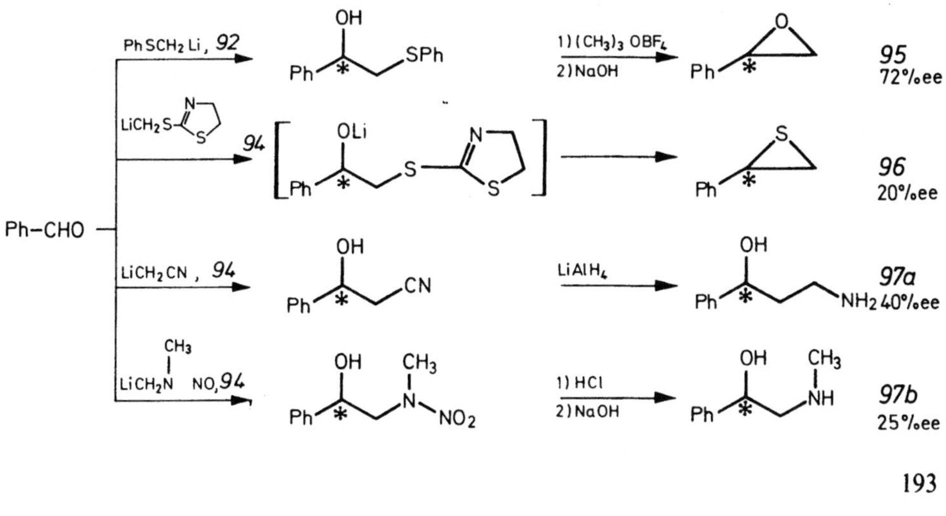

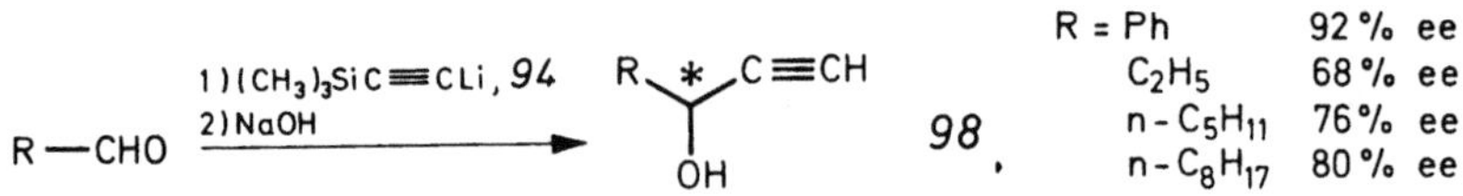

Some of the chiral acetylenic alcohols (98) were successfully converted to γ-ethyl-γ-butyrolactones, insect pheromones of *Trogoderma*. In addition, they are important intermediates for the synthesis of products with antibacterial activity [113].

Aside from alkyllithium compounds, only dialkylmagnesium compounds gave high optical yields of the alcohols using ligand (94) as chiral auxiliary [109c, 114].

The diamine (99) was prepared from (S)-proline [90b] or (S)-glutamic acid [115] maintaining the asymmetric center. Racemic 2-(anilinomethyl)pyrrolidine, prepared from (RS)-5-oxopyrrolidine-2-carboxylic acid, was effectively resolved into a pair of enantiomers by fractional crystallization of its mandelic acid salt [116]. Moreover, the preferential crystallization of its 4-hydrobenzoic acid salt was found to produce both enantiomers in high optical purities by alternate seeding [116].

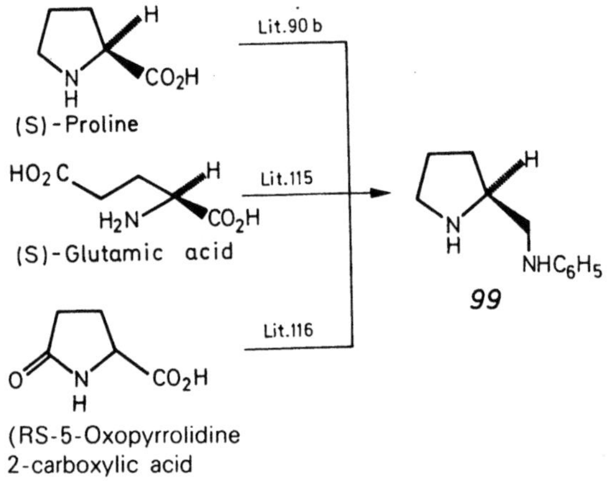

The diamine (99) was applied [117] to the synthesis of chiral α-hydroxyaldehydes. Thus, treatment of the aminal (100), prepared from the chiral diamine (99) and phenylglyoxal, with the Grignard reagent affords the hydroxyaminal, which in turn was hydrolyzed to yield α-alkyl-α-hydroxyphenylacetaldehyde (101). The chiral auxiliary was recovered [117].

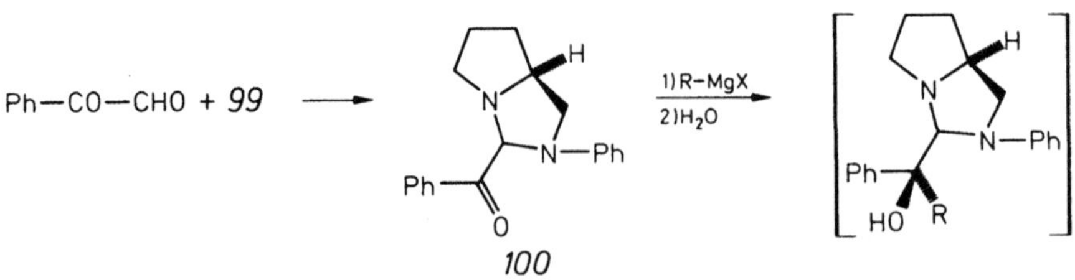

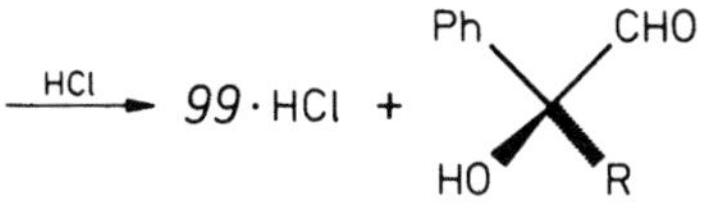

101

R = CH_3 (34% ee), C_2H_5 (92% ee), i-C_3H_7 (40% ee), n-C_4H_9 (88% ee), sec.-C_4H_9 (42% ee)

Mukaiyama et al. developed a rather general and versatile method for the preparation of optically active α-hydroxyaldehydes by using the diamine (99) as chiral adjuvant. Thus, one Grignard reagent (R^1MgX) is reacted with the aminal (102) of methyl glyoxylate. In the next step a second kind of Grignard reagent (R^2-MgX) is diastereoselectively added to the ketoaminal, and the desired chiral α-hydroxyaldehyde (103) is obtained by hydrolysis [117,118].

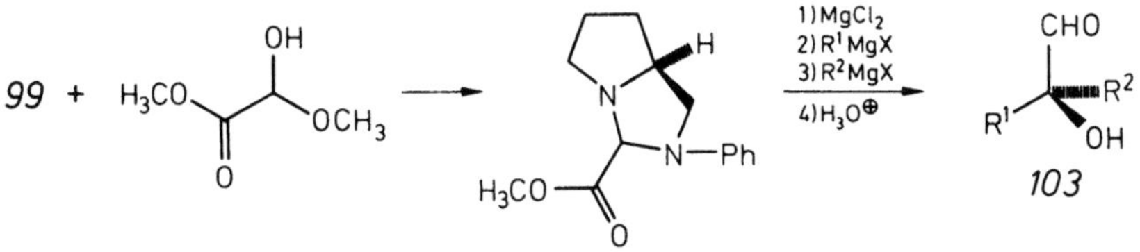

102 103

If, according to the CIP rules, R^1 has a higher priority than R^2, (103) has an (S)-configuration; otherwise it has an (R)-configuration.

This synthetic principle was used by Mukaiyama et al. in the asymmetric total synthesis of
frontalin (104) [119], a pheromone of several species of beetles belonging to the genus *Dendroctonus*,
a marine antibiotic, (—)-malyngolide (105) [120], discovered in the marine blue-green alga, *Lyngbya majuscula Gomot*, and
an α-tocopherol precursor (106) [121].

The aminal (102) was always a key intermediate in the synthesis:

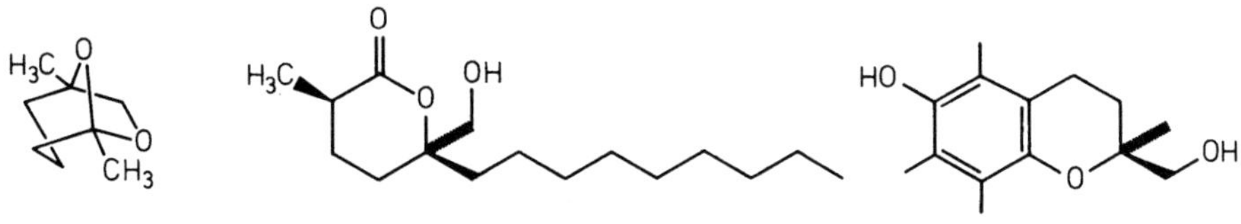

(+)-Frontalin 104 (-)-Malyngolide 105 Vitamin E Precursor 106

Chiral β-formyl-β-hydroxycarboxylic esters were also obtained by the employment of either lithium or zinc enolate of ethyl acetate in place of Grignard reagents in the above-mentioned reaction in moderate to excellent optical purity (62 to 92% e.e.) [122].

Asymmetric synthesis of 2-oxaindane is also achieved by utilizing the aminal (107), prepared from 2-bromobenzaldehyde and the chiral auxiliary (99). Various chiral

3-alkyl-1-hydroxy-2-oxaindanes (108) were obtained by the reaction of the lithium salt, prepared by treatment of the aminal (107) with n-butyllithium, to aldehydes and subsequent hydrolysis.

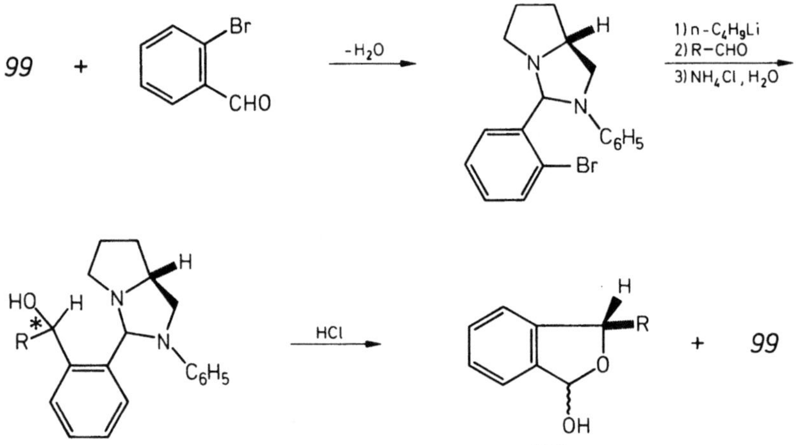

108

$R = C_2H_5$ (88% ee), i-C_3H_7 (90% ee), n-C_4H_9 (87% ee), n-C_8H_7 (90% ee), CH_3–CH=CH (20% ee)

Recently, M. Asami and T. Mukaiyama [124] synthesized α-benzyloxyaldehydes (109) having a chiral tertiary center at the α-carbon atom in high enantiomeric excess by successive treatment of the aminal (102) with diisobutylaluminium hydride (DIBAL-H) and Grignard reagents. The asymmetric reaction is applied to the total synthesis of *exo*-(+)-brevicomin (110), the principal aggregation pheromone in the frass of the female western pine beetle (*Dendroctonus brevicomis*).

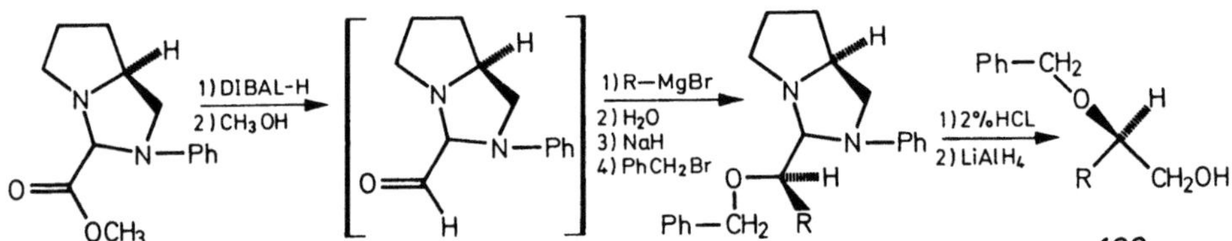

102

109

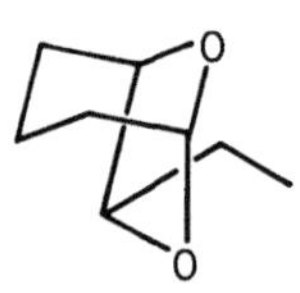

exo-(+)-Brevicomin
110

The chiral azomethine compounds (111) were synthesized by condensation of (S)-valinol with benzaldehyde and substituted benzaldehydes, respectively [125]. The products of 1. these reactions consisted solely of the E isomers due to the difference of bulkiness between the hydrogen atom and aryl group. The reaction of these chiral azomethines (111) with benzylmagnesium chloride produced α-substituted phenethylamines (112) in good yields. No other diastereomer was detected by TLC, GC, and NMR in any case [125].

On the other hand, azomethine (113) was synthesized by condensation of (S)-valinol with phenylacetaldehyde in good yield. The reaction of this chiral azomethine (113) with aryllithium afforded again chiral α-substituted phenethylamines (114). The relationship between (112) and (114) was diastereomeric, due to the different configurations at the newly created chiral center at the 1-position. The diastereoselectivity is more than 98%, because the other diastereomer was not detectable in any case [125].

It seems that these asymmetric reactions include two important stereoselective steps:

a) Condensation of (S)-valinol with aldehydes leads to the E isomer at the C=N bond.
b) Complexation of the chiral azomethines (111) or (113) with the magnesium or lithium reagents leads to a highly stereoselective attack.

Similar results [126] were obtained when related optically active azomethines were synthesized from (S)-alanine, (S)-leucine, and (S)-isoleucine *via* (S)-2-aminoalkanols. On the other hand, the aminoalkanols having (R)-configuration were synthesized from (R)-amino acids, i.e., (R)-alanine and (R)leucine [126]. Again, the asymmetric reaction of the chiral azomethines with Grignard reagents was found to be extremely highly stereoselective. However, the reactions of the azomethines derived from (S)- and (R)-alanine with benzylmagnesium chloride afforded only poor stereoselectivity.

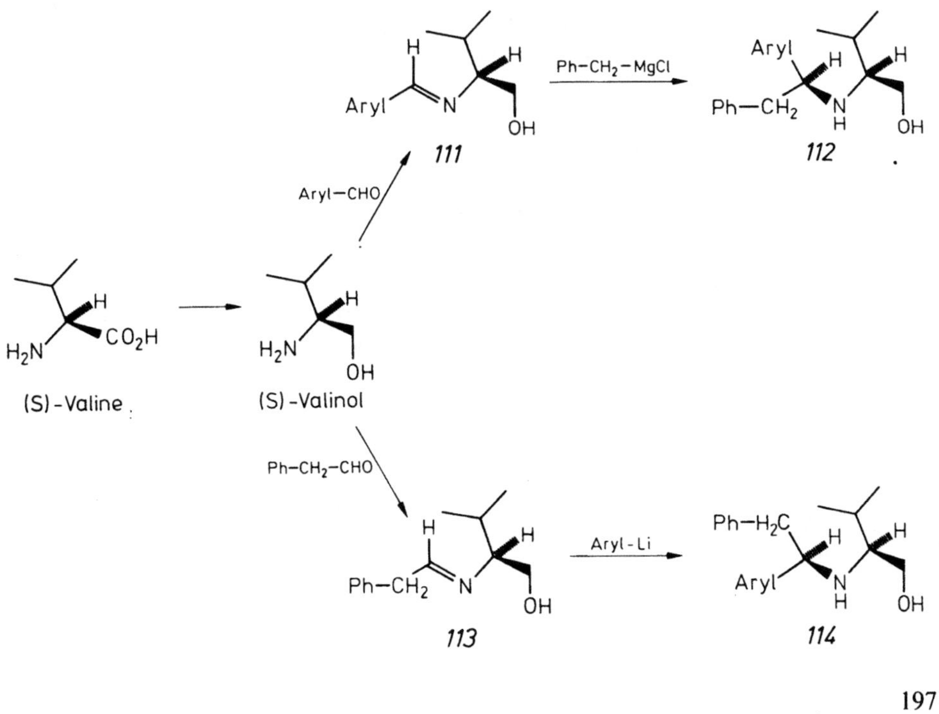

Recently, new chiral oxazolidines (115a–c) were synthesized by the condensation of (S)-N-methylvalinol with some aldehydes [127]. According to NMR-spectra, only one diastereomer was obtained in this reaction, suggesting again that the isopropyl group of optically active (S)-valine produces very high asymmetric induction. The reactions of (115a–c) with ethylmagnesium bromide produced optically active amines (116 and 116') in good yields [127]. These reaction products were elucidated to consist of a mixture of two diastereomers. The major component (116) was formed with 58 to 80% ds.

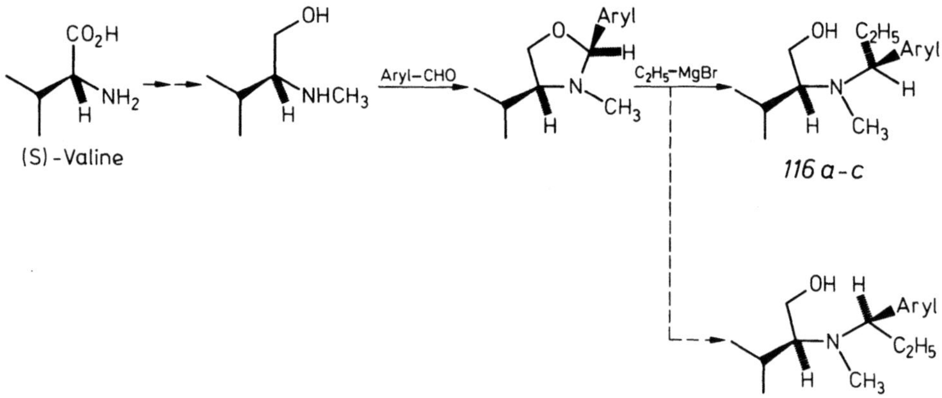

Aryl = Phenyl, 4-Chloro-phenyl, p-Anisyl

On the other hand, the oxazolidine (115d) was synthesized by the condensation of (S)-N-methylvalinol with propionaldehyde in a stereospecific manner. Reaction of (115d) with phenyllithium produced (116'a) as the major product, while (116a), the diastereomer, was the minor product. The ratio of major to minor product was estimated to be 89:11 [127].

The mechanism of this interesting asymmetric synthesis may be assumed to be as follows:

First, the magnesium atom of the Grignard reagent attaches to the lone pair of 1-position oxygen from the less sterically hindered side. Then, the ethyl anion attacks the 2-position carbon from the same side, because the reaction of (115a–c) with ethylmagnesium bromide produced the (1S, 1'S)-configuration amines (116a–c) as the major products. On the other hand, the minor component (116'a–c) is considered to be formed by attack of ethyl anion at 2-position from behind.

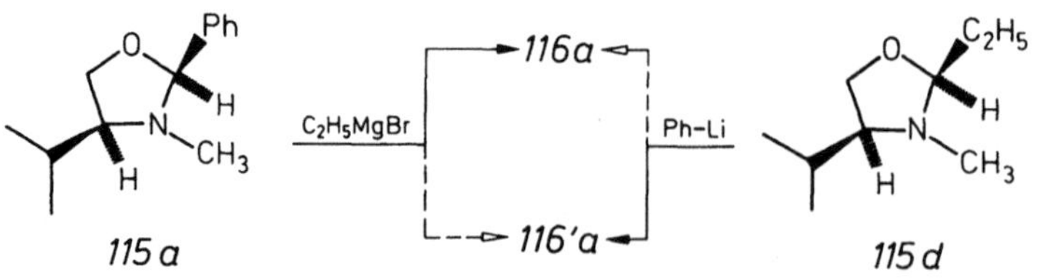

Reactions of the Pictet-Spengler Type:

Various symptoms of ethanol intoxication, dependence and withdrawal may be caused by the reaction of acetaldehyde, the primary metabolite of ethanol, with compounds possessing a β-arylethylamine structure to produce pharmacologically active tetrahydroisoquinolines. Thus, phenylalanine, tryptophane and histidine can react with aldehydes in a reaction of the Pictet-Spengler type. The reaction products are tetrahydroisoquinolinecarboxylic acid derivatives, tetrahydro-β-carbolines, and tetrahydroimidazopyridines, respectively. Such carbolines were synthesized as early as 1948, but from racemic tryptophane [128]. Later, Brossi et al. [129] reported a Pictet-Spengler condensation of (S)-Dopa (3) and of its 0-alkyl derivatives with acetyldehyde. This reaction occurred with asymmetric induction and gave rise predominantly to the tetrahydroisoquinolinecarboxylic acids (117a) (ds = 80%) in which the carboxyl and methyl groups are in the cis position to one another. Thus, the asymmetric center of the amino acid (3) induces the configuration at C-1 with high stereoselectivity.

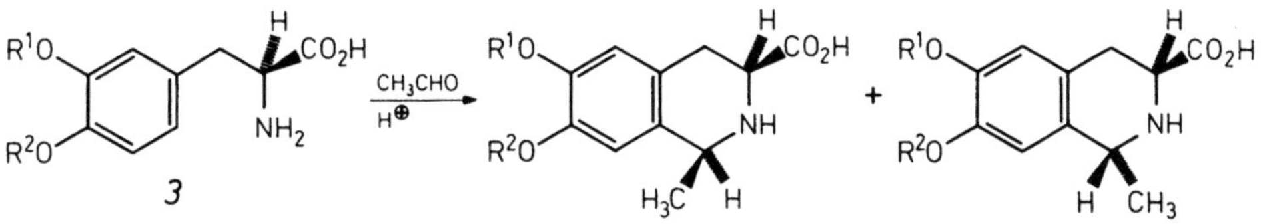

3 117a 117b

R¹ = R² = H, CH₃

S. Yamada and co-workers [130] obtained tetrahydroisoquinoline (120) by the condensation of (S)-dopa methyl ester hydrochloride (118) with the carboxylate (119), a precursor of 3,4-dimethoxybenzaldehyde, in high stereoselectivity (ds = 76%). (120) was decarboxylated to yield a natural product named (S)-laudanosine (121) [131]. (R)-Laudanosine [132] and (S)-reticuline [133] have been obtained by the same method. Rapoport et al. [134] synthesized some 8- and 13-methylberberine alkaloids starting from (117a) ($R^1 = R^2 = H$).

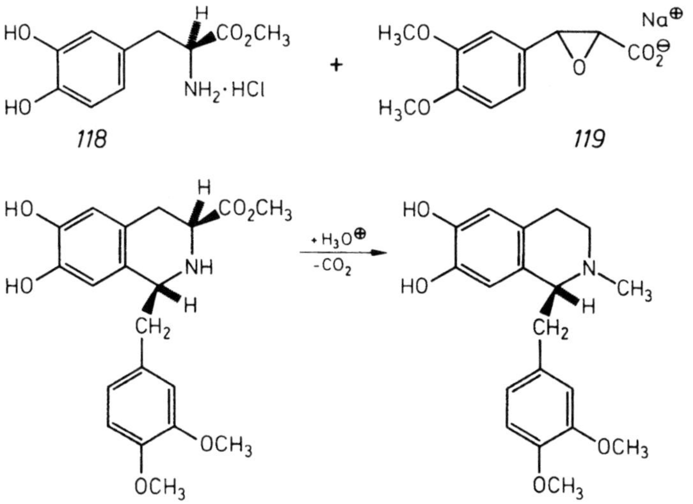

118 119 120 121

In a series of papers, Cook et al. [135] described the Pictet-Spengler reaction of (S)-tryptophan derivatives (122) and the appropriate aldehydes in benzene to form exclusively tetrahydro-β-carbolines (123a). On the other hand, Grigg et al. [136] reported that in the condensation described by Cook et al. [135] the presence of a Bronsted acid is essential. Otherwise the Pictet-Spengler reaction occurs extremely slowly, if at all. Grigg et al. [136] obtained in a model reaction the tetrahydro-β-carbolines (123) as a stereoisomeric mixture of (123a) and (123b) in a ratio of 1:1.2. Several other groups [137] have also investigated the ratio of cis/trans isomers produced in the Pictet-Spengler reaction of tryptophan derivatives with aldehydes. However, in all of the reactions discussed, mixtures of cis and trans diastereomers were reported with the exception of the harman substitution pattern (1-methyl) and, in fact, Brossi et al. [138] have isolated cis and trans isomers (1-methyl-3-carboxyl) in this series. Only Cook et al. [135] found that the Pictet-Spengler reaction of N_b-benzyltryptophan methyl ester with aldehydes occurs in a stereospecific fashion.

A base-catalyzed Pictet-Spengler reaction of (S)-histidine with benzaldehyde was recently reported to yield a mixture of diastereomers, 59% trans and 22% cis [139].

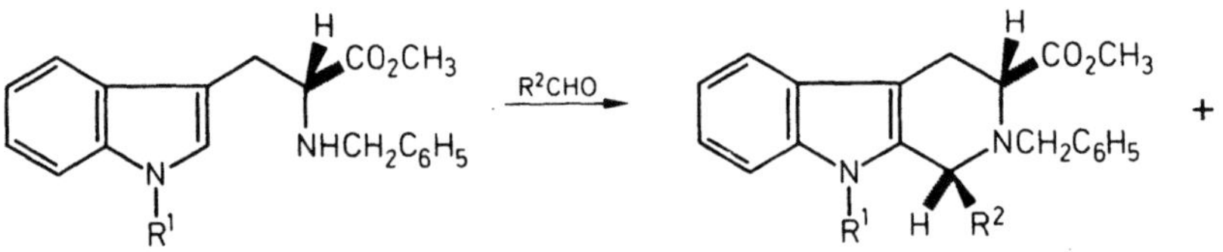

122 123a

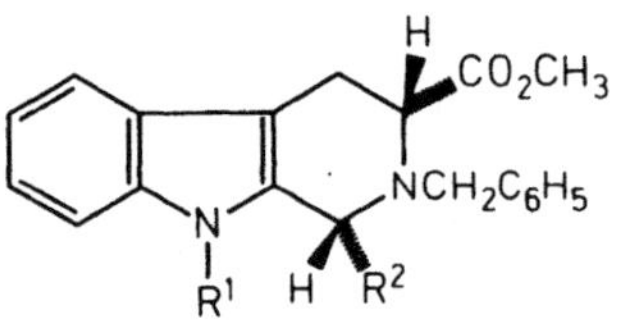

123b

R^1 = H, CH_3

R = C_2H_5, cyclo-C_6H_{11}, 2-OH—C_6H_4, C_6H_5 .

Asymmetric Strecker Syntheses:

When an aldehyde is allowed to react with an optically active amine and hydrocyanic acid, one of the two diastereomeric amino nitriles, (124a) or (124b), may be formed in excess. To prepare the chiral amino acids (125a) or (125b), the nitriles (124a) and (124b), respectively, are hydrolyzed with mineral acids, whereupon R* is split off. However, this asymmetric synthesis of amino acids has no industrial significance.

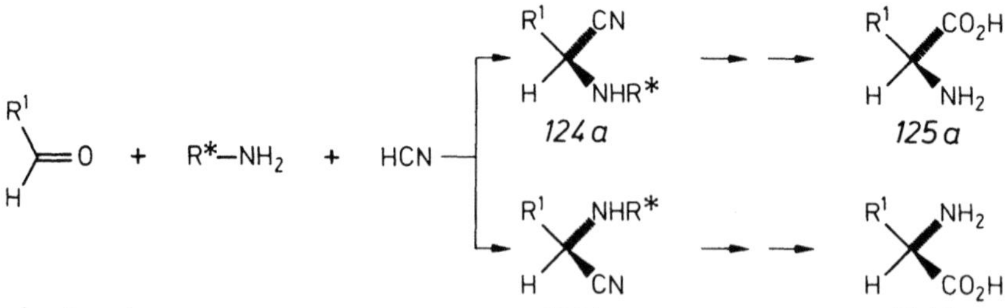

124a 125a

124b 124b

R^1 = CH_3, C_6H_5

Most commonly, α-phenylethylamine is used as chiral amine in the asymmetric Strecker synthesis. Amino acid derivatives have also been used quite successfully as chiral amines. Especially (S)-tert.-butylglycine tert.-butyl ester was proven to be a powerful chiral inductor [140]. The optical yields were as high as 96.5%. The best results were obtained in nonpolar solvents such as n-hexane [140].

The asymmetric Strecker synthesis was also applied in the preparation of other chiral products. In these reactions japanese chemists [141] always used amino acid derivatives as the chiral amine component which is responsible for the induction of asymmetry.

3.2.2 Reactions at the α-Carbon Atom of Carbonyl Groups

Alkylation of Imino- and Enamino Compounds:

Methods which allow the construction of carbon-carbon bonds α to the carbonyl group and simultaneously generate a new asymmetric center at the α-position, belong

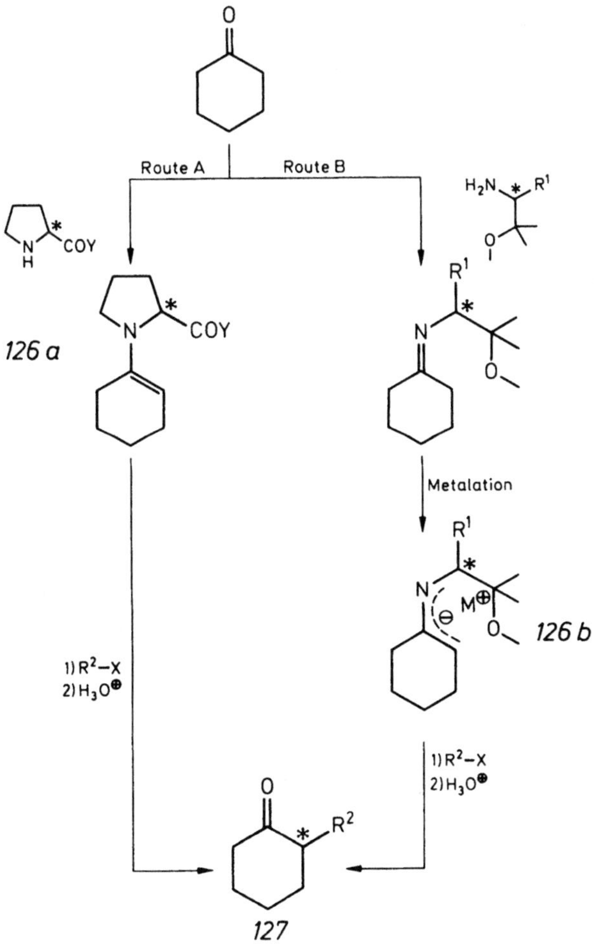

to the most important synthetic operations. A milestone in this field was the introduction of metalated imines as reactive enolate equivalents by Wittig et al. [142] and Stork et al. [143]. Metalated hydrazones have proven to be of similar synthetic utility in organic synthesis.

To obtain an asymmetric induction during C–C bond formation, one needs an enantiomerically pure amine compound.

In 1968, Horeau et al. [144] reported an enantioselective methylation of cyclohexanone via an optically active terpenylimine to yield 2-methylcyclohexanone with 72% enantiomeric excess. The following deals only with those chiral amino compounds that are derived directly from amino acids [145].

The alkylation of caclohexanone has been studied as a model reaction in detail. Generally, enamino compounds (126) are allowed to react with alkyl halides or α,β-unsaturated carbonyl compounds. The enamine (126a) is prepared directly from the ketone and a chiral secondary amine (route A). A "metalloenamine" (126b) can be synthesized from chiral azomethine, derived from the model ketone and a primary chiral amine (route B). The primary amine used for the formation of (126b) must possess an oxygen function. This oxygen function plays a key role in the coordination of the lithium ion in the complex (126b).
2-Alkylated cyclohexanones (127) have been obtained by this procedure with 72–98% enantiomeric excess [146]. If cyclohexanone is replaced in this model reaction by acyclic aldehydes, the e.e. value drops significantly, to 42–54% [146].

A series of chiral amines derived from amino acids has been used; the best optical yields have been obtained with phenylalanine derivatives (128) ($R^1 = CH_2C_6H_5$) and tert-leucine tert-butyl ester (129) [146].

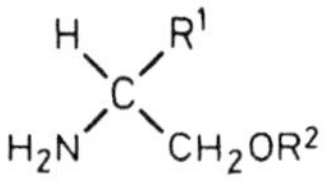

128

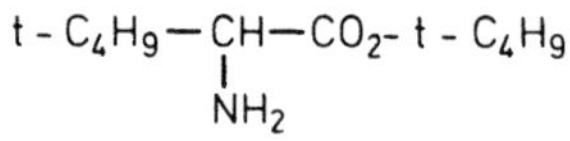

129

$R^1 = C_6H_5, CH_2C_6H_5, i\text{-}C_3H_7, CH_2CH(CH_3)_2$, cyclo-$C_6H_{11}CH_2$, CH_3CH_2

$R^2 = H, CH_3, CH_2CH_2OCH_3, CH_2CH_2N(CH_3)_2, CH_2CH_2OCH_2CH_2OCH_3, CH_2CH_2CH_2OCH_3$

The Stork reaction between methylvinyl ketone and enamine (130) derived from (S)-proline derivatives (131) is of particular interest since chiral cyclohexenones can be obtained. These are useful in many natural product syntheses. Optical yields of 20–50% have been reported [147].

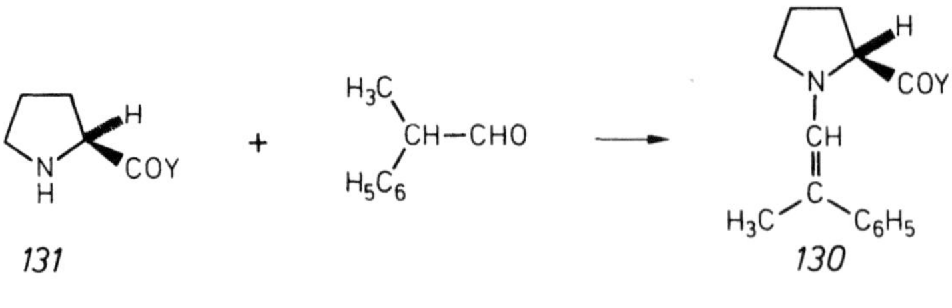

131 *130*

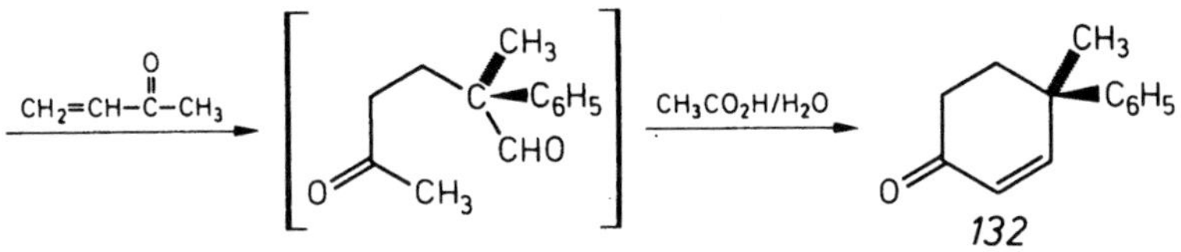

132

Y = O-t-C_4H_9, $N(CH_3)_2$, $N(C_2H_5)_2$, 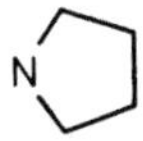

The alkylation of enamines (126a) (Y = OCH_3, OC_2H_5, O-t-C_4H_9) derived from (S)-proline esters was first described by Yamada et al. [148].

Activated olefins (acrylonitrile, methyl acrylate), and halides such as allyl bromide and ethyl bromoacetate were used as electrophiles. In nonpolar solvents, the enamines (126a) were alkylated with high enantioselectivity, but poor chemical yields. In polar solvents, the chemical yields were acceptable, the optical yields poor [148]. A similar reaction sequence has been used successfully for the synthesis of (+)-mesembrine (133) [149].

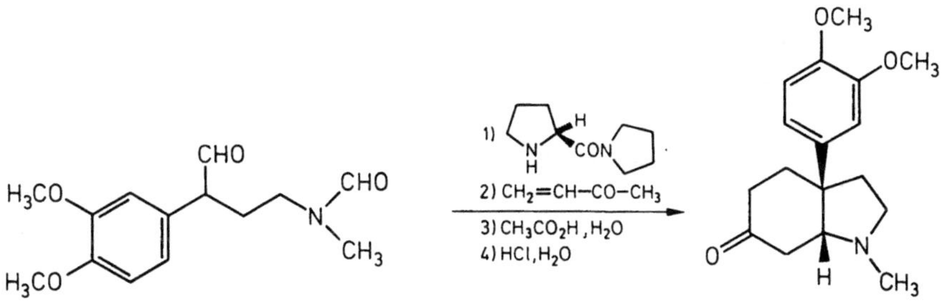

133

In the asymmetric synthesis of 4,4-disubstituted cyclohexenones of the type (132) it was possible to raise the optical yield to a maximum of 54% by varying the structures of the carbonyl compounds [150] and of the proline derivatives (131) [151].

The acid-catalyzed cyclization of the chiral enamine (134) produces (R)-α-cyclocitral (135) and hence (R)-trans-α-damascone (136) in 33% enantiomeric excess [152].

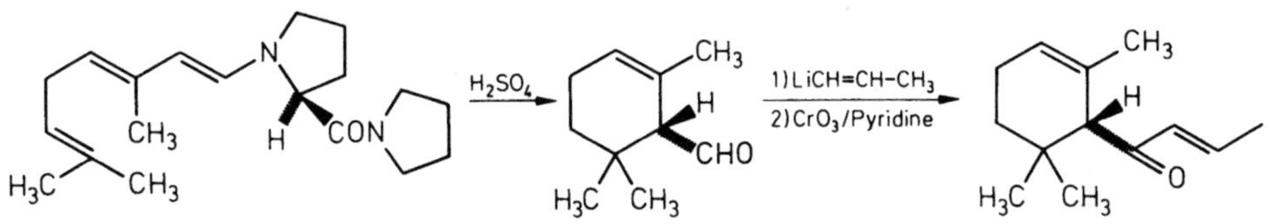

134 *135* *136*

Other natural products in the form of chiral diterpenes and steroids, such as podocarpic acid [153], (R)- and (S)-$\Delta^{1,9}$-2-octalone [154] and vincamine [155], have been prepared from (132) via the phenanthrone derivatives (137).

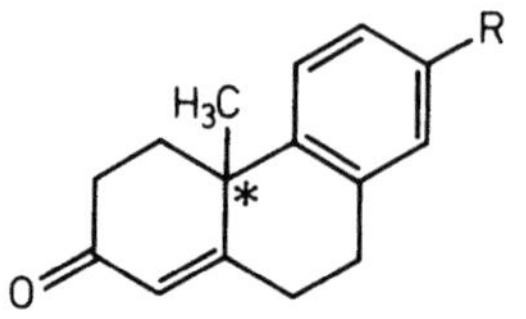

137

In late 1975, Enders et al. [156] started a research project directed towards the development of a new synthetic method for asymmetric carbon-carbon bond formation. A new chiral auxiliary, namely the (S)-proline derivative SAMP (137), was allowed to react with aldehydes and ketones to give the hydrazones (138), which can be alkylated in the α-position in an diastereoselective manner [157,158]. Lithiation [159] of the SAMP hydrazones (138), which are formed in excellent yields, leads to chelate complexes of known configuration [160]. Upon treatment of the chelate complexes with alkyl halogenides the new hydrazones (139) are formed. Cleavage of the product hydrazones (139) leads to 2-alkylated carbonyl compounds (140).

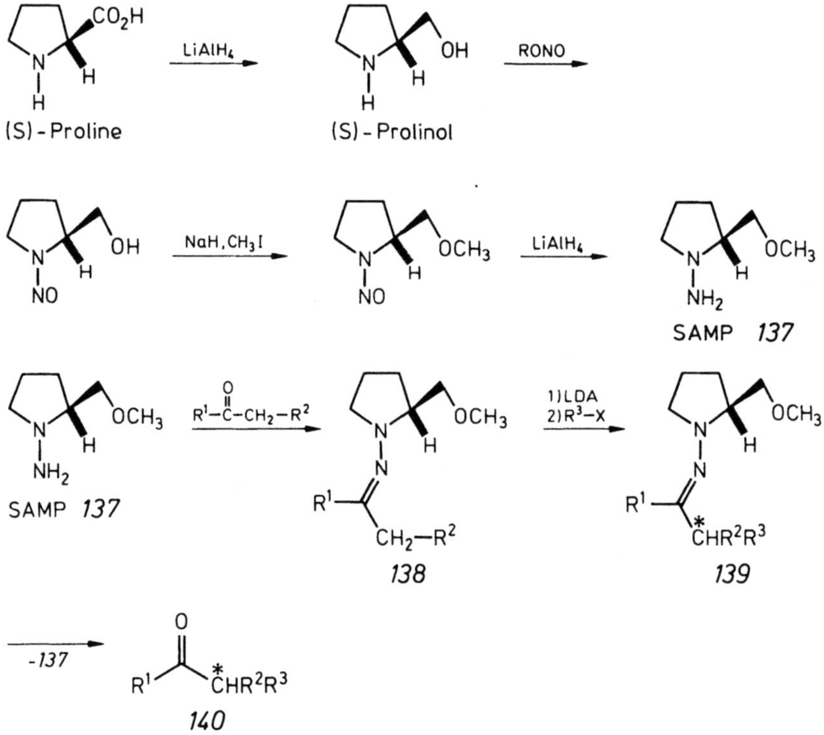

For the cleavage reaction two methods have been described: ozonolysis at —78 °C, which can be used to recycle the chiral information (137), or acid hydrolysis in a two-phase system. No racemization of the product ketone was observed under these conditions.

A number of optically active cyclic ketones have been prepared this way. The overall chemical yields in all cases (cyclopentanones, cyclohexanones, cycloheptanones, and cyclooctanones) were good; the enantiomeric excess in some cases exceeded 95% and was generally satisfying to excellent.

Pennanen [161)] used the SAMP-hydrazone method in a total synthesis of the sesquiterpene (+)-eremophilenolide (143). The key step in the synthetic scheme was the enantioselective α'-alkylation of cyclohexanone. Thus, with cyclohexanone SAMP (137) readily gave the hydrazone (141), which was subjected to lithiation with lithium di(isopropyl)amide (LDA), followed by treatment with 4-bromo-1-butene. The regeneration of the keto group was performed with methyl iodide/HCl producing the enone (142) which was converted to (143) in several steps. The stereochemistry of all intermediates was controlled by the asymmetric center in (142). In other words: the absolute configuration of (143) having *four* asymmetric carbon atoms is derived from the chiral information incorporated in (142), and this was induced by the influence of the chiral amino acid (S)-proline via SAMP (137). The optical purity of the sesquiterpene (143) which was obtained by the SAMP-method was found to be high (89% e.e.).

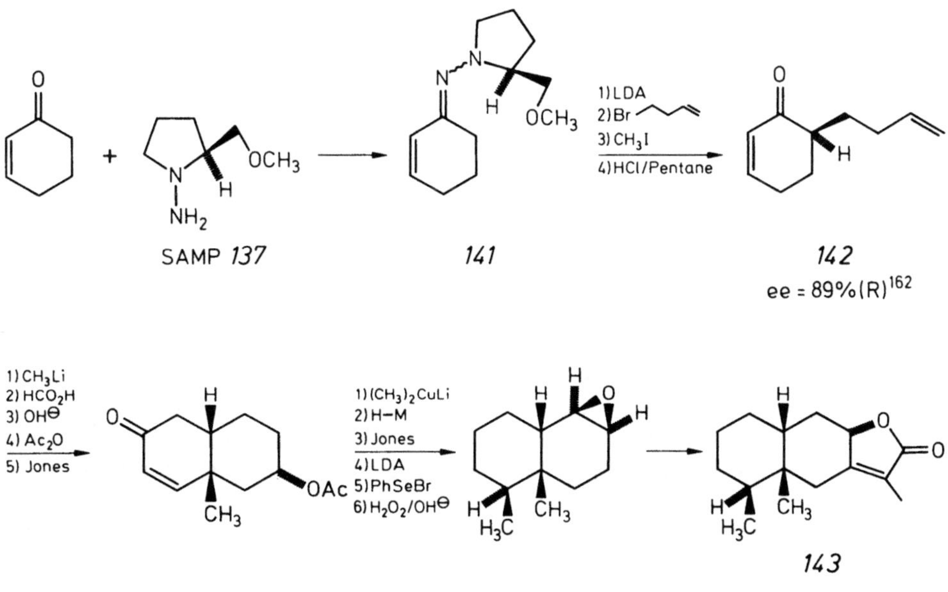

In the case of the asymmetric syntheses of conformationally much more flexible aldehydes and acyclic ketones applying the SAMP-method, the control of both metalation and alkylation selectivity is necessary to reach an excellent overall stereoselectivity. A variety of acyclic ketones can be prepared via the related SAMP-hydrazones in good chemical yields and routinely with overall enantioselectivities of 94–99.5% e.e.

(S)-4-methyl-3-heptanone (146), the alarm pheromone of the leafcutting ant *Atta texana* is about 400 times more active than its optical isomer. This acyclic ketone (146) could be prepared by simply starting from diethyl ketone and propyliodide.

Transformation of diethyl ketone to the corresponding SAMP-hydrazone (144) followed by metalation with LDA and trapping with propyliodide produced the (ZSS)-diastereomer of the product hydrazone (145). This was then cleaved to produce the pheromone (146) in 60% chemical yield with 99.5% enantiomeric excess.

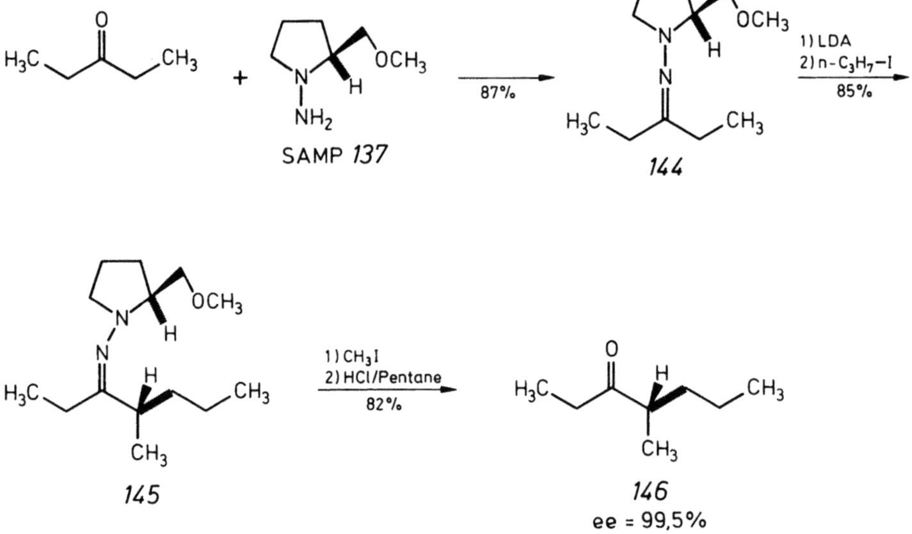

Thus, even in this flexible acyclic back-bone case, the electrophilic substitution occurs with almost complete asymmetric induction, which means that both deprotonation selectivity and alkylation selectivity are almost 100%.

The (R)- and (S)-enantiomers of (E)-4.6-dimethyl-6-octene-3-one (147), a defense substance of spiders (known commonly as "daddy longlegs" *Leiobunum vittatum* and *L. calcar*) were recently synthesized by Enders and Baus [163] using the (R)-proline derivative RAMP and the (S)-proline derivative SAMP (137) as chiral auxiliary, respectively. (S)- and (R)-enantiomers of (147) have been obtained in an overall chemical yield of 70% and in very high stereoselectivities of ≧95% e.e., respectively.

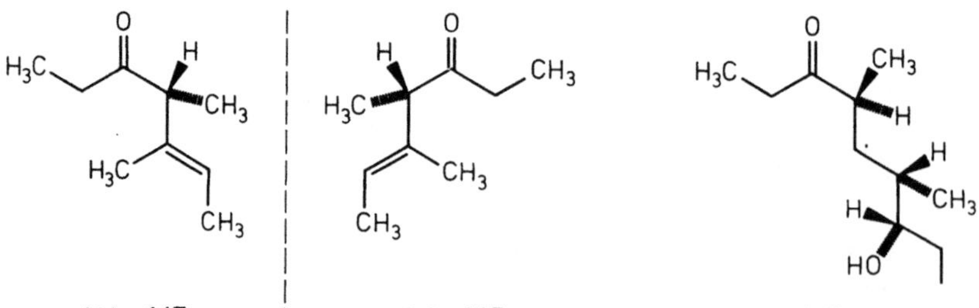

(R)-*147* (S)-*147* *148*

In an elegant asymmetric synthesis of natural serricornin (148) by Mori et al. [164], an overall enantioselective alkylation of diethyl ketone via its SAMP-hydrazone (144) was again the key step. (148) is the sex pheromone of the female cigarette beetle, *Lasioderma serricorne F.*

As expected for the SAMP-alkylation, the (4S)-configuration was generated in excess. Thus, the absolute stereochemistry of (—)-serricornin (148) could be determined to be (4S,6S,7S). Other synthetic stereoisomers of serricornin prepared by Mori et al. [165] were almost devoid of pheromone activity.

Bestmann et al. [166] used the SAMP/RAMP-hydrazone method very successfully in their synthesis of both enantiomers of methylsubstituted pheromone analogues (150) of *Lepidopetra* species *Manestra* brassicae (R = n-C_4H_9), *Argyrotaenia velutinana*, *Ostrinia nubilalis* and *Tortrix viridana* (R = C_2H_5).

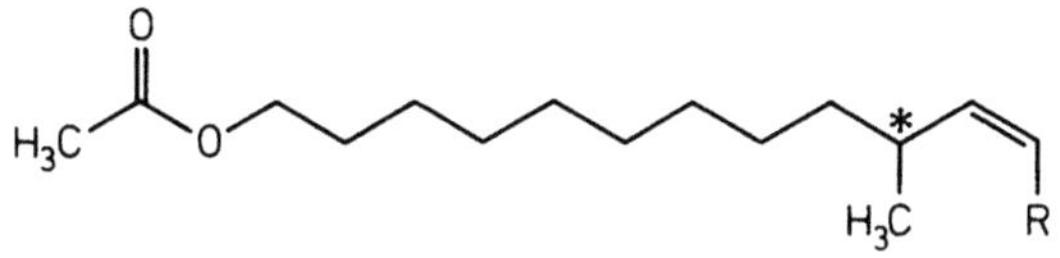

150

Recently, Nicolaou et al. [167] reported an elegant asymmetric synthesis of the ionophore antibiotic X-14547 A (153) isolated at Hoffmann-La Roche from *Streptomyces antibioticus NRRL 8167*. The key step in this synthesis was an enantioselective α-alkylation of n-butanal via its SAMP-hydrazone (151) to produce the intermediate (152). The asymmetric induction as determined by NMR-spectra of the product SAMP-hydrazone, was 95% e.e.

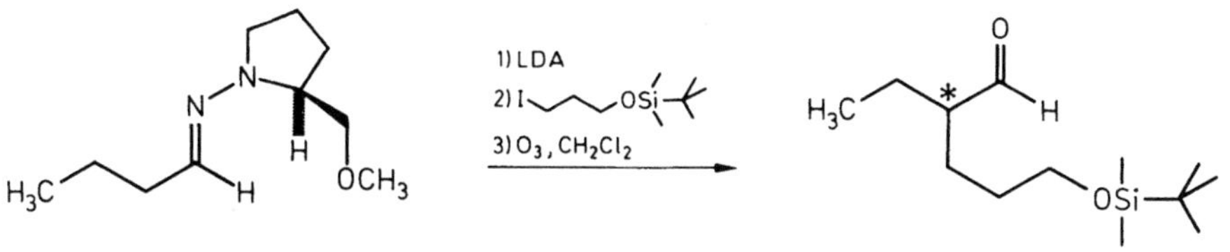

151 *152*

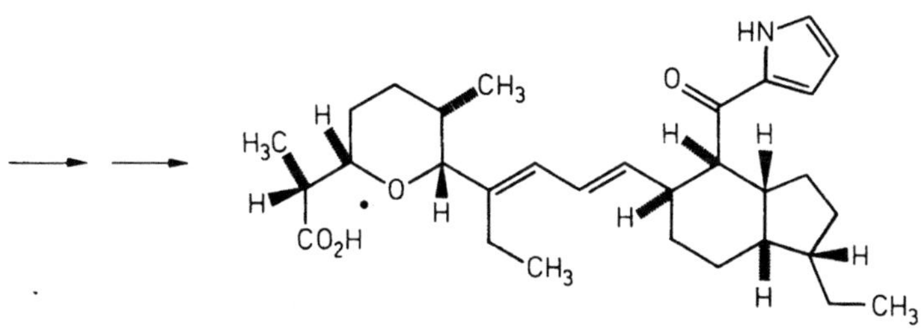

153

Regiospecific and enantioselective aldol reactions [168] were also performed with SAMP (137). Lithiated hydrazones obtained from ketones (154) as described above were alkylated with carbonyl compounds and the adducts then treated with chlorotrimethylsilane. The resulting trimethylsilylethers (155) were finally oxidatively hydrolyzed to yield the chiral β-hydroxyketones (156) (e.e. = 31–62%) [168].

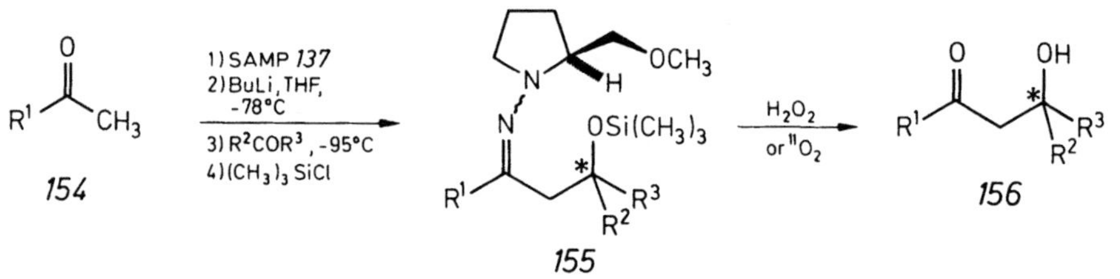

In cases involving unsymmetrical ketones (154) ($R^1 \neq CH_3$), the C–C bond formation occurs regiospecifically at the less substituted alkyl group in α-position to the carbonyl function.

[6]-Gingerol (157), the pungent principal of ginger, was prepared by this method from the cheap starting materials vanillin, acetone, and n-hexanal. 36% e.e. of the (R)-compound [using SAMP (137)] and 39% e.e. of the naturally configurated (S)-gingerol [using the mirror image of SAMP (137), e.g. RAMP] were reached [169]. Yashibushiketol (158), occurring in the tree *Alnus firma*, was obtained only with a low e.e. value via SAMP (137) [170].

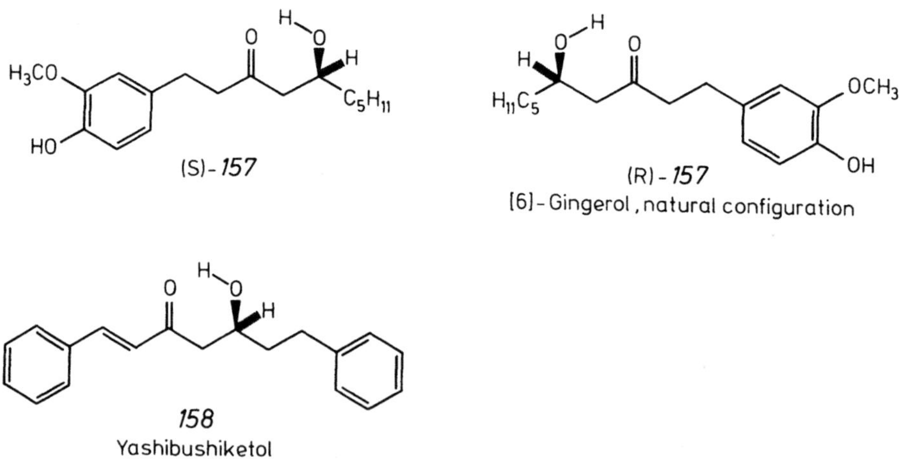

One way to achieve a higher stereoselectivity in these aldol reactions could obviously be the variation of the alkoxy group on the pyrrolidine sidechain of the chiral auxiliary. Thus, Enders and co-workers synthesized the SAMP-analogue (159). While acetone-SAMP-hydrazone leads to a (+)-β-hydroxyketone in 47% e.e., the corresponding acetone-(159)-hydrazone, bearing the sterically demanding trityl group, gave rise to its enantiomer (−)-β-hydroxyketone in only 38% e.e. This result is not easy to explain. Further investigations are necessary to achieve a better understanding of this phenomenon.

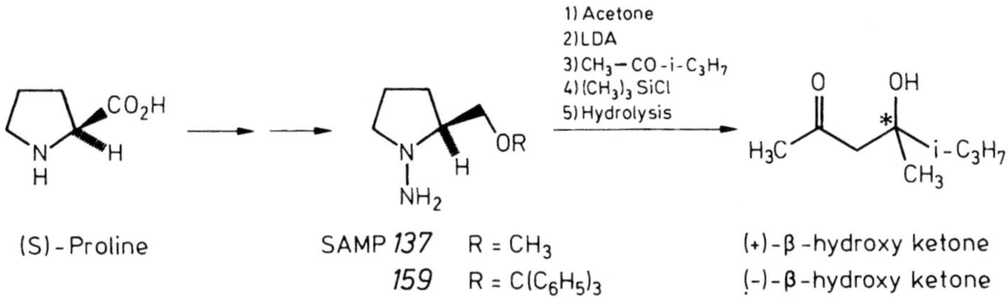

A highly enantioselective aldol-type reaction forming various β-hydroxycarbonyl compounds (162) from 3-acetylthiazolidine-2-thione (160) and achiral aldehydes was achieved via divalent tin enolate employing a chiral diamine (161) derived from naturally occurring (S)-proline as ligand [171]. Thus, stannous triflate was treated with (160) in the presence of N-ethylpiperidine as a base. (S)-1-Methyl-2-[(piperidin-1-yl)-methyl] pyrrolidine (161), the chiral auxiliary, was added and the resulting reaction mixture was cooled. Then, an aldehyde was added and, after the usual work up, the reaction mixture afforded the corresponding aldol-type product (162) in good yield and stereoselectivity. In some cases, optical purities of more than 90% enantiomeric excess were achieved by Iwasa and Mukaiyama [171].

Table 3. Enantioselective Aldol-Type Reaction of (160) to yield (162)

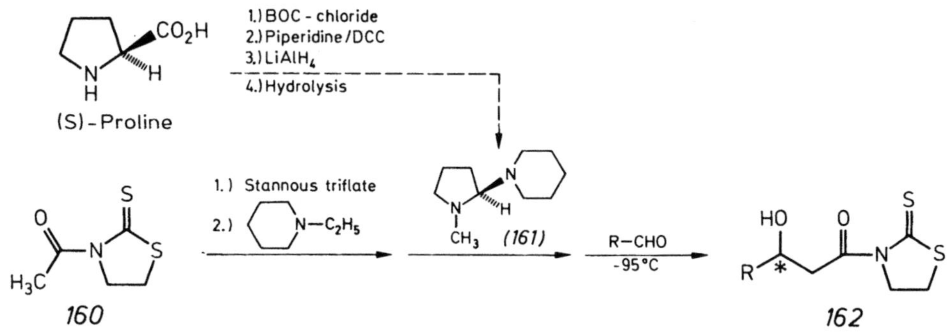

R−CHO	162 Yield (%)	162 Optical purity (%)
C_6H_5−CHO	79	65
i-C_3H_7−CHO	63	>90
$C_6H_5CH_2CH_2$−CHO	76	>90
cyclo-C_6H_{11}−CHO	81	88
C_2H_5−CHO	70	90
n-C_5H_{11}−CHO	65	90

Iwasawa and Mukaiyama have previously reported the first example of forming highly optically active aldols from aromatic ketones and various aldehydes, again via divalent tin enolates employing chiral diamines derived from (S)-proline as ligands [172].

Enders et al. [173] transformed open chain and cyclic β-ketoesters into the corresponding SAMP-hydrazones. Metalation with n-butyllithium, followed by trapping of the intermediate "anions" with alkyhalides generates esterhydrazones which upon cleavage by ozonolysis finally leads to optically active β-ketoesters. While the overall chemical yields are good, the enantiomeric excesses of 18–60% are relatively low.

Enders and Lotter [174] developed an asymmetric synthesis of α-hydroxyketones and vicinal diols using the (S)-proline derivative (S)-1-formyl-2-methoxymethyl-pyrrolidine as chiral auxiliary. However, the α-hydroxyketones and vicinal diols, respectively, were only obtained with low stereoselectivity.

Alkylation of Enolates: Asymmetric syntheses involving enolate reactions such as alkylations, aldol additions and acylations in which the chiral auxiliary A*-H is both readily obtained and easily recoverable after the desired bond construction had been achieved by Evans et al. [175].

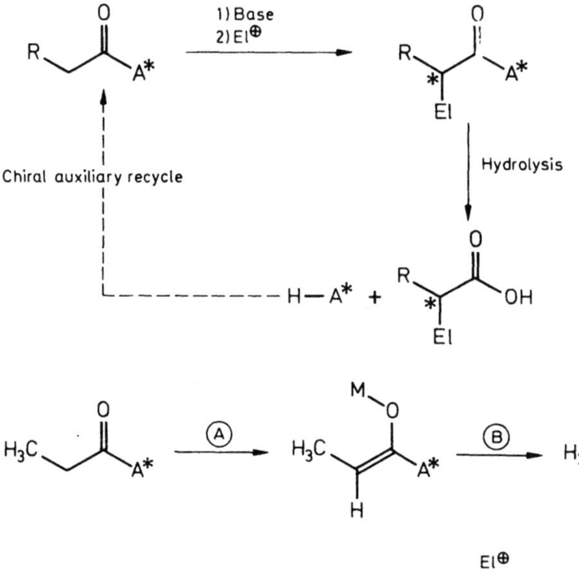

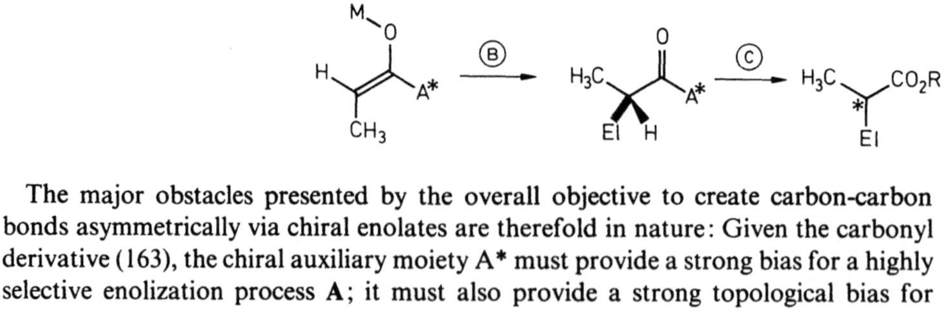

The major obstacles presented by the overall objective to create carbon-carbon bonds asymmetrically via chiral enolates are therefold in nature: Given the carbonyl derivative (163), the chiral auxiliary moiety A* must provide a strong bias for a highly selective enolization process **A**; it must also provide a strong topological bias for enolate diastereoface selection in the bond construction **B**; and finally its nondestructive removal **C** must be carried out with minimal racemization under mild conditions.

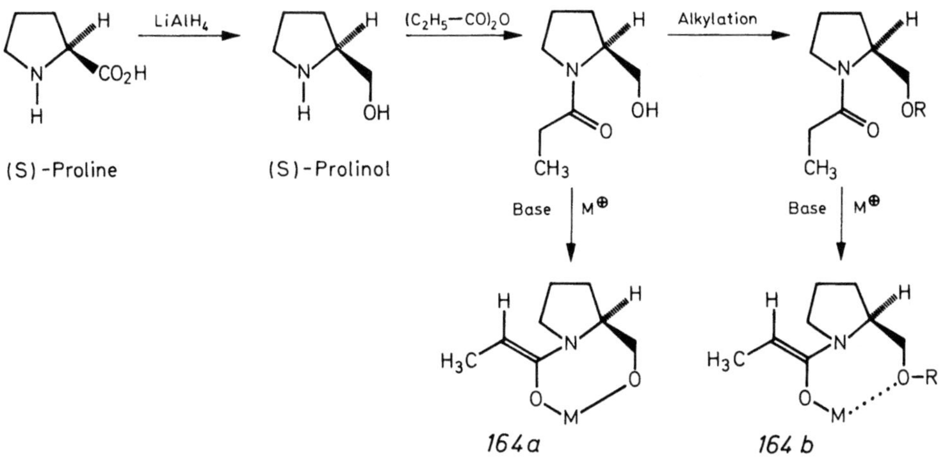

The chiral auxiliaries H-A* developed by Evans et al.[176] were derivatives of naturally occurring amino acids. The (S)-proline-derived amide enolates (164) as well as the (S)-valine-derived amide enolates (166) and imide enolates (165) have proven to be exceptionally versatile chiral nucleophiles.

The highly nucleophilic (S)-prolinol amide enolate (164) (M = Li) was alkylated employing a range of alkyl halides. The carboxylic acids (167) were obtained in chemical yields of 78–96% and outstanding optical yields[177].

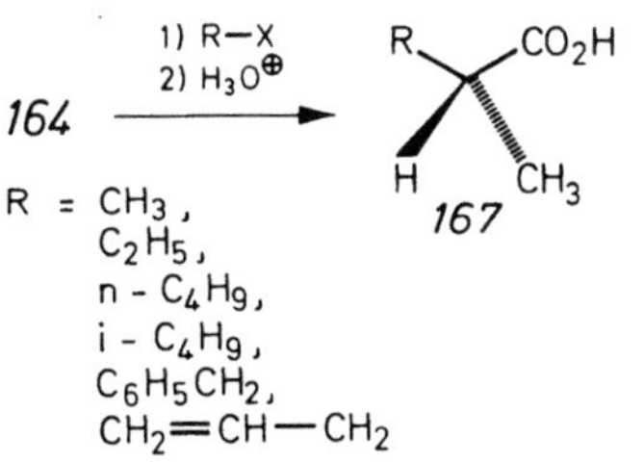

Recently, both enantiomeric forms of callosobruchusic acid (170), a pheromone of the azuki bean weevil, *Callosobruchus chinensis* L., which induces the male to extrude his genital organ and to attempt copulation, were synthesized by Mori et al.[178], applying Evan's alkylation method in natural product synthesis as the key step. Thus, (S)-prolinol propionamide was converted to its enolate (164) by treatment with LDA.

Then it was alkylated with iodide (168) in the presence of HMPA. Finally the reaction was quenched to produce (169) in 46% yield with 96.6% diastereomeric purity. The amide (169) was hydrolyzed to produce (R)-(170) with 93% e.e. In the same manner, by alkylating (R)-prolinol propionamide with (168), the amide (169′) was obtained. Acid hydrolysis of (169′) produced (S)-(170) with 92% e.e.

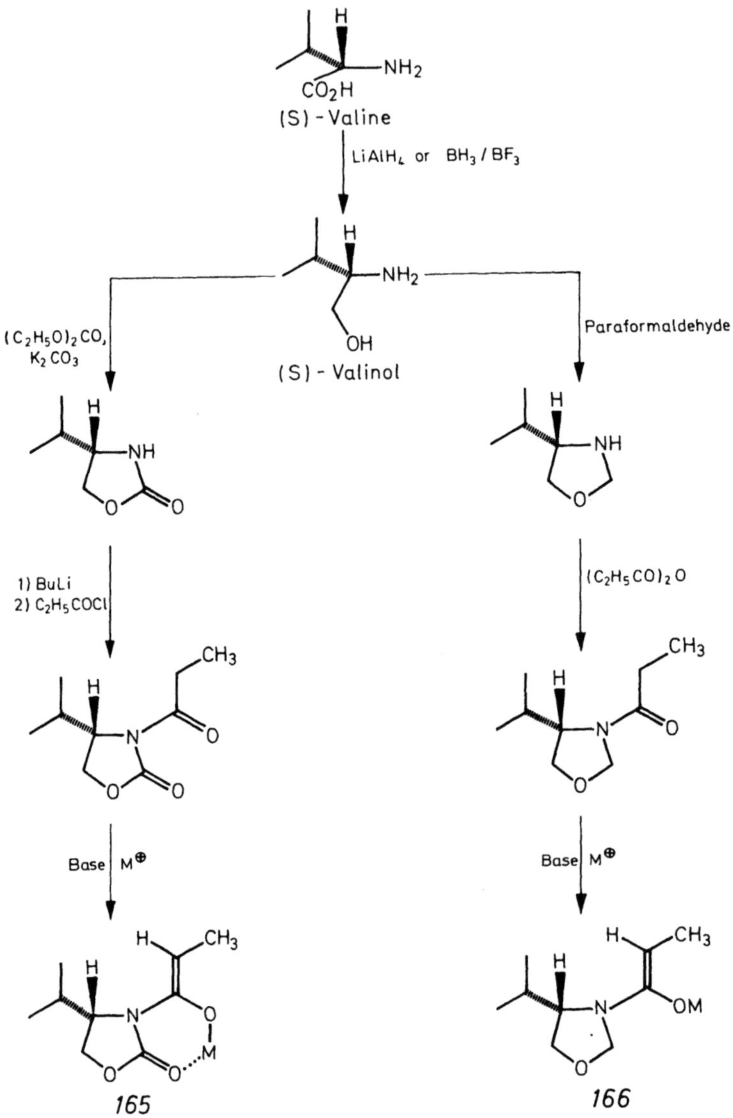

Not surprinsingly, the aldol addition of the lithium enolates derived from these systems proved to be unsatisfactory. However, the derived zirkonium enolates in these and related systems have proven to be exceptional [176]. The amides (171) and (172), each of which is readily derived from (S)-proline and (S)-valine respectively, exhibit good stereoselectivity with a range of aldehydes. The optical purity of the β-hydroxy amides (173) was very good (>95% e.e.). However, this method has a limitation which has been associated with the acidic conditions that are required to hydrolize these chiral amides (173) to their derived carboxylic acids (174). While

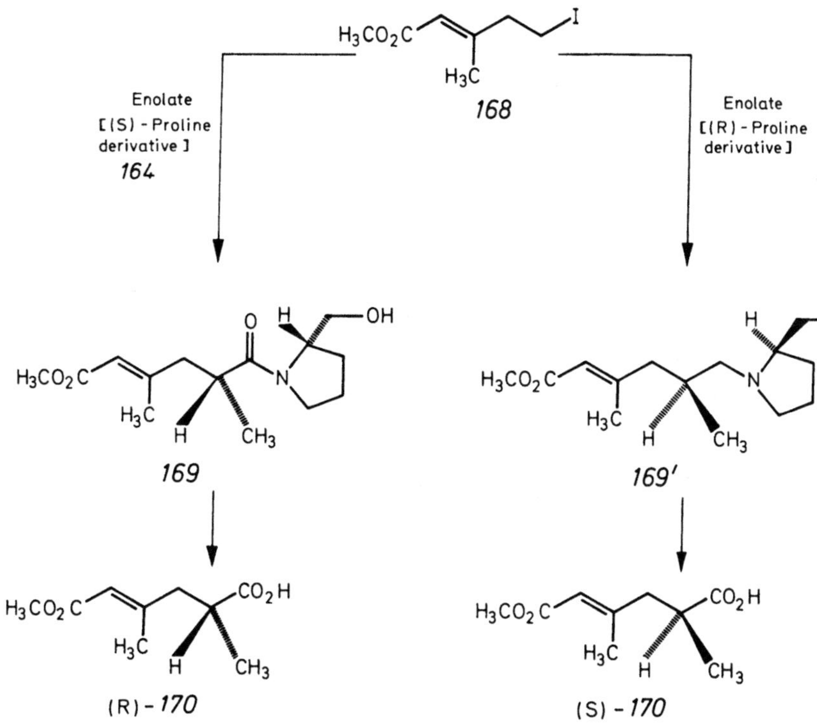

this is not a problem in simple systems, in more complex cases where acid labile protecting groups are present, these hydrolytic conditions have proven to greatly limit these enolate systems [175].

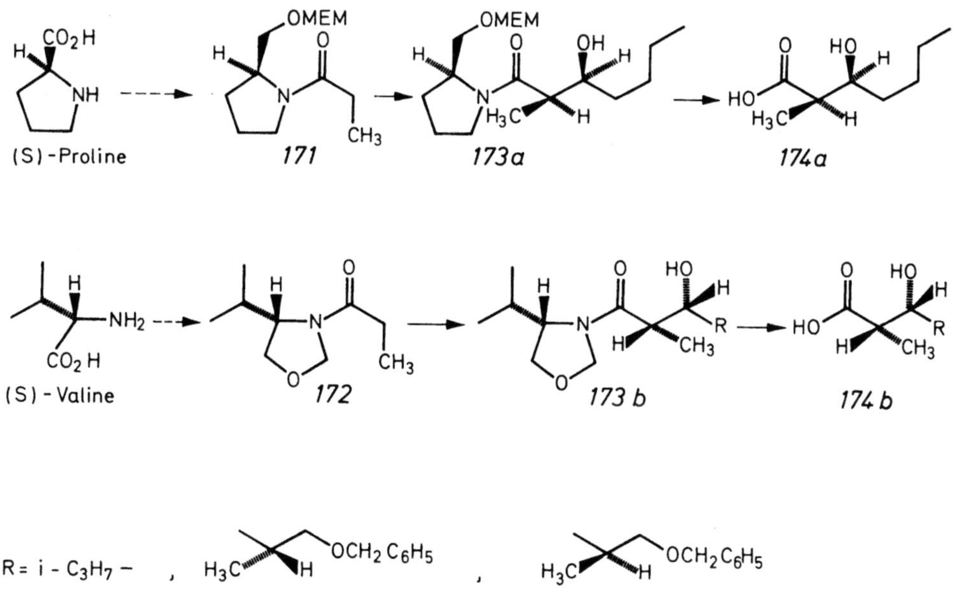

Table 4. Stereoselective Alkylation of the Chiral 2-Oxazolidones (175)

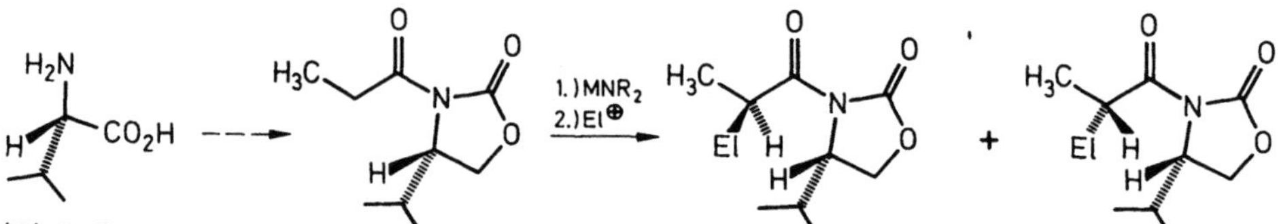

(S)-Valine — *175* — *176a* — *176b*

Electrophile	Metal	Ratio *176a* : *176b*	Yield (%)[a]
$C_6H_5CH_2Br$	Li	120 : 1	75
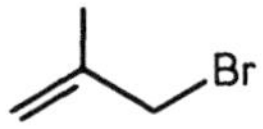	Li	98 : 2	62
	Li	98 : 2	71
	Na	99 : 1	-
$C_6H_5CH_2OCH_2Br$	Li	98 : 2	77

a) *176a* : *176b* ≫ 99:1

Due to these limitations Evans et al. focussed on the exploration of imide-derived enolates (165). They expected these systems to react stereoselective in carbon-carbon bond formation and that the derived imides might be readily hydrolized or reduced under the mild conditions required for the construction of complex products. One of the two chiral 2-oxazolidones (175) chosen for study by Evans et al. [179] is derived from (S)-valine and was readily prepared from this inexpensive commercially available α-amino acid having an optical purity exceeding 99%. The preparation of the related imide-derived enolate (165) is shown in the next scheme. Alkylation reactions employing (175) resulted in excellent diastereoface selection, as summarized in Table 4 [179].

To date, Evans and co-workers have examined a series of transformations that result in the mild, nondestructive removal of the oxazolidone auxiliaries from (176). Basic hydrolysis, transesterification, and reduction of (176) are all viable chemical operations in these systems.

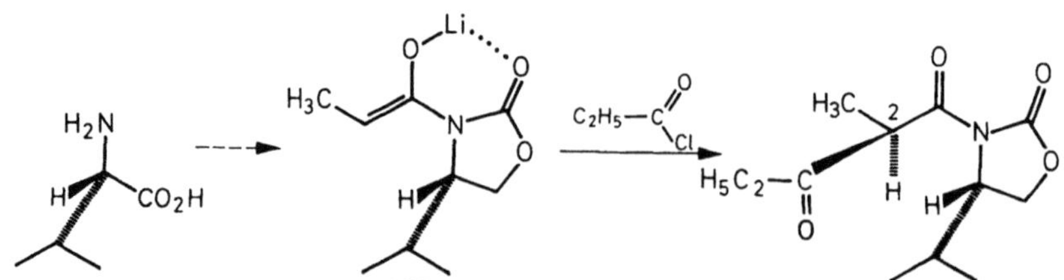

(S)-Valine — *165* — *177*

2R/2S > 96/4

Besides alkyl halides, other electrophiles have been allowed to react with the imide-derived enolate (165). For example acylation of (165) with propionylchloride afforded the chiral β-keto imide (177) in 95% yield with high stereoselectivity [175].

From the illustrated (S)-valinol imide (175), the derived dibutylboryl enolates undergo condensation with a broad range of aldehydes in greater than 99% asymmetric induction for both newly formed asymmetric centers [180]. Evans et al. have shown that the propionyl sidely chain in (175) may be replaced by other alkanoyl substituents without loss of stereoselectivity in the aldol type reaction [180].

The chiral lactone alcohol derivative (178) [181] can be readily prepared from natural (S)-glutamic acid, the cheapest chiral α-amino acid. Lactone (178) was alkylated to yield optically active 3-substituted lactone alcohol derivatives, (179) and (180), which were intermediates in the stereoselective synthesis of various natural products [182].

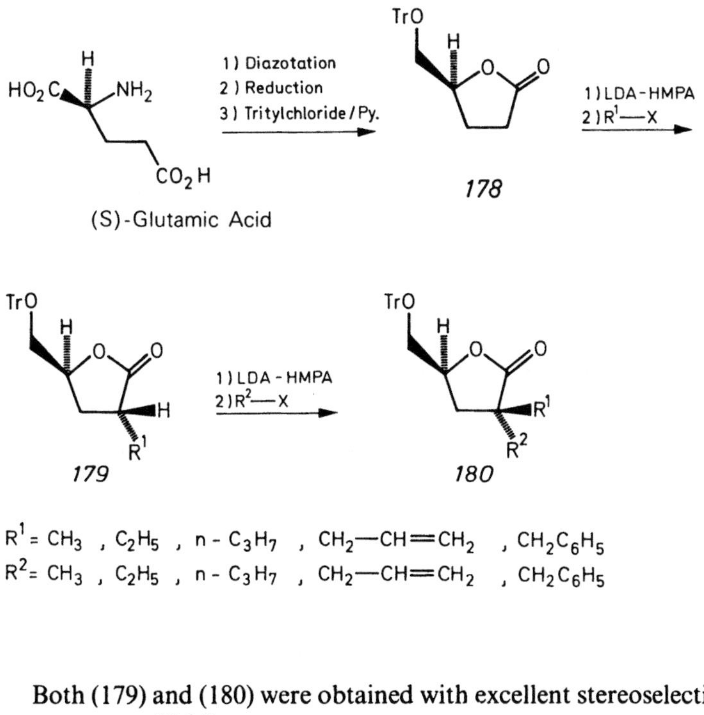

R^1 = CH_3 , C_2H_5 , n - C_3H_7 , CH_2—CH=CH_2 , $CH_2C_6H_5$

R^2 = CH_3 , C_2H_5 , n - C_3H_7 , CH_2—CH=CH_2 , $CH_2C_6H_5$

Both (179) and (180) were obtained with excellent stereoselectivity [183] (d.s. ≧92%; most cases >98%).

The chiral lactone (178) has been used for the synthesis of a variety of natural products, such as sugars, lignans, terpenes, alkaloids, and β-lactams as a chiral building block [182c, 184]. The use of (178) as a powerful inductor of asymmetry was mainly established by Takano et al. [181–184]; one can expect more highly interesting reports from this group.

(S)-Aspartic acid has been converted to derivatives (181) and (182) which undergo alkylation reactions with stereoselectivities that are enantiomerically complementary [185].

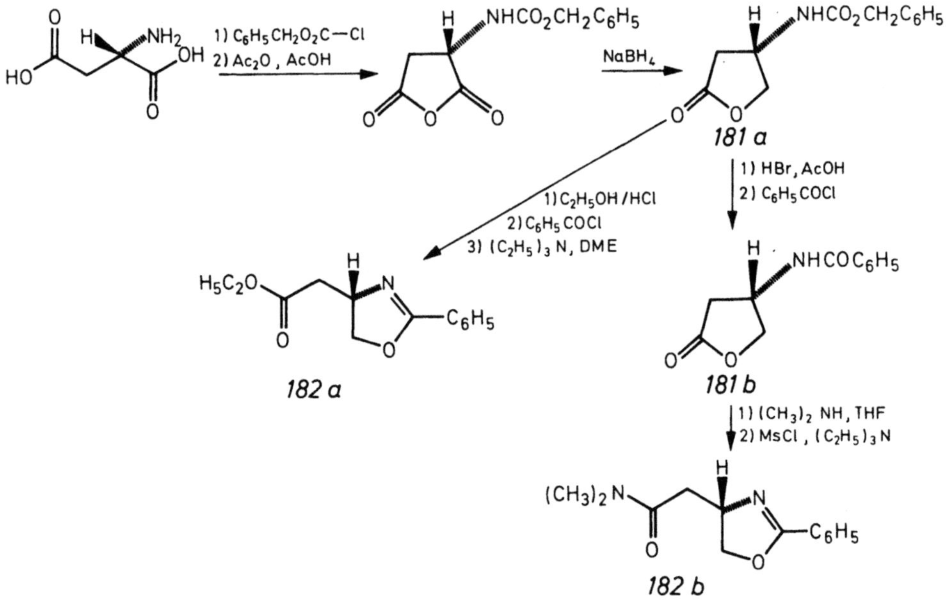

With multigram quantities of compounds (181) and (182) in hand, attention was turned to the stereoselective alkylation of their derived enolates. The lactones (181) were smoothly transformed to their corresponding dianions which subsequently suffered alkylation favoring the expected trans product (183a) [185].

Table 5. Enantioselective Alkylation of (S)-Aspartic Acid Derivatives (181a, b)

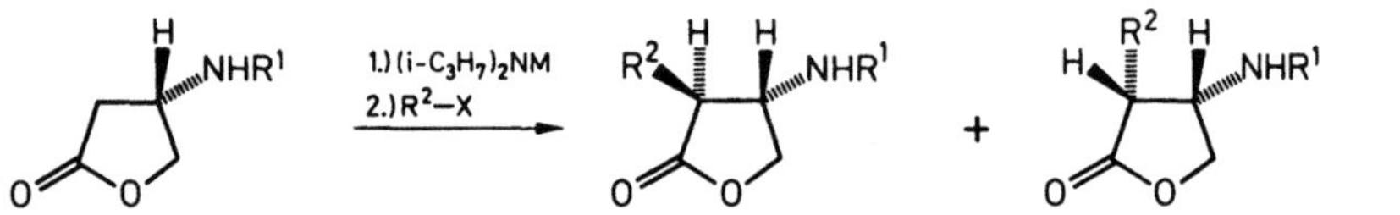

181 183a 183b

R^1	M	R^2—X	Ratio *183a* : *183b*	Yield (%)
$CO_2CH_2C_6H_5$	Li	CH_3I	88 : 12	97
$CO_2CH_2C_6H_5$	MgCl	CH_3I	88 : 12	88
$CO_2CH_2C_6H_5$	Li	CH_2=$CHCH_2Br$	88 : 12	80
COC_6H_5	Li	CH_3I	91 : 9	77

Alternatively, it was expected that chelation with the β-nitrogen in the enolates resulting from oxazolines (182) would impose a diastereofacial bias opposite to that found in the lactones (181). Deprotonation of (182) afforded enolates which were

Table 6. Enantioselective Alkylation of Oxazolines (182), derived from (S)-Aspartic Acid

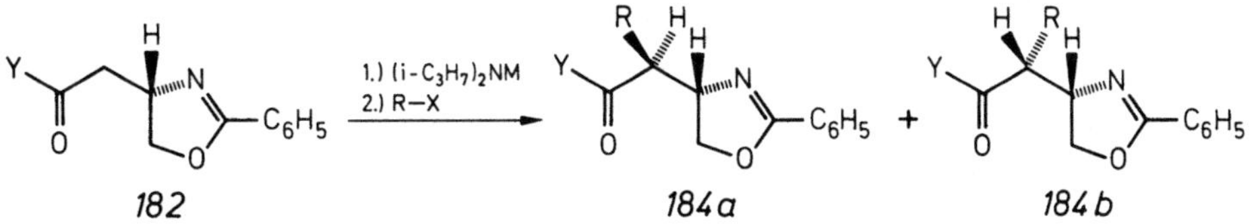

Y	M	R—X	Ratio *184a* : *184b*	Yield (%)
OC_2H_5	MgCl	CH_3I	66 : 34	84
OC_2H_5	Li	CH_3I	67 : 33	81
$N(CH_3)_2$	MgCl	CH_3I	76 : 24	100 (crude)
$N(CH_3)_2$	Li	CH_3I	83 : 17	78
$N(CH_3)_2$	Li	$C_6H_5CH_2Br$	92 : 8	87
$N(CH_3)_2$	Li	$CH_2{=}CH{-}CH_2Br$	94 : 6	67

sufficiently reactive to undergo alkylation reactions at low temperatures without activating agents such as HMPA. As illustrated in Table 6, the expectations of McGarvey et al. [185] were realized as the antiisomer (184a) predominated. The best asymmetric induction was exhibited by the lithium enolate of amide (182b) [185]. Since diastereomeric alkylation products (184) may often be resolved by simple column chromatography, one can easily isolate isomerically pure compounds.

A related approach was recently successful for the stereocontrolled synthesis of an antibiotic. In this synthesis the Japanese chemists used (S)-3-[(benzyloxycarbonyl)-amino]-4-(methoxycarbonyl)butyric acid as chiral auxiliary [186].

α-Alkylations of α-Amino Acid Derivatives: Optically active α-alkyl-α-amino acids deserve attention because of their documented or potential biological activity. Some are valuable pharmaceuticals, such as (S)-α-methyldopa. Others are interesting as enzyme inhibitors. In fact, enzyme inhibition studies with non-proteinogen amino acids have furnished valuable information about the mode of action of certain enzymes. Obviously, there is a demand for chiral non-proteinogen amino acids both for pure and applied organic or bioorganic chemistry.

Schöllkopf et al. [187] synthesized α-alkyl-α-amino acids (186) by the alkylation of chiral 1-substituted 2-imidazolin-5-ones (185), which can be prepared from α-amino acid (S)-phenylethylamides and orthoformic esters. The optical yields of the products (186) were in many cases higher than 95%.

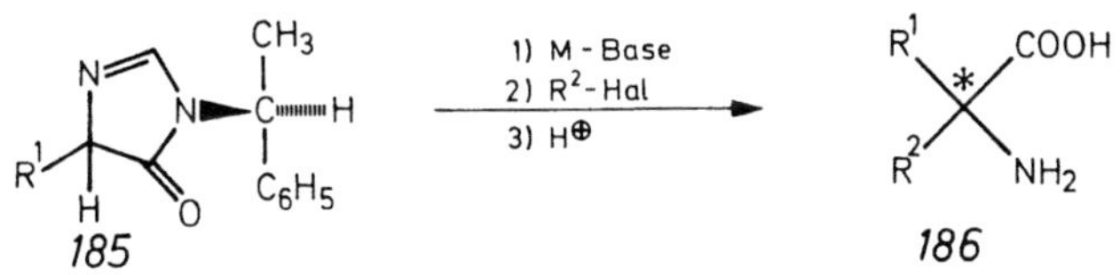

$R^1 = CH_3, C_2H_5$, n-C_3H_7, i-C_3H_7, n-C_4H_9, $CH_2CH{=}CH_2$, CH_2-Aryl
$R^2 = CH_3$, i-C_3H_7, n-C_4H_9, $CH_2CH{=}CH_2$, $CH_2CO_2C_2H_5$, CH_2-Aryl

Excellent reviews covering the enantioselective synthesis of non-proteinogenic amino acids via metallated *bis*-lactim ethers of cyclic dipeptides (2,5-diketopiperazines) (187) have been compiled by Schöllkopf [188].

Bis-lactim ethers (187) contain a chiral inducing center, an acidic carbon-hydrogen bond, and two sites susceptible to hydrolysis. They may react with BuLi to give lithium compounds of type (188), which possess a prochiral carbon atom. Lithium compounds, like (188), smoothly add electrophiles (such as alkylating agents or carbonyl compounds) with outstanding diastereoface differentiation. In many cases the d.s.-value of the adduct exceeds 95%. On hydrolysis, the adducts were cleaved, liberating the chiral auxiliary and the desired products, the optically active amino acids (189) or amino acid methyl esters. The recovered chiral auxiliary (also an amino acid or an amino acid methyl ester) and the target molecule (189) are separable either by fractional distillation (esters) or by chromatography (free acids). In Scheme 93 the enantioselective synthesis of α-methyl amino acids is given as an example for a successful application of Schöllkopf's now well-established *bis*-lactim ether method.

"Symmetrical" *bis*-lactim ethers of type (187) — built up from two identical amino acids — do have one disadvantage, inherent in the system: only 50% of the chiral auxiliary — in this case (S)-alanine — is recovered; the other 50% is first racemized via (188) and finally incorporated in the product (189). To avoid this disadvantage Schöllkopf et al. have developed methods to synthesize "mixed" *bis*-lactim ethers, starting from two different amino acids, e.g. (S)-valine and (R,S)-alanine. Thus, the authors obtained cyclo [(S)-val-(R,S)-ala] and prepared the related *bis*-lactim ether (190).

The *bis*-lactim ether (190) was metallated with BuLi regiospecifically in the alanine part of the molecule to produce, after the reaction with alkyl halides, the (3R)-adducts (191) with ds > 95%. Upon hydrolysis the (R)-α-methyl amino acid esters (192) were obtained with e.e. values >95% [189].

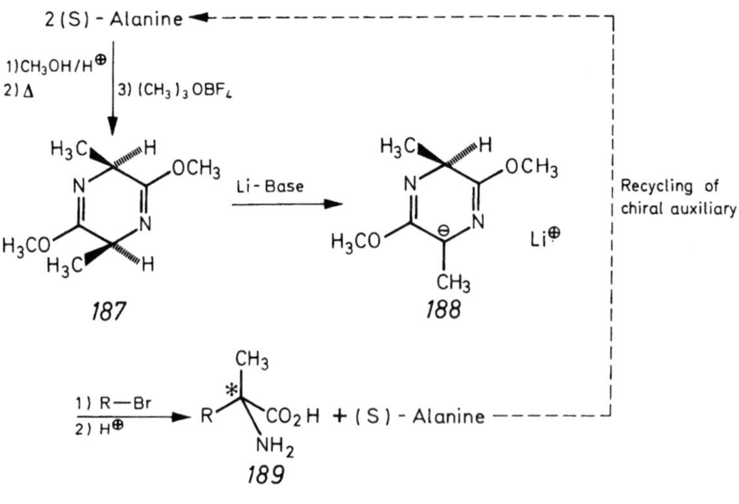

R = i - C_3H_7, $CH_2CH{=}CH_2$, CH_2-Aryl, $C(OH)(C_6H_5)_2$, n - C_8H_{17}, Cinnamyl

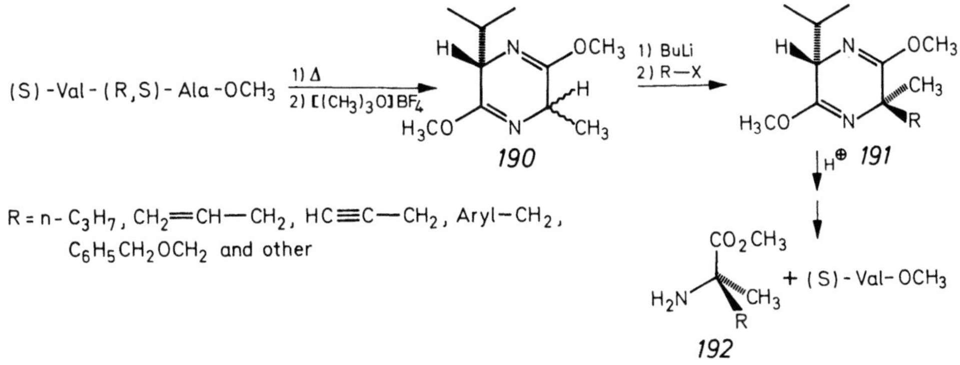

Applying the *bis*-lactim ether method, Schöllkopf and co-workers synthesized enantioselectively

- α-unsubstituted amino acids from cyclo [(S)-val-gly] [188],
- (S)-cystein derivatives [188,190] with the *bis*-lactim ether of cyclo [(S)-val-gly] [188],
- (S)-α-methyl-cystein derivatives [190,191] using (190) as chiral auxiliary reagent,
- α-methyl serines and α-alkenyl serines by allowing (188) to react with ketones and aldehydes [188,192], respectively,
- (2R)-serines starting with the *bis*-lactim ether of cyclo [(S)-val-gly] [193].

3,6-Dihydro-3-phenyl-2*H*-1,4-oxazin-2-ones (194) [194] were synthesized from (R,S)-phenylglycine and (S)-2-hydroxyalkanoic acids (193), the latter derived from (S)-amino acids. The heterocycles (194) contain an endocyclic chiral center at C-6. Metallation of (194) and subsequent treatment with alkyl halides produced the adducts (195) in good chemical yields and with d.s. values from 50 up to more than 95% due to an asymmetric induction. On hydrolysis of the heterocycles (195), the 2-hydroxy-alkanoic acids (193) and the optically active (S)-α-alkyl-α-phenylglycines (196) were liberated [194].

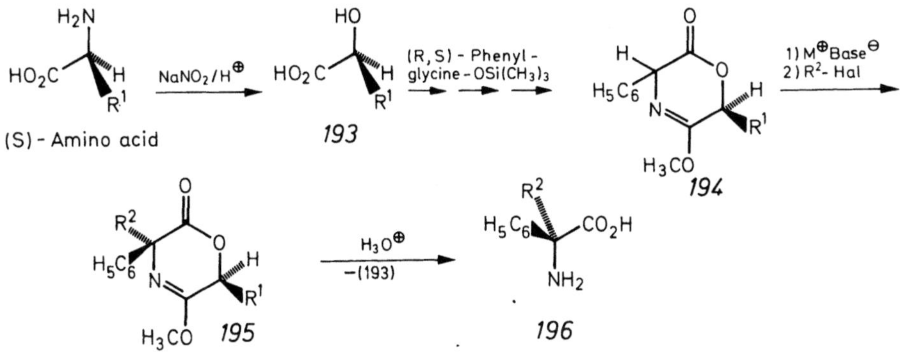

R^1 = i-C_3H_7, t-C_4H_9, C_6H_5, $CH_2C_6H_5$ R^2 = CH_3, $CH_2{=}CH{-}CH_2$, $C_6H_5CH_2$

The asymmetric methylation of N-benzylidene-(R,S)-phenylalanine methyl ester was carried out [195] in the presence of lithium salts of secondary amines derived from

(S)-proline. The lithium amides of *poly*-(imino-1-isobutylethylene) and its corresponding low-molecular-weight model compound, derived from (S)-leucine, were similarly used in order to examine the polymer effects with regard to the stereoselectivity. After acetylation, N-acetyl-α-methylphenylalanine was obtained in max. 31% optical yield [195].

Seebach and Naef [196] generated chiral enolates with asymmetric induction from α-heterosubstituted carboxylic acids. Reactions of these enolates with alkyl halides were found to be highly diastereoselective. Thus, the overall enantioselective α-alkylation of chiral, non-racemic α-heterosubstituted carboxylic acids was realized. No "external" chiral auxiliary was necessary in order to produce the α-alkylated target molecules. Thus, (S)-proline was refluxed in a pentane solution of pivalaldehyde in the presence of an acid catalyst, with azeotropic removal of water. (197) was isolated as a single diastereomer by distillation. The enolate generated from (197) was allylated and produced (198) with a d.s. value >98%. The substitution (197) → (198) probably takes place with retention of configuration [196].

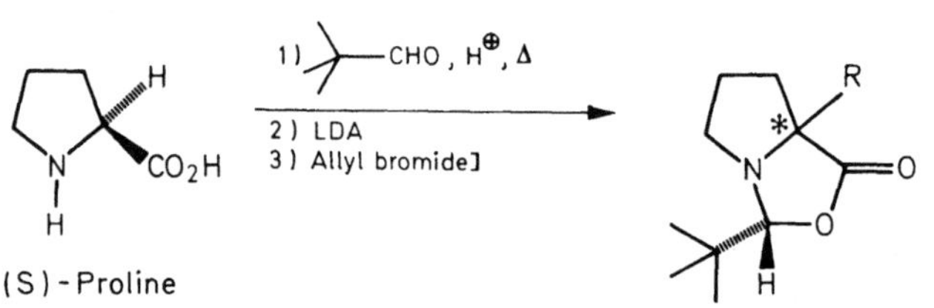

(S)-Proline

197 R = H
198 R = $CH_2—CH=CH_2$

A group at the Academy of Sciences in Moscow [197] has synthesized chiral threonine. Derivatives of cyclic imino acids form copper complexes with glacine and carbonyl compounds. Hydroxyethylation with acetaldehyde and decomposition of the resulting complexes produced threonine with an optical purity of up to 97–100% and with *threo/allo* ratios of up to 19:1 [197]. The chiral reagents could be recovered and re-used without loss of stereoselectivity. The mechanism of this asymmetric synthesis of amino acids *via* glacine Schiff base/metal complexes was also discussed [197].

3.2.3 Reactions at the β-Carbon Atom of Carbonyl Groups

Michael-Additions: The most important application of this type of reaction is the carbon-carbon bond formation β to a carbonyl function by the addition of carbanion. The asymmetric version of the Michael-addition has been known since 1973. An optically active vinyl sulfoxide served as chiral auxiliary [198].

The first Michael-addition with induction of asymmetry under the influence of an chiral amino acid derivative was reported by Koga et al. [199] in 1976. Thus, the 1,4-addition of Grignard reagents [199a] and potassium diethyl malonate [199b] to chiral α,β-unsaturated aldimines (199) was realized. It was suggested that acting as a base to give a metallodinamine, the Grignard reagent produced a chelated complex (200) which transfers an alkyl group to the adjacent electrophilic position of the aldimine to produce (201). Hydrolysis produced the (R)- or (S)- aldehyde (202) in high optical yield (63–98 %) depending upon the configuration of the chiral α-amino acid auxiliary. Alternatively, the enantiomer of (202) could be obtained by simply exchanging the R^1 group in aldimine (199) for the R^3 group in the Grignard reagent while retaining the same chiral amino acid reagent.

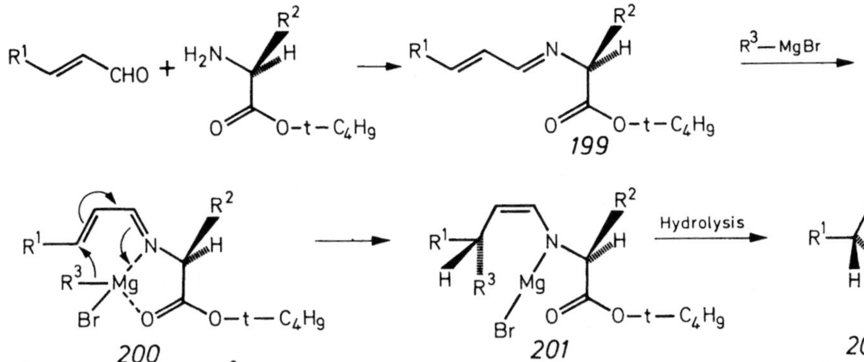

$R^1 = CH_3, C_6H_5$ $R^2 = i\text{-}C_3H_7$, $i\text{-}C_4H_9$, $t\text{-}C_4H_9$

$R^3 = C_2H_5$, $n\text{-}C_4H_9$, cyclo-C_6H_{11}, $CH(CO_2C_2H_5)_2$, C_6H_5, $(CH_3)_2C{=}CHCH_2CH_2$

A useful method for the diastereoselective and enantioselective synthesis of *trans*- and *cis*-1,2-disubstituted cycloalkanecarboxaldehydes was devised by Koga et al. [199f)] starting from cycloalkanecarboxaldehydes. (S)-*tert.*-Leucine *tert.*-butyl ester, a highly effective chiral auxiliary reagent, could be recovered for recycling without any loss of optical purity in a reaction sequence similar to that in the acyclic synthesis of (202).

Mukaiyama et al. [200)] synthesized optically active 3-substituted succinialdehyde acid esters (204) via a Michael-addition. The methyl ester of fumaraldehydic acid was converted into the corresponding aminal (203) by treatment with the (S)-proline-derived chiral diamine (99). The Michael-addition of Grignard reagent to the aminal, followed by hydrolysis produced stereoselectivily 3-substituted succinaldehydic acid ester (204) in good yield.

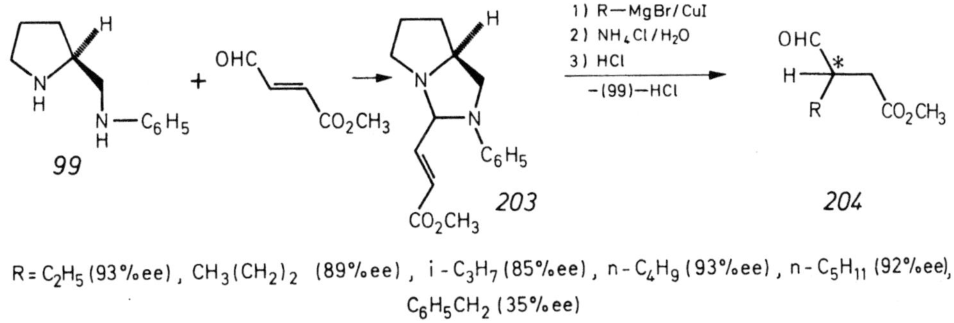

R = C_2H_5 (93% ee), $CH_3(CH_2)_2$ (89% ee), $i\text{-}C_3H_7$ (85% ee), $n\text{-}C_4H_9$ (93% ee), $n\text{-}C_5H_{11}$ (92% ee), $C_6H_5CH_2$ (35% ee)

In the reaction of 2-cyclohexenone and of 4-phenylbut-3-en-2-one with chiral organocuprates of type (205), optical yields of only 5% and 15%, respectively, were obtained [201]. Methylation of 1,3-diphenyl-2-propene-1-one with (205) ($R = CH_3$) produced (S)-1,3-diphenyl-1-butanone in 68% enantiomeric excess [202].

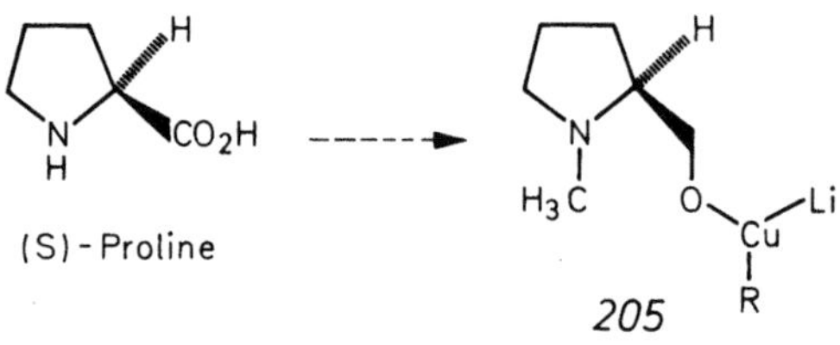

When (2S)-1-(1-cyclohexene-1-yl)-2-(methoxymethyl)pyrrolidine (206), enamine from cyclohexanone, and (S)-proline-derived (2S)-(methoxymethyl)pyrrolidine is added to the Knoevenagel condensation products (207), mainly one of the possible four diastereomers is formed. The diastereomeric purity was found to be excellent (d.s. > 90%) [203]. The stereochemical course of this highly effective asymmetric synthesis allowed the synthesis of the optically active target molecules (208). A possible mechanism discussed by Blarer and Seebach [203].

In related asymmetric Michael-additions of enamine (206) and 2-aryl-1-nitroethylenes, only one of the four possible enantiomerically pure diastereomers was formed [204]. Hydrolysis of the crude primary products furnished α-alkylated cyclohexanones of >90% enantiomeric excess [204].

Instead of introducing the (S)-proline-derived chiral auxiliary (206), its enantiomer in the Michael-addition, the authors obtained the enantiomeric product (208′) having opposite optical rotation compared to (208).

Table 7. Asymmetric Michael-Addition of Knoevenagel Condensation Products

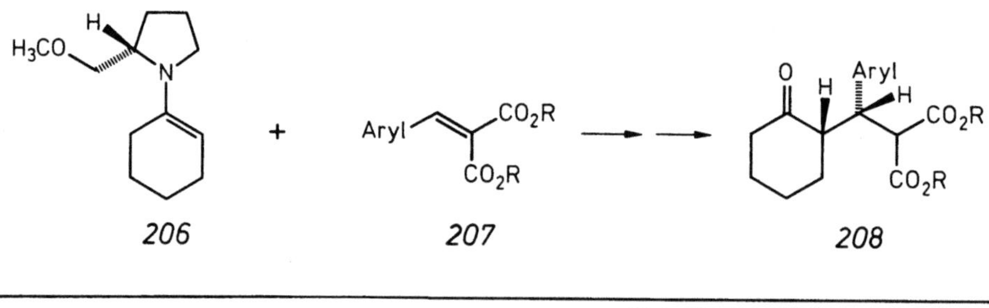

		Ketodiester *208*, crude		
Aryl	R	Yield (%)	ds (%)	ee (%)
C_6H_5	C_2H_5	70	>95	95
C_6H_5	CH_3	76	>95	92
$4\text{-}Cl\text{-}C_6H_4$	CH_3	53	88	80
$4\text{-}NO_2\text{-}C_6H_4$	CH_3	35	88	83
$3,4\text{-}(OCH_2O)\text{-}C_6H_3$	CH_3	35	95	82

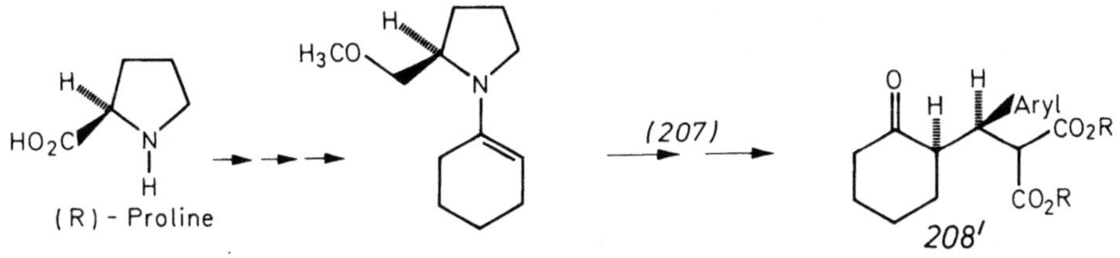

Enders et al.[205] metalated the (S)-proline-derived chiral allylamines (209). The resulting homoenolate (210) was subsequently alkylated. Upon hydrolysis β-substituted aldehydes (211) were obtained (e.e. = 64–67%).

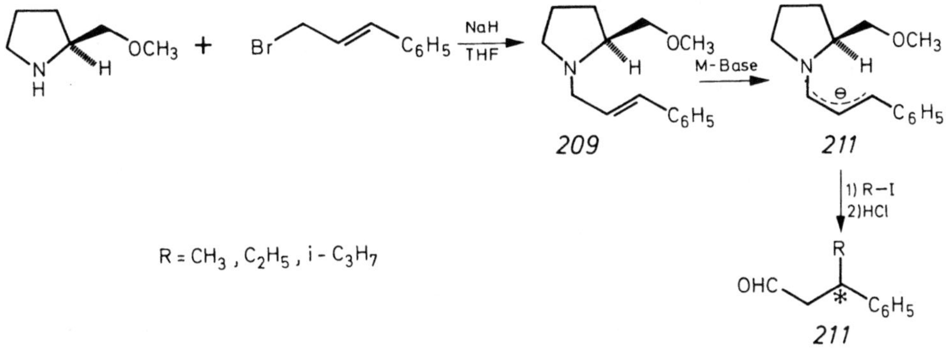

The bacteriostatic gliotoxin was prepared[206] by a total synthesis involving an asymmetric Michael-addition. A chiral amino acid derivative served as chiral auxiliary in the key step.

3.2.4 Cycloadditions

The use of a chiral dienophile or enophile in the Diels-Alder reaction effects asymmetric induction. This asymmetric Diels-Alder chemistry, pioneered by Korolew and

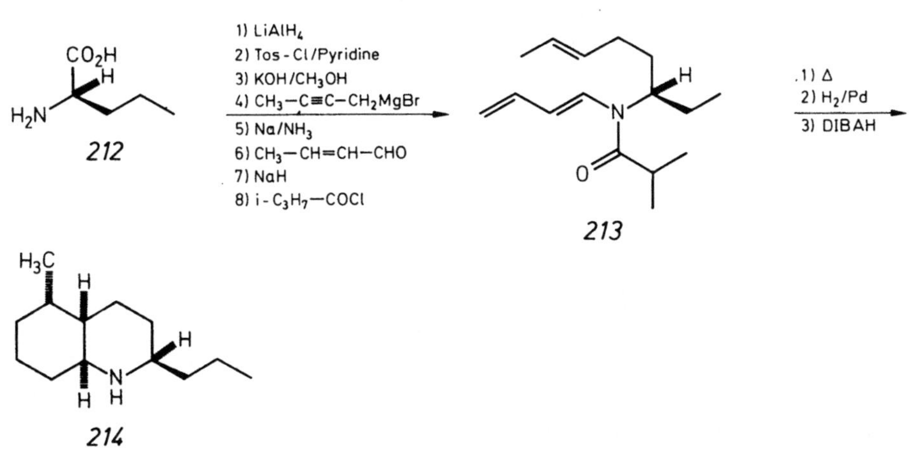

Mur [207], has received renewed interest in recent years. A fine review covering the intermolecular asymmetric Diels-Alder reaction was compiled by Mori [208]. In this article the use of terpenes and carbohydrates as chiral auxiliaries is discussed; no amino acid derivatives are mentioned in this context. A chiral α-hydroxycarboxylic acid derivative was also used to achieve an asymmetric Diels-Alder reaction [209]. High asymmetric induction could be detected in the intramolecular Diels-Alder reaction of chiral molecules.

Thus, (2R)-pumiliotoxin C (214) has been prepared from (R)-norvaline (212). The asymmetric center in the triene (213) controls the configuration at three carbon atoms [210]. α-Kainic acid, isolated from the algae *Digena simplex* and *Centrocerus clavulatum*, was prepared by total synthesis. Its enantioselective synthesis involved a stereocontrolled intramolecular cycloaddition of a (S)-glutamic acid [211]. Asymmetric cycloadditions also play a decisive role in the synthesis of chiral cytochalasins. In this case [212] the primary chiral information was carried by (S)-alanine and (S)-phenylalanine, respectively.

The (S)-leucine derivative (215) was allowed to react with diene (216) to afford the *threo* isomer (217) as the major product (d.s. = 80%) [213]. Mukaiyama et al. [215] have reported the total synthesis of the sesquiterpene (+)-farnesiferol, starting from (R)-phenylglycinol, a derivative of the amino acid (R)-phenylglcine. They key step of this synthesis was an asymmetric Diels-Alder reaction.

Baggiolini et al. [216] have succeeded in synthesizing the vitamin d-biotin from (RR)-cystine *via* intramolecular [3 + 2] cycloaddition with a d.s. value of 80%. Danishefsky et al. [217] synthesized pretyrosine, the biosynthetic precursor of (S)-tyrosine, by a [4 + 2]-cycloaddition applying an (S)-glutamic acid derivative as chiral auxiliary. Phenylglycine was used as the source of the chiral information in an asymmetric variation of a 1,3-dipolar cycloaddition, as reported by Grigg and Kemp et al. [218].

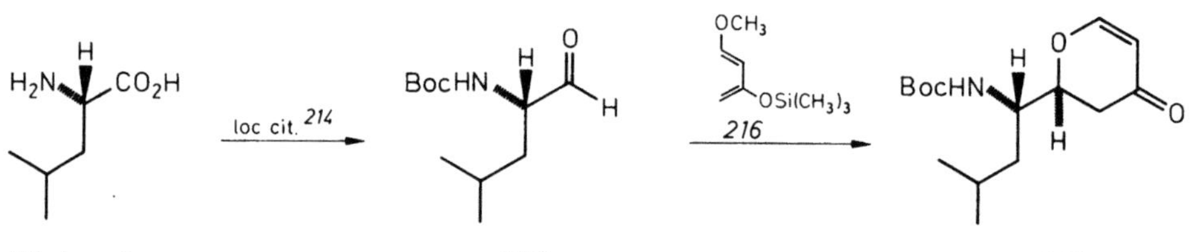

(S)-Leucine 215 217

3.2.5 [2,3]-Sigmatropic Rearrangements *via* Chiral Ammonium Ylides

The [2,3]-sigmatropic rearrangement of (E)-(218a), a derivative of the chiral cyclic α-amino acid (S)-proline, produced the aminonitrile (219) in a stereoselective manner. Saponification of (219) yielded (+)-2-methyl-2-phenyl-3-butenal (220) with an enantiomeric excess of 90% [219]. In replacing the benzyloxymethyl moiety in (218a) by a methyl group, the optical purity of the chiral aldehyde (220) obtained in the corresponding reaction sequence decreases considerably [219].

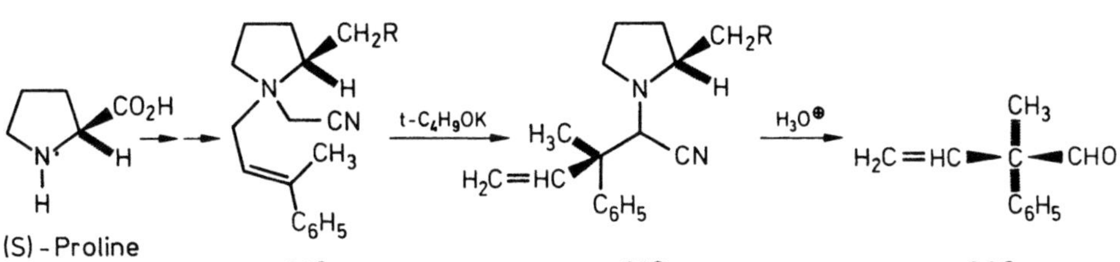

(S)-Proline

218 219 220

a R = $OCH_2C_6H_5$
b R = H

3.2.6 Stereoselective Photochemical Syntheses

Asymmetric synthesis, either enantioselective or diastereoselective, has seldom been performed by photochemical reactions. One of the first examples that may be classified as a photochemical asymmetric synthesis is the photoalkylation of the most simple amino acid, glycine. Elad and Sperling [220] demonstrated that, if glycine is part of a polypeptide chain, there is good control (up to 40% e.e.) in the creation of the new chiral center. A radical mechanism operates after the first step of photoinitiation of the process.

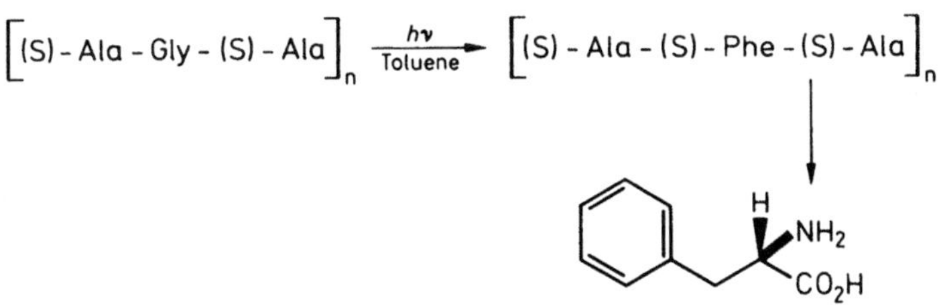

(S)-Phenylalanine

Thus, glycine, as part of a peptide chain, was stereocontrol-alkylated by irradiation in the presence of toluene. The new asymmetric center was induced by (S)-alanine moieties in the peptide chain.

The protoberberine alkaloid, xylopinine, has been synthesized in an optically active form by Kametani et al. [221]. A key reaction in this synthesis was the photochemical cyclization of the optically active amino acid derivative 1,2,3,4-tetrahydro-6,7-dimethoxy-3-methoxycarbonyl-1-methylene-2-veratroylisoquinoline with 1,3 asymmetric induction (d.s. <50%). Eschenmoser et al. [222] discovered a photochemically-induced secocorrin → corrin cycloisomerization; an (S)-glutamic acid derivative served as the source of chiral information in this transformation.

3.2.7 β-Lactam Syntheses

The asymmetric synthesis of β-lactames is still of interest to many organic and pharmaceutical chemists because of the great importance of these antibiotics. A detailed discussion of the numerous β-lactam syntheses which involves the induction of asymmetry by chiral amino acids is beyond the scope of this review article. Therefore, the reader is referred to reviews [223] and some more recent original publications describing the asymmetric synthesis of β-lactames starting from chiral amino acids, particularly serine [224, 225], threonine [225, 226], asparagin [227], and aspartic acid [228].

3.2.8 α-Alkylation of Amines

Kolb and Barth [229] synthesized α-substituted optically active amines or amino acids (223). Again the authors employed a derivative of naturally occurring (S)-proline, namely (—)-(S)-1-dimethoxymethyl-2-methoxymethyl-pyrrolidine (221) as chiral auxiliary agent. The metalation of the amidines (160) leads to azaallyl anions homologous with (222). After alkylation and hydrolysis, the desired α-substituted amines and amino acids, respectively, are obtained with some stereoselectivity.

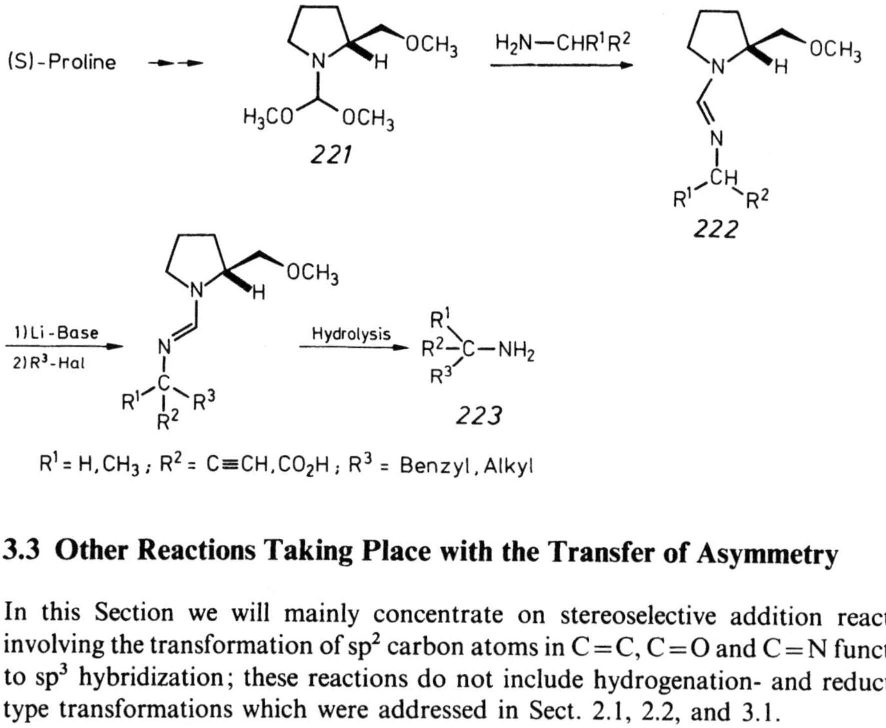

3.3 Other Reactions Taking Place with the Transfer of Asymmetry

In this Section we will mainly concentrate on stereoselective addition reactions involving the transformation of sp^2 carbon atoms in C=C, C=O and C=N functions to sp^3 hybridization; these reactions do not include hydrogenation- and reduction-type transformations which were addressed in Sect. 2.1, 2.2, and 3.1.

3.3.1 Addition of XH-Compounds to C=C Double Bonds (X = O,S,Halogen)

The naturally occurring chiral amino acid (S)-baikiain (224) is an attractive substrate for the preparation of 5- or 4-substituted pipecolinic acids. (S)-Baikiain (224) is a non-proteinogenic imino acid which can be isolated from in Rhodesian teak wood (*Baikiea plurijuga*). After protecting the nitrogen function of (224), Callens, Anteunis, and Reyniers [230] have oxymercurated (225) using $Hg(OAc)_2$. After reductive work-up with $NaBH_4$/NaOH the authors isolated a 7:3 mixture of Z-(2S,5R)-5-hydroxy-pipecolic acid (226a) and Z-(2S,4R)-4-hydroxy-pipecolic acid (226b).

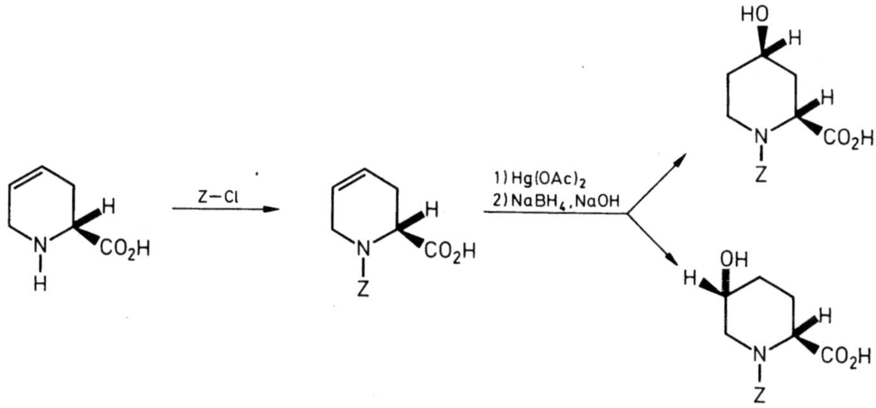

226b

226a

Terashima et al.[231)] reported an asymmetric halolactonization reaction. This highly stereoselective reaction permits the synthesis of intermediates for the preparation of chiral α,α-disubstituted α-hydroxycarboxylic acids (227)[231c)], α-hydroxyketones (228)[231c)], functionalized epoxides (229)[231d,e)] and natural products[231h,j)]. Only amino acids have so far been used as a source of the chiral information in the asymmetric halolactonization reaction. Again, the best results have been obtained by using cyclic imino acid enantiomers, namely proline.

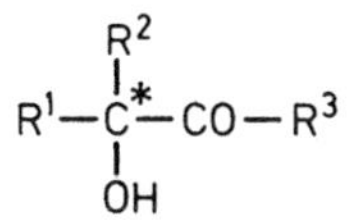

227, R^1, R^2 = Alkyl or Aryl
R^3 = OH

228, R^1, R^2R^3 = Alkyl or Aryl

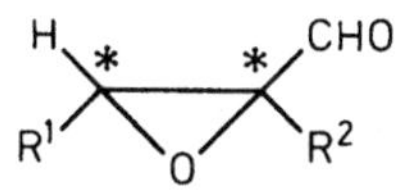

229, R^1, R^2 = Alkyl or Aryl

α,β-Unsaturated N-acyl-(S)-prolines (230) were treated with N-bromo-succinimide (NBS). The halolactonization leads to the formation of a mixture of diastereomeric lactones (231a) and (231b), (231a) being the major product (d.s. > 90%)[231)].

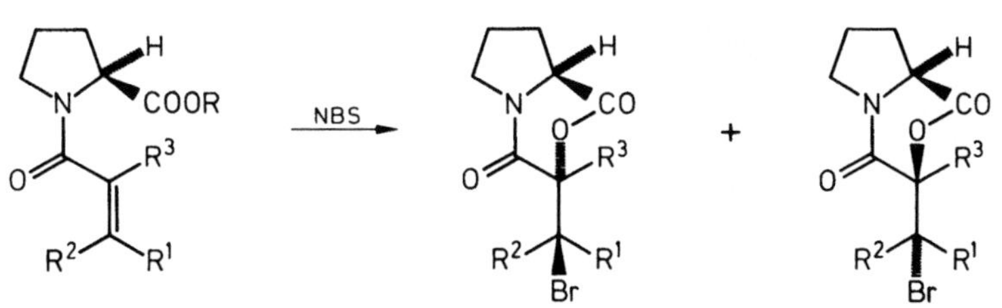

230 *231a* *231b*

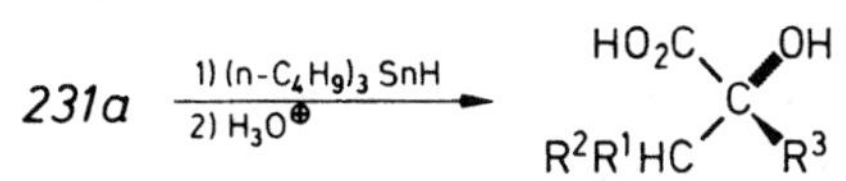

R^1 = H, CH_3, n-C_6H_{13}, C_6H_5; R^2 = H, CH_3; R^3 = CH_3

The asymmetric induction in the formation of (231) proceeds *via* a bromonium ion [231c)]. Debromination of (231a) with tri-*n*-butyltin hydride followed by saponification gave the chiral α-hydroxycarboxylic acid (232) in high optical purity. (S)-proline was recovered for recycling.

In the total synthesis of an anthracycline antibiotic, the key step was an asymmetric halolactonization reaction. The corresponding bromolactones were formed with high stereoselectivity (d.s. > 90%). (S)-Proline was used as chiral auxiliary.

The 4-hydroxy-(S)-proline-derived acid (232) was subjected to electrophilic lactonization either with J_2-KJ-$NaHCO_3$ to yield the iodolactone (233a), or benzeneselenyl chloride to give the phenylselenolactone (23b). Reductive removal of X from these products was achieved with tri-*n*-butyl- or triphenyltin hydride, followed by hydrogenolysis to yield (234) with at least 99% optical purity [231j)].

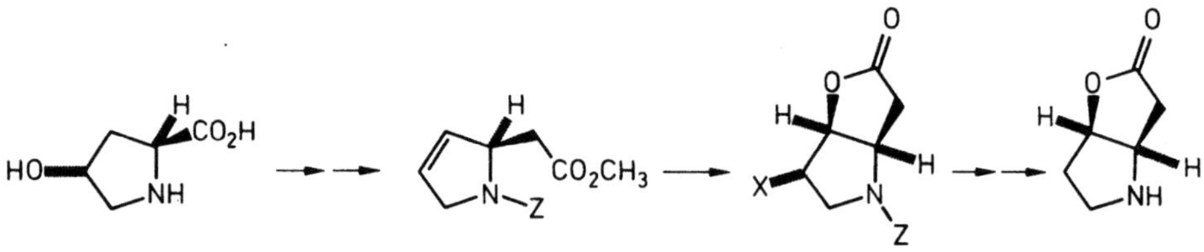

4-Hydroxy-(S)-proline 232 233 234

Stereoselective formation of 3-alkyl-6-methoxy-2,5-piperazine-dione derivatives by the addition of methanol in the presence of NBS to 3-alkyl-6-alkylidene-2,5-piperazinediones was recently reported by Shin et al. [232)]. The asymmetric induction in this reaction was accomplished by the chiral center of a derivative of the natural proteinogenic chiral amino acid threonine.

While investigating the biomimetic formation of cysteine, Schmidt et al. [233)] added thiolates to N-protected chiral α-aminoacrylic acid derivatives (dehydropeptides). (235) was obtained in optical yields up to 90%.

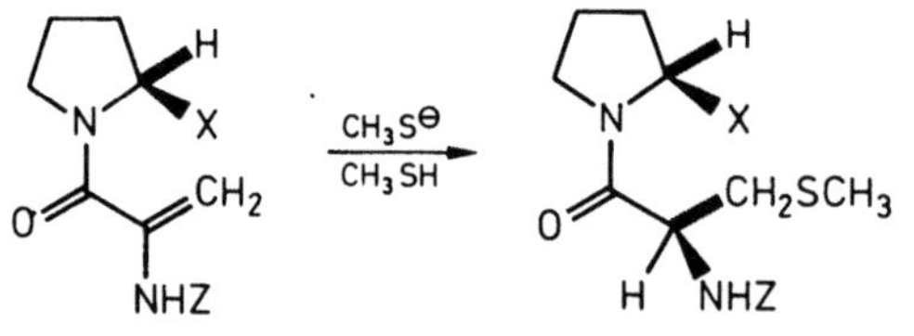

235

X = $CONHCH_3$, $CONH_2$, $CON(CH_3)_2$; Z = $C_6H_5CH_2OCO$

An analogue of the drug captopril (1) was prepared by Vasella et al. [234)] by a similar approach. In this case the diastereoselectivity was very low.

The asymmetric addition of thiolates to 2-cyclohexenone was induced in an intermolecular induction process by catalytic amounts of the 4-hydroxy-(S)-proline derivative (236) [235)].

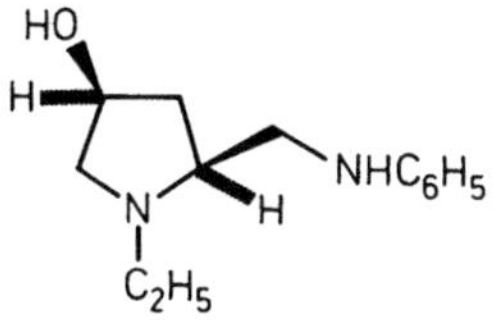

236

Recently, tryptophane has been converted to the methyl ester of lysergic acid in ten steps, involving a stereoselective HBr-addition to a C=C double bond [236].

3.3.2 Epoxidation

A stereoselective total synthesis of dendrobatide toxin 251 D was developed by Overman et al. [237] involving an epoxidation of the (S)-proline derivative (237) to furnish the oxirane (238) as major product. In their approach towards the total synthesis of the same natural product Thomas et al. [238] investigated the stereoselectivity of the epoxide formation from (S)-5-acetylpyrrolidin-2-one and dimethyloxosulfonium methylide. A diastereoselectivity of d.s. 50–60% was achieved [238].

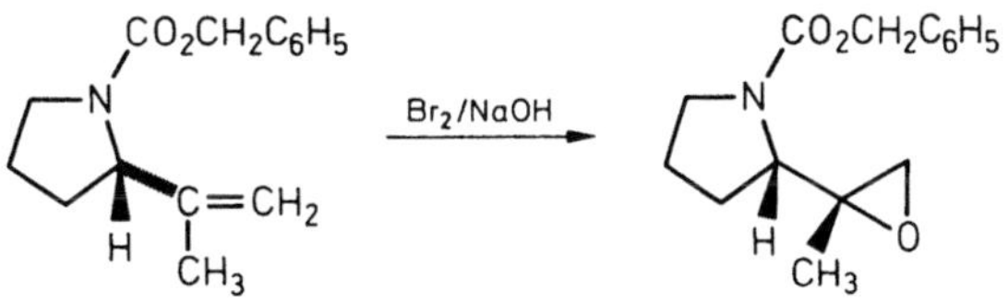

237 238

3.3.3 Carbene Addition to a C=C Double Bond

The observation that *cis*-3,4-methylene-(S)-proline (240a) isolated from *Aesculus parviflora* (the buckeye chestnut of the USA) had significant effects on pollen viability, has initiated renewed interest in the chemistry of this cyclic imino acid. Thus, 4-hydroxy-(S)-proline was converted [239] into (S)-3,4-diadehydroproline (239), and its N-trifuoroacetyl methylester was successfully reacted with diazomethane to yield, after deprotection, the desired target molecule (240a) as the major product. A minor product was *trans*-3,4-methylene-(S)-proline (240b) [240].

3.3.4 Addition of X—H-Compounds to C=O Double Bonds (X = S,N,O)

Reactions of carbonyl compounds with vicinal aminothiols, particularly with (R)-cysteine and (S)-penicillamine (4), are of chemical, biochemical, and pharmacological interest and have been investigated extensively. When (R)-cysteine was condensed with benzaldehydes, the corresponding thiazolidine compound (241) was obtained in almost quantitative yield [241]. This reaction occurs stereospecifically [241].

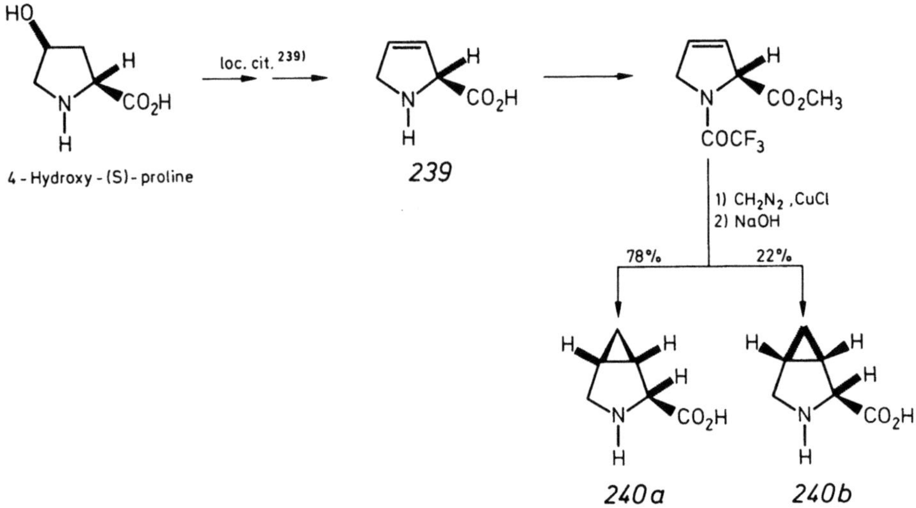

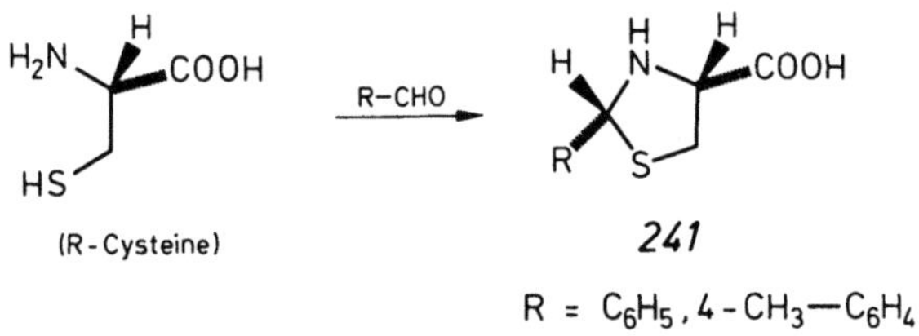

The increasingly accepted hypothesis that acetaldehyde may be the causative agent in initiating the multitude of acute pharmacological and chronic pathophysiological effects of alcohol prompted Nagasawa et al. to seek methods to reduce its blood levels. One possibility would be the administration of (S)-penicillamine (4), a compound related to cysteine. The condensation of this amino acid with acetaldehyde produced 2,5,5-trimethylthiazolidine-4-carboxylic acid [242]. The chirality of this compound was deducted by NMR analysis to be 72% 2S, 4S and 28% 2R, 4S. Thus, this result is consistent with the configuration found previously for the thiazolidines formed from (R)-cysteine and aldehydes [241].

An asymmetric synthesis of the vitamin (+)-biotin has been reported [241a] using thiazoline (241) as substrate.

Addition of (S)-α-amino acids to β,β-dibromo-α-imino acids proceeded to produce the corresponding optically active α,α-diamino acid derivatives with d.s. = 33 to 85% [243].

The diastereoselective condensation of pivalaldehyde and (S)-proline was described by Seebach and Naef [196] to yield (197).

3.3.5 Equilibration and Stereoselective Protonation

In the synthesis of optically active 2-phenylalkanoic acids (244), a derivative of (S)-phenylalanine was used as a chiral auxiliary [244]. The carboxylic acids (244) were obtained in optical yields of up to 53%.

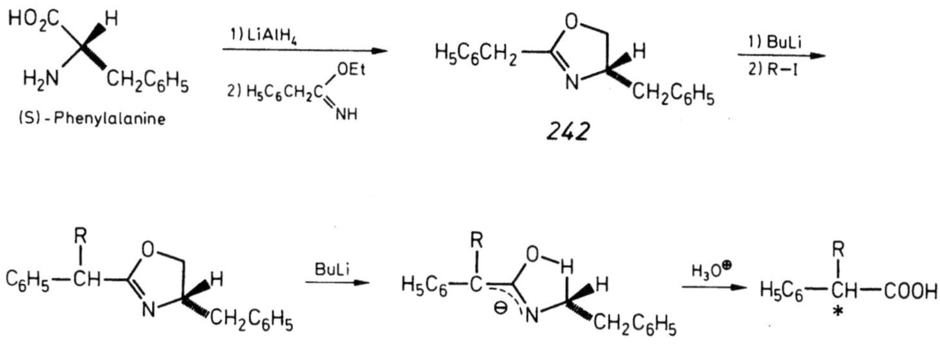

242 243 244

R = CH_3, C_2H_5, i-C_3H_7, n-C_4H_9, n-C_5H_{11}

Chiral oxazolines (242), which had been originally used by Meyers et al. for asymmetric synthesis, were also applied for the asymmetric transformation of 2-chloroalkanoic acids to produce chiral 2-chloroalkanenecarboxylic acids in 45–73% optical yield [245].

Imidazolines (245) have been prepared from (S)-alanine and (S)-proline. Upon hydrolysis (R)-alanine was obtained. This result can be explained in terms of epimerization and stereoselective protonation with asymmetric induction by the chiral center originating from (S)-proline [246].

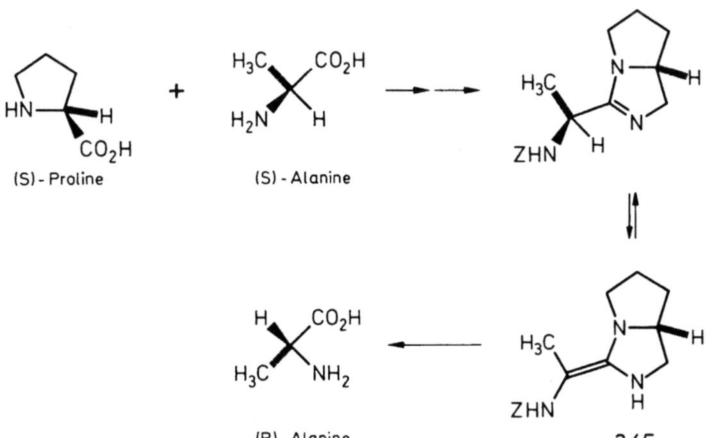

245

Z = $C_6H_5CH_2OCO$

(R)-Alanine was obtained from (245) in an optical yield of up to 94% [246]. Non-natural alanine can, of course, also be obtained from racemic alanine in this reaction sequence.

Racemic 6-benzoyloxycyclocitral (246) has been converted to (S)-6-hydroxycyclocitral (248) *via* the oxazolinone (247). (248) was obtained in 63% enantiomeric excess [247].

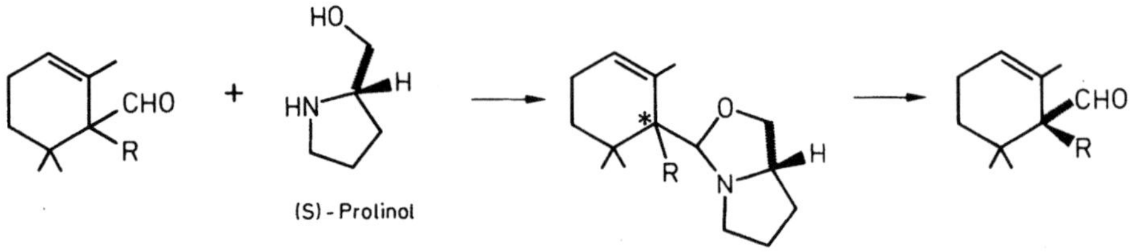

(S)-Prolinol

246 R = H, $OCOC_6H_5$ | *247* R = H, OH | *248* R = H, OH

By heating the racemic Robinson annelation product (249) with an equimolar amount of (S)-proline pyrrolidide (250), the related dienamine (251) was obtained, which, upon hydrolysis was converted to (S)-(249) [248].

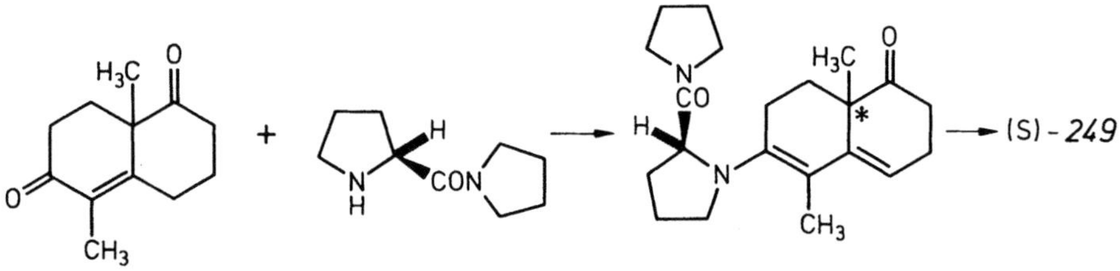

249 *250* *251*

Kinetics and stereochemistry of deuterium exchange of the α-hydrogen of an amino acid moiety in metal complexes of amino acid Schiff bases with *ortho*-hydroxyacetophenone have been studied by Belokon et al. [249].

3.3.6 Nitrogen, Phosphorus and Sulfur Atoms as Asymmetric Centers

Chiral N-Oxides [250]: In the reaction of N-benzyl-N-methylamino amino acids with H_2O_2 in alkaline water solution, mixtures of diastereomeric N-oxides, containing new centers of chirality on nitrogen atoms, were obtained. The reaction was performed with the corresponding derivatives of (S)- [or (R)] alanine, (S)- [or (R)] leucine, (S)- [or (R)] phenylalanine, and (S)- [or (R)] proline, respectively. In the reaction a distinct stereoselectivity could be observed: for alanine, leucine, and phenylalanine derivatives the formation of $N_{(S)}C_{(S)}$, or correspondingly $N_{(R)}C_{(R)}$ diastereomer is favoured. The reaction of (S)-proline derivatives leads, however, exclusively to the $N_{(R)}C_{(S)}$ stereoisomer (252); on the other hand, (R)-proline yielded stereoselectively the $N_{(S)}C_{(R)}$ compound (253).

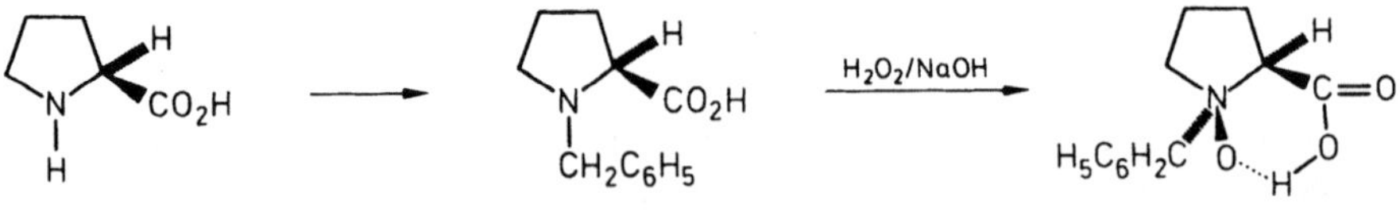

(S)-Proline

252

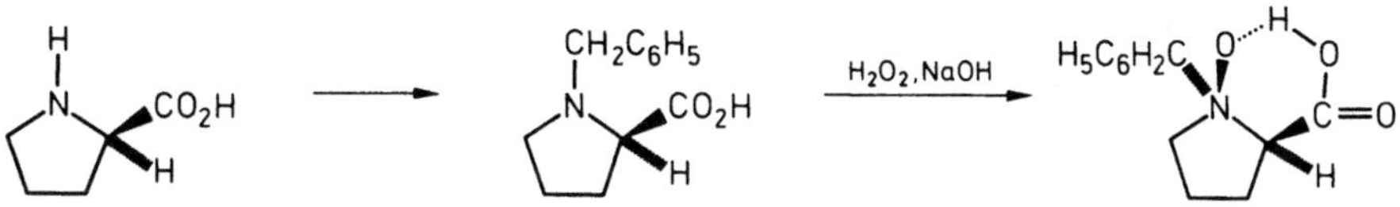

(R)-Proline

253

Chiral Phosphorus Compounds: Koizumi et al. [251] have prepared a series of chiral organophosphorus compounds (256) in which the phosphorus atom is the asymmetric center, whereby amino acid derivatives were used as chiral auxiliary reagents.

The reaction of phosphonic acid chloride (254) with (S)-proline ethyl ester afforded a mixture of diasteromeric amides (255) in high diastereoselectivity. The diastereomers (255) can easily be purified by chromatography. The chiral, practically optical pure organophosphorus compound (256) was obtained from purified (255) by acid alcoholysis.

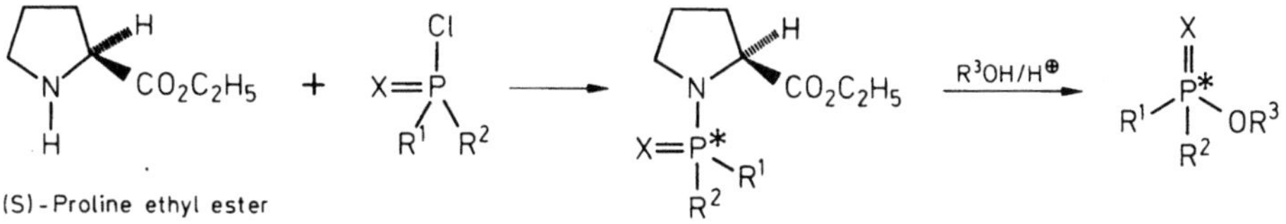

(S)-Proline ethyl ester

254 255 256

X = O, S

R^1 = CH_3, C_6H_5, OC_6H_5

R^2 = C_2H_5, n-C_4H_9, C_6H_5, OC_6H_5

R^3 = CH_3, C_2H_5, n-C_3H_7, i-C_3H_7

Some diastereoselectivity was also observed in the reaction of (S)-prolinol with the dichloride (257) leading to the chiral organophosphorus compounds (258).

(S)-Prolinol

257 258a 258b

X = O, S

Chiral Sulfur Compounds: Dehydromethionine was first prepared by Lavine [252], who correctly assigned its structure (without regard to relative stereochemistry). An analysis of the crystal structure of the racemic compound revealed the carboxy-group to be on the opposite side of the ring to the S-methyl group. The structure of the dehydromethionine derived from (S)-methionine is therefore that of compound (259), having the (1R,3S)-configuration [253]. (259) is readily obtained by oxidizing (S)-methionine with iodine in methanol, and although this oxidation normally produces sulfoxides from sulfides, (259) is formed because of intramolecular attack by the amino-group in a sulfonium iodide intermediate. The stereochemistry of the conversion [253b] of dehydromethionine (259) into methionine sulfoxide (260) was also studied. Dehydromethionine (259) was treated with alkali hydroxide in water and, after neutralization, the addition of acetone led to the precipitation of (260). This produced methionine sulfoxide (260) in yields of 80–85% and with $[\alpha]_D^{24} = +120°$ (C = 1.8 in 1N hydroxychloric acid), compared with the highest reported value for (S)-methionine-(S)-sulfoxide (260) of $+131°$. Thus, the preparation of (260) from (S)-methionine *via* (1R,3S)-dehydromethionine (259) is not as stereospecific as the oxidation of (S)-methionine with H_2O_2 [252] or gold (III) chloride [254]. However, it has the advantages of being rapid, experimentally simple, and of using radily available, cheap starting materials [253b].

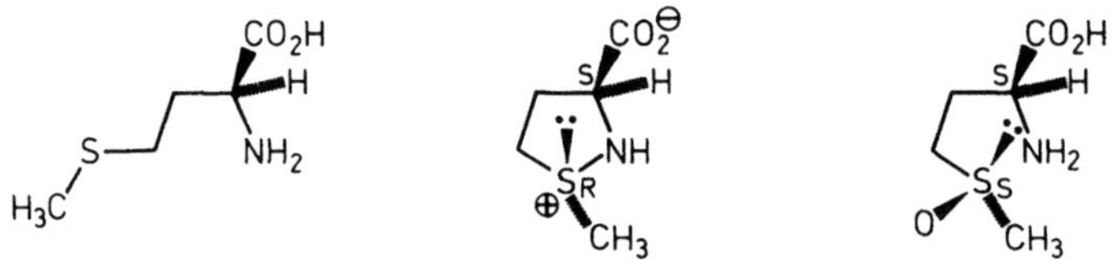

(S)-Methionine   *259*   *260*

Other sulfoxides of high optical purity have been derived from enantiomers of cysteine [255] and methionine [256].

3.3.7 Miscellaneous

Recent developments regarding the utility of chiral amino acids in asymmetric synthesis of natural products were reported. Examples of such syntheses are the preparation of carbohydrates from (S)-glutamic acid [257], (S)-alanine [258], or (S)-threonine [259], and syntheses of alkaloids [260], terpenes [261], peptide [262] derivatives, and toxines [263].

A chiral recognition was observed in aminolysis of 3-acyl-4(R)-methoxycarbonyl-1,3-thiazolidine-2-thione, a derivative of (R)-cysteine, by racemic amines to give an optically active amide [(S)-excess] and amine [(R)-excess] [264]. In the reaction of cyclic *meso*-1,3-diols with chiral N-protected phenylalanyl chlorides, Yamada et al. [265] observed the preferential formation of one of the two possible diastereomeric monoesters, which has been used for the synthesis of optically active steroids [266] and prostaglandins [267].

The asymmetric synthesis of (1R,3S)-*cis*-chrysanthemic acid was reported by Mukaiyama et al. [268]; (R)-phenylglycinol was used as chiral auxiliary.

Asymmetric transformations of α-amino acids promoted by optically active metal complexes have been reported by several groups [269]. The control of the stereoselective hydrolysis reactions of racemic esters by chiral micellar compounds prepared from amino acids has been intensively investigated [270].

4 Concluding Remarks

Asymmetric synthesis has evolved rapidly during recent years. Most of the progress is registered in synthetic methods; less emphasis has been given to theoretical concepts and mechanistic studies. Methods have been devised for achieving optical yields exceeding 95%. A number of stochiometric reactions with respect to the chiral auxiliary moiety are now highly efficient.

A trend in asymmetric synthesis is the utilization of cheap chiral products such as amino acids, hydroxy acids, sugars, or terpenes as starting materials or catalysts. Nature provides us with a wide range of these optically active natural compounds ("chiral pool"). Particularly frequent use is made of sugars and naturally occurring hydroxycarboxylic acids such as tartaric acid and malic acid. In recent years, however, increasing use has been made of amino acids as educts in asymmetric syntheses. Until 1981, the induction of asymmetry by amino acids was the subject of roughly 250 original papers, most of which were published in the last ten years. Surveying the existing literature to date (September 1983) it is notable that by now 400 research reports have been published which deal with the topic of this review. Thus, drastically increasing use has been made of amino acids as educts in asymmetric syntheses. One reason for this development is the growing interest in asymmetric synthesis as such. Another reason should also be noted: amino acids are particularly versatile chiral compounds, most are available in significant quantities and — other than in the case of sugars and hydroxycarboxylic acid — both enantiomers of many amino acids are being produced or could be produced on an industrial scale.

Strikingly high stereoselectivities have been achieved in asymmetric syntheses with optically pure proline or proline derivatives, probably due to the rigidity of the five-membered ring. Other preferably used chiral auxiliaries include (S)-phenylalanine, (S)-valine and tert.-(S)-leucine.

From an industrial chemist's point of view the use of proline, phenylalanine, valine, and other commercially available amino acids, is fine. To date, however, tert.-(S)-leucine is still an exotic compound. It should also be noted that the recycling of the chiral amino acid moiety is of importance for possible technical processes. On the other hand, the recovery of the chiral auxiliary sometimes does not make sense, especially in syntheses which the require the use of stochiometric amounts of expensive reagents, e.g. LDA.

Traditionally, amino acids have been utilized in the mainstream of organic chemistry primarily as building blocks for peptide syntheses. One may expect that in the future the use of amino acids as starting materials for non-peptide compounds will be a subject of ever-increasing interest. Many chiral target molecules with widely variable structures will be prepared from amino acids in the future.

5 References

1. Cushman, D. W. et al.: Biochemistry *16*, 5484 (1977)
2. a) Bestmann, H. J., Vostrowsky, O.: Naturwissensch. *69*, 457 (1982)
   b) Light, D. M., Birch, M. C.: ibid. *69*, 243 (1982)
3. Moser, H., Rihs, G., Sauter, H.: Z. Naturforsch. *37b*, 451 (1982)
4. Markwald, W.: Ber. dtsch. Chem. Ges. *37*, 1368 (1904)

5. Morrison, J. D., Mosher, H. S.: Asymmetric Organic Reactions, Englewood Cliffs, New Jersey, Prentice-Hall 1971
6. For a definition of the term "prochirality" see: Eliel, E. L.: Topics in Current Chemistry *105*, 1 (1982)
7. For a recent review see: Whitesides, G. M., Wong, C.-H.: Aldrichimica Acta *16*, 27 (1983)
8. Chibata, S.: Applications of Immobilized Enzymes for Asymmetric Reactions, Am. Chem. Soc. Symp. Series *185*, Washington, DC, 1982
9. a) Scott, J. W., Valentine, D.: Science *184*, 943 (1974)
   b) Valentine, D., Scott, J. W.: Synthesis *1978*, 329
10. a) Meyers, A. I.: Acc. Chem. Res. *11*, 375 (1978)
    b) Meyers, A. I., Mihelic, E. D.: New Synthetic Methods *5*, 105 (1979)
11. Kagan, H. B., Fiaud, J. C.: Top. Stereochem. *10*, 175 (1978)
12. a) Otsuka, S.: Kagaku Zokan (Kyoto) *97*, 3 (1982) [Chem. Abstr. *98*, 15999 e (1983)]
    b) Otsuka, S., Tani, K.: Kagaku Zokan (Kyoto) *97*, 127 (1982) [Chem. Abstr. *98*, 15997 c (1983)]
13. Apsimon, J. W., Sequin, R. P.: Tetrahedron *35*, 2797 (1979)
14. Fischli, A.: Chimia *30*, 4 (1976)
15. Wynberg, H.: Recl. Trav. Chim. Pays-Bas *100*, 393 (1981)
16. Masamune, S., Choy, W.: Aldrichimica Acta *15*, 47 (1982)
17. Heathcock, C. H.: Science *214*, 395 (1981)
18. a) Fraser-Reid, B., Anderson, R. C.: Fortschr. Chem. Org. Naturst. *39*, 1 (1980)
    b) Cohen, N.: Accounts Chem. Res. *9*, 412 (1976)
    c) Fraser-Reid, B., Mo Sun, K., Tam, T. F.: Bull. Chem. Soc. Fr. *1981*, II-230
    d) Fraser-Reid, B. et al.: Carbohydrate derivatives in the asymmetric synthesis of natural products, in: Organic synthesis today and tomorrow (Trost, B. M., Hutchinson, C. R. ed.), p. 319, Oxford, New York 1981
    e) Hanessian, S. et al.: Can. J. Chem. *60*, 2292 (1982)
    f) Hoppe, I., Schöllkopf, U.: Liebigs Ann. Chem. *1982*, 1548
    g) Kinoshita, M. et al.: Carbohydr. Res. *109*, 5 (1982)
    h) Sherk, A. E., Fraser-Reid, B.: J. Org. Chem. *47*, 932 (1982)
    i) Fraser-Reid, B. et al.: ibid. *47*, 935 (1982)
    j) Fraser-Reid, B. et al.: ibid. *47*, 941 (1982)
    k) Pochet, S., Huynh-Dinh, T.: ibid. *47*, 193 (1982)
    l) Mubarak, A. M., Brown, D. M.: J. Chem. Soc. Perkin I *1982*, 809
    m) Redlich, H. et al.: Liebigs Ann. Chem. *1983*, 393
    n) Redlich, H. et al.: ibid. *1983*, 412
    o) Schmidt, R. R., Maier, M.: Synthesis *1982*, 747
    p) Redlich, H., Xiang-jun, J.: Liebigs Ann. Chem. *1982*, 717
    q) Redlich, H.: ibid. *1982*, 708
    r) Paulsen, H., Stubbe, M.: ibid. *1983*, 535
    s) Prugh, J. D., Deana, A. A.: Tetrahedron Lett. *1982*, 281
    t) Vasella, A., Voeffray, R.: J. Chem. Soc. Chem. Commun. *1981*, 97
    u) Takano, S. et al.: Tetrahedron Lett. *1983*, 401
    v) Kakinuma, K.: ibid. *1982*, 1697
    w) Paulsen, H., Stubbe, M.: ibid. *1982*, 3171
19. a) Brown, H. C. et al.: Tetrahedron *37*, 3547 (1981)
    b) Brown, H. C. et al.: J. Org. Chem. *47*, 5065 (1982)
    c) Brown, H. C. et al.: ibid. *47*, 5074 (1982)
    d) Whitesell, J. K. et al.: J. Chem. Soc. Chem. Commun. *1982*, 988
    e) Whitesell, J. K. et al.: ibid. *1982*, 989
    f) Sugita, T. et al.: Bull. Chem. Soc. Jpn. *55*, 1677 (1982)
    g) Noe, C. R.: Chem. Ber. *115*, 1607 (1982)
    h) Noe, C. R.: ibid. *115*, 1591 (1982)
    i) Noe, C. R.: ibid. *115*, 1576 (1982)
    j) Consighio, G., Pino, P.: Advances in asymmetric hydrocarbonylation, Am. Chem. Soc. Symp. Series *196*, p. 371, Washington, D.C., 1982
    k) Brandänge, S. et al.: Acta Chem. Scand. *B35*, 273 (1981)
    l) Oppolzer, W. et al.: Helv. Chim. Acta *64*, 2802 (1981)

m) Oppolzer, W., Löher, H. J.: ibid. *64*, 2808 (1981)
n) Annuziata, R., Cinquini, M.: Synthesis *1982*, 767
20. a) Bosnich, B., Mackenzie, P. B.: Pure Appl. Chem. *54*, 189 (1982)
b) Teramoto, T. et al.: Tetrahedron Lett. *1981*, 1109
c) Duhamel, L., Plaquevent, J.-C.: Bull. Soc. Chim. Fr. *1982*, II-75
d) Sharpless, K. B. et al.: Pure Appl. Chem. *55*, 589 (1983)
e) Osawa, T., Harada, T.: Chem. Lett. *1982*, 315
f) Törös, S. et al.: J. Organomet. Chem. *232*, C 17 (1982)
g) Pittmann, C. U. et al.: J. Chem. Soc. Chem. Commun. *1982*, 473
h) Bosnich, B. et al.: ibid. *1981*, 1278
i) Frater, G. et al.: Tetrahedron Lett. *1981*, 4221
j) Yoshikawa, S. et al.: Chem. Lett. *1982*, 1179
k) Seebach, D., Hungerbühler, E.: Asymmetric synthesis with tartaric acid, in: Modern synthetic methods, Vol. 2, p. 91 (Scheffold, R. ed.), Frankfurt am Main, Salle/Sauerländer 1980
21. a) Matsumoto, K., Uchida, T.: Chem. Lett. *1981*, 1673
b) Wynberg, H., Staring, E. G. J.: J. Am. Chem. Soc. *104*, 166 (1982)
c) Kunieda, N., Kinoshita, M.: Phosphorus Sulfur *1981*, 383
22. a) Takahashi, H., Inagaki, H.: Chem. Pharm. Bull. *30*, 922 (1982)
b) Takahashi, H. et al.: ibid. *29*, 3387 (1981)
c) Bates, H. A.: J. Org. Chem. *46*, 4931 (1981)
d) Mangeney, P. et al.: Tatrahedron Lett. *1983*, 373
e) Beck, W. et al.: Z. Naturforsch. *36b*, 578 (1981)
23. Drauz, K., Kleemann, A., Martens, J.: Angew. Chem. *94*, 590 (1982), Angew. Chem. Int. Ed. Engl. *21*, 584 (1982)
24. Jones, J. B., Sih, C. J., Perlmann, D.: Techniques of Chemistry, Application of Biochemical Systems in Organic Chemistry, Part I and II, New York, Wiley 1976
25. Laymann, P. L.: Chem. Eng. News, Jan. 3, 1983, p. 18
26. a) Hoppe, B., Martens, J.: Chemie in unserer Zeit *17*, 41 (1983)
b) Sadovnikova, M. S., Belikov, V. M.: Russ. Chem. Rev. *47*, 199 (1978)
c) Kleemann, A.: Chemiker-Z. *106*, 151 (1982)
27. a) Izumi, Y., Chibata, I., Itoh, T.: Angew. Chem. *90*, 187 (1978), Angew. Chem. Int. Ed. Engl. *17*, 176 (1978)
b) Shpak, V. S. et al.: Vestn. Akad. Nauk SSSR *1983*, 107
c) Hoppe, B., Martens, J.: Chemie in unserer Zeit *18*, 73 (1984)
28. Yamada, H., Kumagai, H.: Pure Appl. Chem. *50*, 1117 (1978)
29. a) Dinelli, D., Marconi, W., Cecere, F., Galli, G., Morisi, F.: Enzyme Engineering, Vol. 3 (Pye, E. K., Weetall, H. H. ed.), New York, Plenum 1978, p. 477
b) Guivarch, M., Gillonnier, C., Brunie, J.-C.: Bull. Soc. Chim. Fr. *1980*, II-91
30. Čaplar, V., Comisso, G., Šunjić, V.: Synthesis *1981*, 85
31. Akashi, T.: Review given at CHEMRAWN II in Manila (1982)
32. a) Harmon, R. E., Gupta, S. K., Brown, D. J.: Chem. Rev. *1973*, 7321
b) Benes, J., Hetflejs, J.: Chem. Listy *68*, 916 (1974)
c) Brown, J. M.: Chem. Ind. *1982*, 737
d) Knowles, W. S. et al.: Studies of Asymmetric Homogeneous Catalysts (Alyea, E. C., Meek, D. v. W. ed.), Advances in Chemistry Series No. 196, p. 325, Washington, D.C., 1982
e) Halpern, J.: Inorg. Chim. Acta *50*, 11 (1981)
f) Brunner, H.: Kontakte (Merck) *1981*, 3
g) Halpern, J.: Pure Appl. Chem. *55*, 99 (1983)
33. a) Bourson, J., Oliveros, L.: J. Organometal. Chem. *229*, 77 (1982)
b) Matteoli, U. et al.: J. Mol. Catal. *12*, 265 (1981)
34. a) Amma, J. P., Stille, J. K.: J. Org. Chem. *47*, 468 (1982)
b) Yamashita, M. et al.: Bull. Chem. Soc. Jpn. *55*, 2917 (1982)
35. Gladiali, S. et al.: J. Organomet. Chem. *244*, 289 (1983)
36. Hayashi, T., Kumada, M.: Acc. Chem. Res. *15*, 395 (1982)
37. Horner, L., Winkler, H., Rapp, A., Mentrup, A., Hoffmann, H., Beck, P.: Tetrahedron Lett. *1961*, 161
38. a) Knowles, W. S., Sabacky, M. J.: J. Chem. Soc. Chem. Commun. *1968*, 1445

b) Knowles, W. S., Sabacky, M. J., Vineyard, B. D., Weinkauf, D. J.: J. Am. Chem. Soc. *97*, 2567 (1975)
39. Horner, L.: Pure Appl. Chem. *52*, 843 (1980)
40. a) Achiwa, K.: J. Am. Chem. Soc. *98*, 8265 (1976)
b) Baker, G. L., Fritschel, S. J., Stille, J. R., Stille, J. K.: J. Org. Chem. *46*, 2954 (1981)
41. a) Achiwa, K.: Tetrahedron Lett. *1977*, 3755
b) Achiwa, K.: Chem. Lett. *1977*, 777
c) Achiwa, K.: ibid. *1978*, 561
d) Achiwa, K., Soga, T.: Tetrahedron Lett. *1978*, 1119
e) Achiwa, K., Chaloner, P. A., Parker, D.: J. Organomet. Chem. *218*, 249 (1981)
f) Ojima, I., Yoda, N.: Tetrahedron Lett. *1980*, 1051
42. Kellner, K., Tzschach, A., Nagy-Magos, Z., Markó, L.: J. Organomet. Chem. *193*, 307 (1980)
43. Bergstein, W., Kleemann, A., Martens, J.: Synthesis *1981*, 76
44. Joó, F., Trócsányi, E.: J. Organomet. Chem. *231*, 63 (1982)
45. Osakada, K. et al.: Chem. Lett. *1981*, 1691
46. Takeuchi, S., Ohgo, Y.: Bull. Chem. Soc. Jpn. *54*, 2136 (1981)
47. Brown, J. M. et al.: Rational Approaches to Asymmetric Hydrogenation, ACS Symp. Series *119*, 169 (1980)
48. Scott, J. W. et al.: J. Org. Chem. *46*, 5086 (1981)
49. a) Ojima, I., Suzuki, T.: Tetrahedron Lett. *1980*, 1239
b) Ojima, I., Yoda, N.: ibid. *1982*, 3913
c) Ojima, I., Yoda, N., Yatabe, M.: ibid. *1982*, 3917
d) Levine-Pinto, H. et al.: Tetrahedron *38*, 119 (1982)
e) Ojima, I. et al.: J. Org. Chem. *47*, 1329 (1982)
50. Ojima, I., Yoda, N.: Tetrahedron Lett. *1982*, 3913
51. a) Achiwa, K.: Heterocycles *8*, 247 (1977)
b) Valentine, D., Sun, R. C., Toth, K.: J. Org. Chem. *45*, 3703 (1980)
c) Koenig, K. E., Bachman, G. L., Vineyard, B. D.: ibid. *45*, 2362 (1980)
52. Schrock, R. R., Osborn, J. A.: J. Chem. Soc. Chem. Commun. *1970*, 567
53. Ojima, I., Kogure, I., Achiwa, K.: ibid. *1977*, 428
54. a) Ojima, I., Kogure, T., Terasaki, T., Achiwa, K.: J. Org. Chem. *43*, 3444 (1978)
b) Achiwa, K., Kogure, T., Ojima, I.: Tetrahedron Lett. *1977*, 4431
c) Ojima, I., Kogure, T.: J. Organomet. Chem. *195*, 239 (1980)
55. a) Woerde, H. W., Bostelaar, L. J., Hoek, A., Sachtler, W. M. H.: J. Catal. *76*, 316 (1982)
b) Nakahata, M. et al.: Bull. Chem. Soc. Jpn. *55*, 2186 (1982)
c) Kawaken Fine Chemicals Co., Ltd. (Inventors: Harada, T. et al.) Germ. Pat. 2917752 (1982)
56. Brunner, H.: Chemie in unserer Zeit *14*, 177 (1980)
57. a) Izumi, Y.: Angew. Chem. *83*, 956 (1971), Angew. Chem. Int. Ed. Engl. *10*, 871 (1971)
b) Klabunovskii, E. I. et al.: Izv. Akad. Nauk SSSR, Ser. Khim. *1980*, 121 [Chem. Abstr. *93*, 25560 r (1980)]
58. Isoda, T., Ichikawa, A., Shimamoto, T.: J. Sci. Res. Inst. Tokyo *34*, 143 (1958)
59. Gourley, R. N., Grimshaw, J., Millar, P. G.: J. Chem. Soc. C *1970*, 2318
60. Kopilov, J., Kariv, E., Miller, L. L.: Electroanal. Chem. *99*, 3450 (1977)
61. a) Watkins, B. W., Behling, J. B., Kariv, E., Miller, L. L.: ibid. *97*, 3549 (1975)
b) Firth, B. E., Miller, L. L., Mitani, M., Rogers, T., Lennox, J., Murray, R. W.: ibid. *98*, 8271 (1976)
c) Firth, B. E., Miller, L. L.: ibid. *98*, 8272 (1976)
62. Nonaka, T., Kumai, Y., Ito, M., Fuchigami, T.: Rep. Asahi Glass Found. Ind. Technol. *32*, 155 (1978)
63. Abe, S., Nonaka, T., Fuchigami, T.: J. Am. Chem. Soc. *105*, 3630 (1983)
64. Brunner, H., Riepl, G., Weitzer, H.: Angew. Chem. *95*, 326 (1983), Angew. Chem. Int. Ed. Engl. *22*, 326 (1983)
65. Eder, U., Sauer, G., Wiechert, R.: Angew. Chem. *83*, 492 (1971), Angew. Chem. Int. Ed. Engl. *10*, 496 (1971)
66. Hajos, Z. G., Parrish, D. R.: J. Org. Chem. *39*, 1615 (1974)
67. a) Nagasawa, K., Hiroi, K., Yamada, S.: J. Pharm. Soc. Jpn.: *95*, 46 (1975)

b) Micheli, R. A. et al.: J. Org. Chem. *40*, 675 (1975)
c) Cohen, N. et al.: ibid. *40*, 681 (1975)
d) Eder, U. et al.: Helv. Chim. Acta *59*, 999 (1976)
e) Eder, U. et al.: Chem. Ber. *109*, 2948 (1976)
f) Wang, K. C., Lee, J.-Y.: J. Taiwan Pharm. Ass. *33*, 55 (1981)
g) Wang, K. C., Lee, J.-Y.: ibid. *33*, 45 (1981)
h) Wang, K. C., Lai, C. F., Lee, J.-Y.: ibid. *31*, 63 (1979)
i) Wang, K. C., Lai, C. F.: ibid. *32*, 44 (1980)
j) Wang, K. C., Lai, C. F.: ibid. *31*, 71 (1979)

68. Cohen, N.: Acc. Chem. Res. *9*, 412 (1976)
69. Takano, S., Kasahara, C., Ogasawara, K.: J. Chem. Soc. Chem. Commun. *1981*, 635
70. Woodward, R. B.: Perspectives in Organic Chemistry (Todd, A. ed.), New York, Interscience Publ. 1956
71. a) Woodward, R. B. et al.: J. Am. Chem. Soc. *103*, 3210 (1981)
b) Woodward, R. B. et al.: ibid. *103*, 3213 (1981)
c) Woodward, R. B. et al.: ibid. *103*, 3215 (1981)
72. a) Coisne, J.-M. et al.: Bull. Soc. Chim. Belg. *90*, 481 (1981)
b) Tamai, Y. et al.: J. Chem. Soc. Chem. Commun. *1983*, 114
73. More recent review: Hayashi, T.: Asymmetric Grignard Cross-Coupling Catalyzed by Chiral Phosphine-Nickel and Phosphine-Palladium Complexes, in Asymmetric Reactions and Processes in Chemistry (Eliel, E. L., Otsuka, S. ed.), American Chemical Soc. Symp. Series Vol. 185, p. 177 (1982)
74. a) Hayashi, T. et al.: Tetrahedron Lett. *1980*, 4623
b) Hayashi, T. et al.: ibid. *1981*, 137
75. a) Tsuboyama, T. S.: Bull. Chem. Soc. Jpn. *35*, 1004 (1962)
b) Ueyanagi, K., Inoue, S.: Makromol. Chem. *178*, 235 (1977)
c) Oku, J., Ito, N., Inoue, S.: ibid. *180*, 1089 (1979)
d) Oku, J., Inoue, S.: J. Chem. Soc. Chem. Commun. *1981*, 229
e) Oku, J., Ito, N., Inoue, S.: Makromol. Chem. *183*, 579 (1982)
76. Becker, W., Pfeil, E.: J. Am. Chem. Soc. *88*, 4299 (1966)
77. a) Juliá, S., Masana, J., Vega, J. C.: Angew. Chem. *92*, 968 (1980), Angew. Chem. Int. Ed. Engl. *19*, 929 (1980), and references cited therein
b) Juliá, S. et al.: J. Chem. Soc. Perkin I *1982*, 1317
c) Colonna, S. et al.: Tetrahedron *39*, 1635 (1983)
d) Juliá, S. et al.: Anal Quim *1983*, in press
78. Yamada, S., Mashiko, T., Terashima, S.: J. Am. Chem. Soc. *99*, 1980 (1977)
79. Michaelson, R. C., Palermo, R. E., Sharpless, K. B.: ibid. *99*, 1990 (1977)
80. Coleman-Kammula, S., Duim-Koolstra, E. Th.: J. Organometal. Chem. *246*, 53 (1983)
81. Mashiko, T., Terashima, S., Yamada, S.: Yakugaku Zasshi *100*, 319 (1980)
82. a) Mukaiyama, T., Ikegawa, A., Suzuki, K.: Chem. Lett. *1981*, 165
b) Suzuki, K., Ikegawa, A., Mukaiyama, T.: Bull. Chem. Soc. Jpn. *55*, 3277 (1982), for related Michael additions catalyzed by alkaloids see: a) Ahuja, R. R. et al.: Ind. J. Chem. *21 B*, 299 (1982), b) Bhongle, N. N.: Ind. J. Chem. *21 B*, 304 (1982), and papers cited therein
83. Suzuki, K., Ikegawa, A., Mukaiyama, T.: Chem. Lett. *1982*, 899
84. a) Onuma, K., Itoh, T., Nakamura, A.: ibid. *1980*, 481
b) Meyer, D. et al.: J. Org. Chem. *45*, 4680 (1980)
c) Poulin, J.-C., Meyer, D., Kagan, H. B.: C.R. Acad. Sci., Ser. C *1980*, 69
d) Izumiya, N. et al.: J. Am. Chem. Soc. *99*, 8346 (1977)
e) Bycroft, B. W., Lee, G. R.: J. Chem. Soc. Chem. Commun. *1975*, 988
f) Bycroft, B. W., Lee, G. R.: ibid. *1976*, 616
g) Davies, J. S., Ibrahim, M. N.: Tetrahedron Lett. *1977*, 1453
h) Poisel, H., Schmidt, U.: Chem. Ber. *106*, 3408 (1973), cf. Pieroni, O. et al.: Int. J. Peptide Protein Res. *10*, 107 (1977)
i) Nakayama, M. et al.: Bull. Chem. Soc. Jpn. *44*, 1150 (1971)
j) Davies, J. S., Eaton, M. C., Ibrahim, M. N.: J. Heterocycl. Chem. *17*, 1813 (1980)
k) Ojima, J., Yatabe, M.: Chem. Lett. *1982*, 1335
l) Akabori, S. et al.: Proc. Jpn. Acad. Sci. *27*, 7 (1951)

m) Bergmann, M., Tietzman, J. E.: J. Biol. Chem. *155*, 535 (1944)
n) Hashimoto, Y. et al.: Int. J. Pept. Protein Res. *21*, 11 (1983)
85. Compagnia di Ricerca Chimia (Inventors: Decorte, E. et al.) Germ. Pat. Appl. (DOS) 32 19 985 (1982)
86. Vincent, M. et al.: Tetrahedron Lett. *1982*, 1677
87. Fujimoto, R., Kishi, Y.: ibid. *1981*, 4197
88. a) Robins, D. J., Sakdarat, S.: J. Chem. Soc. Chem. Commun. *1979*, 1181
b) Robins, D. J., Sakdarat, S.: J. Chem. Soc. Perkin I *1981*, 909
89. Tamura, M., Harada, K.: Bull. Chem. Soc. Jpn. *53*, 561 (1980)
90. a) Mukaiyama, T. et al.: Chem. Lett. *1977*, 783
b) Asami, M. et al.: Bull. Chem. Soc. Jpn. *51*, 1869 (1978)
c) Asami, M., Mukaiyama, T.: Heterocycles *12*, 499 (1979), for similar $NaBH_4$ complexes cf Yamada, K. et al.: Tetrahedron Lett. *22*, 3869 (1981)
91. Umino, N. et al.: Chem. Pharm. Bull. *27*, 1479 (1979)
92. a) Baba, N. et al.: J. Chem. Soc. Chem. Commun. *1980*, 815
b) Seki, M. et al.: J. Org. Chem. *48*, 1370 (1983)
c) Baba, N. et al.: Angew. Chem. *94*, 465 (1982), Angew. Chem. Int. Ed. Engl. *21*, 433 (1982)
93. Yamada, K. et al.: Chem. Pharm. Bull. *31*, 70 (1983)
94. Yamada, K., Takeda, M., Iwakuma, T.: J. Chem. Soc. Perkin I *1983*, 265
95. a) Tanno, N., Terashima. S.: Chem. Pharm. Bull. *31*, 837 (1983)
b) Tanno, N., Terashima, S.: ibid. *31*, 821 (1983)
96. Itsuno, S. et al.: J. Chem. Soc. Chem. Commun. *1983*, 469
97. a) Vries, de J. G., Kellog, R. M.: J. Am. Chem. Soc. *101*, 2760 (1979)
b) Katagi, T., Kuriyama, H.: Heterocycles *19*, 1681 (1982)
c) Jolley, S. T., Bradshaw, J. S., Izatt, R. M.: J. Heterocyclic Chem. *19*, 3 (1982)
98. Baraldi, P. G. et al.: J. Chem. Soc. Perkin I *1982*, 2983
99. Buckley, T. F., Rapoport, H.: J. Am. Chem. Soc. *103*, 6157 (1981)
100. a) Rüeger, H., Benn, M.: Heterocycles *19*, 1677 (1982)
b) Rüeger, H., Benn, M.: ibid. *20*, 235 (1983)
101. Budzikiewicz, H. et al.: Liebigs Ann. Chem. *1979*, 1212
102. a) Knoop, F., Martius, C.: Z. Physiol. Chem. *258*, 238 (1939). The authors believed that they had synthesized octopine; it was shown later, however [Herbst, R. M., Swart, E. A.: J. Org. Chem. *11*, 366 (1946)], that they had obtained isooctopine
b) Akasi, S.: J. Biochem. (Tokyo) *25*, 261, 281 (1937)
c) Harada, K.: Nature *212*, 1571 (1966)
d) Harada, K.: J. Org. Chem. *32*, 1790 (1967)
e) Harada, K., Iwasaki, T., Okawara, T.: Bull. Chem. Soc. Jpn. *46*, 1901 (1973)
f) Hiskey, R. G., Northrop, R. C.: J. Am. Chem. Soc. *87*, 1753 (1965)
g) Kitajima, Y., Waki, M., Izumija, N.: Bull. Chem. Soc. Jpn. *55*, 3870 (1982)
103. a) Yamada, S., Hashimoto, S.: Tetrahedron Lett. *1976*, 997
b) Yamada, S. et al.: ibid. *1976*, 1001
c) Harada, K., Kataoka, Y.: ibid. *1978*, 2103
d) Ikota, N., Achiwa, K., Yamada, S.: Chem. Pharm. Bull. *31*, 887 (1983)
104. Harada, K., Tamura, M.: Bull. Chem. Soc. Jpn. *52*, 1227 (1979)
105. Miyazawa, T. et al.: ibid. *55*, 341 (1982)
106. Sakai, K., Yoneda, N.: Chem. Pharm. Bull. *29*, 1554 (1981)
107. Mohr, N., Budzikiewicz, H.: Tetrahedron *38*, 147 (1982)
108. a) Mukaiyama, T.: Pure Appl. Chem. *54*, 2455 (1982)
b) Mukaiyama, T.: Tetrahedron *37*, 4111 (1981)
109. a) Mukaiyama, T., Soai, K., Kobayashi, S.: Chem. Lett. *1978*, 219
b) Soai, K., Mukaiyama, T.: ibid. *1978*, 491
c) Mukaiyama, T. et al.: J. Am. Chem. Soc. *101*, 1455 (1979)
110. Colombo, L. et al.: Tetrahedron *38*, 2725 (1982)
111. Soai, K. Mukaiyama, T.: Bull. Chem. Soc. Jpn. *52*, 3371 (1979)
112. Mukaiyama, T. et al.: Chem. Lett. *1979*, 447
113. Mukaiyama, T., Suzuki, K.: ibid. *1980*, 255
114. Sato, T. et al.: ibid. *1978*, 601

115. Iriuchijima, S.: Synthesis *1978*, 684
116. Saigo, K., Tanaka, J., Nohira, H.: Bull. Chem. Soc. Jpn. *55*, 2299 (1982)
117. Mukaiyama, T., Sakito, Y., Asami, M.: Chem. Lett. *1978*, 1253
118. Mukaiyama, T., Sakito, Y., Asami, M.: ibid. *1979*, 709
119. Sakito, Y., Mukaiyama, T.: ibid. *1979*, 1027
120. Sakito, Y. et al.: ibid. *1980*, 1223
121. Sakito, Y., Suzukamo, G.: Tetrahedron Lett. *1982*, 4953
122. Sakito, Y., Asami, M., Mukaiyama, T.: Chem. Lett. *1980*, 455
123. Asami, M., Mukaiyama, T.: ibid. *1980*, 17
124. Asami, M., Mukaiyama, T.: ibid. *1983*, 93
125. Takahashi, H., Suzuki, Y., Inagaki, H.: Chem. Pharm. Bull. *30*, 3160 (1982)
126. Suzuki, Y., Takahashi, H.: ibid. *31*, 31 (1983)
127. Takahashi, H., Kametani, T., Suzuki, Y.: Heterocycles *20*, 607 (1983)
128. Snyder, H. R. et al.: J. Am. Chem. Soc. *70*, 219 (1948)
129. Brossi, A., Focella, A., Teitel, S.: Helv. Chim. Acta *55*, 15 (1972)
130. Akimoto, H. et al.: Chem. Pharm. Bull. *22*, 2614 (1974)
131. Konda, M., Shioiri, T., Yamada, S.: ibid. *23*, 1025 (1975)
132. Konda, M., Oh-ishi, Z., Yamada, S.: ibid. *25*, 69 (1977)
133. Konda, M., Shioiri, T., Yamada, S.: ibid. *23*, 1063 (1975)
134. Dean, R. T., Rapoport, H.: J. Org. Chem. *43*, 4183 (1978)
135. a) Soerens, D. et al.: ibid. *44*, 535 (1979)
  b) Sandrin, J. et al.: Heterocycles *4*, 1101 (1976)
  c) Sandrin, J. et al.: ibid. *6*, 1133 (1976)
  d) Wu, G., Yamanaka, E., Cook, J. M.: ibid. *9*, 175 (1978)
  e) Ungemach, F. et al.: Tetrahedron Lett. *1979*, 3225
  f) Ungemach, F. et al.: J. Org. Chem. *46*, 164 (1981)
136. Grigg, R., Gunaratne, H. Q. N., McNaghten, E.: J. Chem. Soc. Perkin I *1983*, 185
137. a) DeSilva, K. T. D., King, D., Smith, G. N.: J. Chem. Soc. Chem. Commun. *1971*, 908
  b) Brown, R. T., Chapple, C.: ibid. *1973*, 886
  c) Hamaguchi, F., Nagasaka, T., Ohki, S.: Yakugaku Zasshi *94*, 351 (1974)
  d) Saxena, A. K., Jain, P. C., Anand, N.: Ind. J. Chem. *12*, 892 (1974)
138. Brossi, A., Focella, A., Teitel, S.: J. Med. Chem. *16*, 418 (1973)
139. a) Cain, M. et al.: Heterocycles *19*, 1003 (1982)
  b) Cook, J. M.: pers. communication
140. Yamada, S., Hashimoto, S.: Chem. Lett. *1976*, 921
141. a) Harada, K., Okawara, T., Matsumoto, K.: Bull. Chem. Soc. Jpn. *46*, 1865 (1973)
  b) Achiwa, K., Yamada, S.: Tetrahedron Lett. *1974*, 1799
  c) Oku, J.-I., Inoue, S.: J. Chem. Soc. Chem. Commun. *1981*, 229
  d) Harada, K., Okawara, T.: J. Org. Chem. *38*, 707 (1973)
142. a) Wittig, G., Frommeld, H. D., Suchanek, P.: Angew. Chem. *75*, 978 (1963), Angew. Chem. Int. Ed. Engl. *2*, 683 (1963)
  b) Wittig, G., Reiff, H.: Angew. Chem. *80*, 8 (1968); Angew. Chem. Int. Ed. Engl. *7*, 7 (1968)
  c) Hesse, A.: Org. Synth. *50*, 66 (1970)
143. a) Stork, G., Dowd, S. R.: J. Am. Chem. Soc. *85*, 2178 (1963)
  b) Stork, G., Dowd, S. R.: Org. Synth. *54*, 46 (1974)
144. Mea-Jacheet, D., Horeau, A.: Bull. Soc. Chim. Fr. *1968*, 4571
145. a) Meyers, A. I.: Acc. Chem. Res. *11*, 375 (1978)
  b) Meyers, A. I., Slade, J.: J. Org. Chem. *45*, 2912 (1980)
  c) Meyers, A. I. et al.: ibid. *45*, 2792 (1980)
146. a) Meyers, A. I., Williams, D. R., Druelinger, M.: J. Am. Chem. Soc. *98*, 3032 (1976)
  b) Meyers, A. I., Poindexter, G. S., Brich, Z.: J. Org. Chem. *43*, 892 (1978)
  c) Meyers, A. I. et al.: J. Am. Chem. Soc. *103*, 3081 (1981)
  d) Meyers, A. I. et al.: ibid. *103*, 3088 (1981)
  e) Whitesell, J. K., Whitesell, M. A.: J. Org. Chem. *42*, 377 (1977)
  f) De Jeso, B., Pommier, J.-C.: Tetrahedron Lett. *1980*, 4511
  g) Hashimoto, S., Koga, K.: ibid. *1978*, 573
  h) Hashimoto, S., Koga, K.: Chem. Pharm. Bull. *27*, 2760 (1979)

147. a) Yamada, S., Otani, G.: Tetrahedron Lett. *1969*, 4237
b) Otani, G., Yamada, S.: Chem. Pharm. Bull. *21*, 2112 (1973)
148. a) Yamada, S., Hiroi, K., Achiwa, K.: Tetrahedron Lett. *1969*, 4233
b) Hiroi, K., Achiwa, K., Yamada, S.: Chem. Pharm. Bull. *20*, 246 (1972)
149. a) Yamada, S., Otani, G.: Tetrahedron Lett. *1971*, 1133
b) Otani, G., Yamada, S.: Chem. Pharm. Bull. *21*, 2130 (1973)
150. Otani, G., Yamada, S.: ibid. *21*, 2125 (1973)
151. a) Sone, T., Hiroi, K., Yamada, S.: ibid. *21*, 2331 (1973)
b) Sone, T., Terashima, S., Yamada, S.: ibid. *24*, 1273 (1976)
152. Shibasaki, M., Terashima, S., Yamada, S.: ibid. *23*, 279 (1975), and references cited therein
153. Sone, T., Terashima, S., Yamada, S.: ibid. *24*, 1288 (1976)
154. Igarashi, T., Oda, J., Inouye, Y.: Bull. Inst. Chem. Res. Kyoto Univ. *50*, 222 (1972) [Chem. Abstr. *77*, 164104 (1972)]
155. Szantay, C., Szabo, L., Kalaus, G.: Tetrahedron *33*, 1803 (1977)
156. Review: Enders, D.: Chemtech *1981*, 504
157. Enders, D., Eichenauer, H.: Angew. Chem. *88*, 579 (1976), Angew. Chem. Int. Ed. Engl. *15*, 549 (1976)
158. Enders, D., Eichenauer, H.: Tetrahedron Lett. *1977*, 191
159. a) Enders, D., Eichenauer, H.: Chem. Ber. *112*, 2933 (1979)
b) Enders, D., Eichenauer, H.: Angew. Chem. *91*, 425 (1979), Angew. Chem. Int. Ed. Engl. *18*, 397 (1979)
160. a) Davenport, K. G. et al.: J. Am. Chem. Soc. *101*, 5654 (1979)
b) Ahlbrecht, H. et al.: Tetrahedron Lett. *1978*, 3961
161. Pennanen, S. J.: Scta Chem. Scand. *B35*, 555 (1981)
162. In the original publication [161)] erroneously assigned as (S)
163. Enders, D., Baus, U.: to be published in Liebigs Ann. Chem. *1983*
164. Mori, K. et al.: to be published in Tetrahedron *39* (1983)
165. a) Chuman, T. et al.: Agric. Biol. Chem. *45*, 2019 (1981)
b) Mori, K. et al.: Tetrahedron Lett. *1981*, 1127
166. Bestmann, H. J., Hirsch, H.-L.: University at Erlangen, private communication to Enders, D.
167. a) Nicolaou, K. C. et al.: J. Am. Chem. Soc. *103*, 6967 (1981)
b) Nicolaou, K. C. et al.: ibid. *103*, 6969 (1981)
168. Eichenauer, H., Friedrich, E., Lutz, W., Enders, D.: Angew. Chem. *90*, 219 (1978), Angew. Chem. Int. Ed. Engl. *17*, 206 (1978)
169. Enders, D., Eichenauer, H., Pieter, R.: Chem. Ber. *112*, 3703 (1979)
170. Eichenauer, H.: Dissertation Justus-Liebig-University at Giessen, *1980*
171. Iwasawa, N., Mukaiyama, T.: Chem. Lett. *1983*, 297
172. Iwasawa, N., Mukaiyama, T.: ibid. *1982*, 1441
173. Enders, D.: University at Bonn, *1983*, private communication
174. Enders, D., Lotter, H.: Angew. Chem. *93*, 831 (1981), Angew. Chem. Int. Ed. Engl. *20*, 795 (1981)
175. Review: Evans, D. A., Aldrichimica Acta *15*, 23 (1982)
176. a) Evans, D. A., Takacs, J. M.: Tetrahedron Lett. *1980*, 4233
b) Evans, D. A., McGee, L. R.: J. Am. Chem. Soc. *103*, 2876 (1981)
177. a) Sonnet, P. E., Heath, R. R.: J. Org. Chem. *45*, 3137 (1980)
b) Evans, D. A., Takacs, J. M., McGee, L. R., Ennis, M. D., Mathre, D. J.: Pure Appl. Chem. *53*, 1109 (1981)
178. Mori, K., Ito, T., Tanaka, K., Honda, H., Yamamoto, I.: Tetrahedron *39*, 2303 (1983)
179. Evans, D. A., Ennis, M. D., Mathre, D. J.: J. Am. Chem. Soc. *104*, 1737 (1982)
180. a) Evans, D. A., Bartroli, J., Shih, T. L.: ibid. *103*, 2127 (1981)
b) Evans, D. A., Bartroli, J.: Tetrahedron Lett. *1982*, 807
181. a) Taniguchi, M., Koga, K., Yamada, S.: Tetrahedron *30*, 3547 (1974)
b) Takano, S., Yonaga, M., Ogasawara, K.: Synthesis *1981*, 265
An alternative synthesis of lactone derivatives (178) from (D)-(+)-mannitol was also reported, c.f.:
c) Takano, S., Goto, E., Hirama, M., Ogasawara, K.: Heterocycles *16*, 389 (1981)
d) Takano, S., Goto, E., Hirama, M., Ogasawara, K.: ibid. *16*, 951 (1981)

182. a) see ref. 181d (there loc. cit. 1) for literature published prior to 1980
b) Takano, S., Yonaga, M., Ogasawara, K.: J. Chem. Soc. Chem. Commun. *1981*, 1153
c) Takano, S., Uchida, W., Hatakeyama, S., Ogasawara, K.: Chem. Lett. *1982*, 733, and literature cited therein
d) Takano, S., Tamura, N., Ogasawara, K.: J. Chem. Soc. Chem. Commun. *1981*, 1155
e) Takano, S., Kasahara, C., Ogasawara, K.: ibid. *1981*, 637
f) Takano, S., Tamura, N., Ogasawara, K., Nakagawa, Y., Sakai, T.: Chem. Lett. *1982*, 933
183. Tomioka, K., Cho, Y.-S., Sato, F., Koga, K.: ibid. *1981*, 1621, and ref. cited therein
184. Takano, S., Kasahara, C., Ogasawara, K.: ibid. *1982*, 631, and literature cited therein
185. McGarvey, G. J., Hiner, R. N., Matsubara, Y., Oh, T.: Tetrahedron Lett. *1983*, 2733
186. Iimori, T., Takahashi, Y., Izawa, T., Kokayashi, S., Ohno, M.: J. Am. Chem. Soc. *105*, 1659 (1983)
187. Schöllkopf, U., Hausberg, H.-H., Hoppe, I., Segal, M., Reiter, U.: Angew. Chem. *90*, 136 (1978), Angew. Chem. Int. Ed. Engl. *17*, 117 (1978)
188. a) Schöllkopf, U.: Topics in Current Chemistry *109*, 65 (1983)
b) Schöllkopf, U.: Tetrahedron *39*, 2085 (1983)
189. Schöllkopf, U., Groth, U., Westphalen, K. O., Deng, C.: Synthesis *1981*, 969
190. In the original publications [188b, 191] erroneously assigned as (R)
191. Groth, U., Schöllkopf, U.: Synthesis *1983*, 37
192. Groth, U., Chiang, Y.-C., Schöllkopf, U.: Liebigs Ann. Chem. *1982*, 1756
193. Schöllkopf, U., Groth, U., Gull, M.-R., Nozulak, J.: ibid. *1983*, 1133
194. Hartwig, W., Schöllkopf, U.: ibid. *1982*, 1952
195. Yamashita, T., Mitsui, H., Watanabe, H., Nakamura, N.: Bull. Chem. Soc. Jpn. *55*, 961 (1982)
196. Seebach, D., Naef, R.: Helv. Chim. Acta *64*, 2704 (1981)
197. a) Belokon, Y. N. et al.: J. Chem. Soc. Chem. Commun. *1982*, 180
b) Belokon, Y. N. et al.: J. Am. Chem. Soc. *105*, 2010 (1983)
198. Tsuchihashi, G., Mitamura, S., Inoue, S., Ogura, K.: Tetrahedron Lett. *1973*, 323
199. a) Hashimoto, S., Yamada, S., Koga, K.: J. Am. Chem. Soc. *98*, 7450 (1976)
b) Hashimoto, S., Komeshima, S., Yamada, S., Koga, K.: Tetrahedron Lett. *1977*, 2907
c) Hashimoto, S., Yamada, S., Koga, K.: Chem. Pharm. Bull. *27*, 771 (1979)
d) Hashimoto, S., Komeshima, S., Yamada, S., Koga, K.: ibid. *27*, 2437 (1979)
e) Kogen, H., Tomioka, K., Hashimoto, S., Koga, K.: Tetrahedron Lett. *1980*, 4005
f) Kogen, H., Tomoika, K., Hashimoto, S., Koga, K.: Tetrahedron *37*, 3951 (1981)
200. Asami, M., Mukaiyama, T.: Chem. Lett. *1979*, 569
201. Gustafsson, B., Hallnemo, G., Ullenius, C.: Acta Chem. Scand. *B34*, 443 (1980)
202. Imamoto, T., Mukaiyama, T.: Chem. Lett. *1980*, 45
203. Blarer, S., Seebach, D.: Chem. Ber. *116*, 2250 (1983)
204. a) Blarer, S., Schweizer, W. B., Seebach, D.: Helv. Chim. Acta *65*, 1637 (1982)
b) Seebach, D., Prelog, V.: Angew. Chem. *94*, 696 (1982), Angew. Chem. Int. Ed. Engl. *21*, 567 (1982)
205. Ahlbrecht, H., Bonnet, G., Enders, D., Zimmermann, G.: Tetrahedron Lett. *1980*, 3175
206. Fukuyama, T., Nakatsuka, S., Kishi, Y.: Tetrahedron *37*, 2045 (1981)
207. Korolev, A., Mur, V.: Dokl. Akad. Nauk. SSSR *59*, 251 (1948), [Chem. Abstr. *42*, 6776f (1948)]
208. Mori, Y.: Yuki Gosei Kagaku Kyokai Shi *40*, 321 (1982)
209. Choy, W., Reed, L. A., Masamune, S.: J. Org. Chem. *48*, 1137 (1983)
210. Oppolzer, W., Flaskamp, E.: Helv. Chim. Acta *60*, 204 (1977)
211. Oppolzer, W., Thirring, K.: J. Am. Chem. Soc. *104*, 4978 (1982)
212. a) Stork, G., Nakahara, Y., Greenlee, W. J.: ibid. *100*, 7775 (1978)
b) Pyne, S. G., Hensel, M. J., Byrn, S. R., McKenzie, A. T., Fuchs, P. L.: ibid. *102*, 5960 (1980)
c) Pyne, S. G., Hensel, M. J., Fuchs, P. L.: ibid. *104*, 5719 (1982)
d) Pyne, S. G., Spellmeyer, D. C., Chen, S., Fuchs, P. L.: ibid. *104*, 5728 (1982)
213. Danishefsky, S., Kobayashi, S., Kerwin, J. F.: J. Org. Chem. *47*, 1981 (1982)
214. Stanfield, C. F., Parker, J. E., Kanellis, P.: ibid. *46*, 4797 (1981)
215. Mukaiyama, T., Iwasawa, N.: Chem. Lett. *1981*, 29, and references cited therein
216. a) Baggiolini, E. G., Lee, H. L., Pizzolato, G., Uskoković, M. R.: Bull. Soc. Chim. Belg. *91*, 967 (1982)

b) Baggiolini, E. G., Lee, H. L., Pizzolato, G., Uskoković, M. R.: J. Am. Chem. Soc. *104*, 6460 (1982)
217. Danishefsky, S., Morris, J., Clizbe, L. A.: ibid. *103*, 1602 (1981)
218. a) Grigg, R., Kemp, J., Malone, J., Tangthongkum, A.: J. Chem. Soc. Chem. Commun. *1980*, 648
b) Grigg, R., Kemp, J.: Tetrahedron Lett. *1980*, 2461
219. Hiroi, K., Nakazawa, K.: Chem. Lett. *1980*, 1077
220. Elad, D., Sperling, J.: J. Am. Chem. Soc. *93*, 967 (1971)
221. Kametani, T., Takagi, N., Toyota, M., Honda, T., Fukumoto, K.: J. Chem. Soc. Perkin I *1981*, 2830
222. Pfaltz, A., Bühler, N., Neier, R., Hirai, K., Eschenmoser, A.: Helv. Chim. Acta *60*, 2653 (1977)
223. Mukerjee, A. K., Singh, A. K.: Tetrahedron *34*, 1731 (1978)
224. Townsend, C. A., Nguyen, L. T.: Tetrahedron Lett. *1982*, 4859
225. Floyd, D. M. et al.: J. Org. Chem. *47*, 5160 (1982)
226. a) Shiozaki, M. et al.: Tetrahedron *39*, 2399 (1983)
b) Yanagisawa, H. et al.: Tetrahedron Lett. *1983*, 1037
c) Shiozaki, M. et al.: ibid. *1981*, 5205
227. Ugi, I.: Angew. Chem. *94*, 826 (1982), Angew. Chem. Int. Ed. Engl. *21*, 810 (1982)
228. Reider, P. J., Grabowski, E. J. J.: Tetrahedron Lett. *1982*, 2293
229. a) Kolb, M., Barth, J.: ibid. *1979*, 2999
b) Kolb, M., Barth, J.: Angew. Chem. *92*, 753 (1980), Angew. Chem. Int. Ed. Engl. *19*, 725 (1980)
230. Callens, R. E. A., Anteunis, M. J. O., Reyniers, F.: Bull. Soc. Chim. Belg. *91*, 713 (1982)
231. a) Terashima, S., Jew, S.: Tetrahedron Lett. *1977*, 1005
b) Jew, S., Terashima, S., Koga, K.: Tetrahedron *35*, 2337 (1979)
c) Jew, S., Terashima, S., Koga, K.: ibid. *35*, 2345 (1979)
d) Terashima, S., Hayashi, M., Koga, K.: Tetrahedron Lett. *1980*, 2733
e) Hayashi, M., Terashima, S., Koga, K.: ibid. *37*, 2797 (1981)
f) Terashima, S., Jew, S., Koga, K.: Chem. Lett. *1977*, 1109
g) Terashima, S., Jew, S., Koga, K.: Tetrahedron Lett. *1977*, 4507
h) Jew, S., Terashima, S., Koga, K.: Chem. Pharm. Bull. *27*, 2351 (1979)
i) Terashima, S., Jew, S., Koga, K.: Tetrahedron Lett. *1978*, 4937
j) Rüeger, H., Benn, M.: Heterocycles *19*, 23 (1982)
232. Shin, C., Yonezawa, Y., Sato, Y., Nakano, T., Yoshimura, J.: ibid. *20*, 405 (1983)
233. a) Schmidt, U., Oehler, E.: Angew. Chem. *88*, 54 (1976); Angew. Chem. Int. Ed. Engl. *15*, 42 (1976)
b) Oehler, E., Prantz, E., Schmidt, U.: Chem. Ber. *111*, 1058 (1978)
234. Vasella, A., Voeffray, R., Pless, J., Huguenin, R.: Helv. Chim. Acta *66*, 1241 (1983)
235. Mukaiyama, T., Ikegawa, A., Suzuki, K.: Chem. Lett. *1981*, 165
236. Rebek, J., Tai, D. F.: Tetrahedron Lett. *1983*, 859
237. a) Overman, L. E., Bell, K. L.: J. Am. Chem. Soc. *103*, 1851 (1981)
b) Overman, L. E., McCready, R. J.: Tetrahedron Lett. *1982*, 4887
238. Fray, M. J., Thomas, E. J., Wallis, J. D.: J. Chem. Soc. Perkin I, *1983*, 395
239. a) Dormoy, J.-R.: Synthesis *1982*, 753
b) Dormoy, J.-R., Castro, B., Chappuis, G., Fritschi, U. S., Grogg, P.: Angew. Chem. *92*, 761 (1980), Angew. Chem. Int. Ed. Engl. *19*, 742 (1980)
240. Fujimoto, Y., Irreverre, F., Karle, J. M., Karle, J. L., Witkop, B.: J. Am. Chem. Soc. *93*, 3471 (1971)
241. a) Confalone, P. N., Pizzolato, G., Baggiolini, E. G., Lollar, D., Usković, M. R.: ibid. *99*, 7020 (1977)
b) Parthasarathy, R., Paul, B., Korytnyk, W.: ibid. *98*, 6634 (1976)
242. a) Nagasawa, H. T., Goon, D. J. W., DeMaster, E. G.: J. Med. Chem. *21*, 1274 (1978)
b) Nagasawa, H. T., Goon, D. J. W., Shirota, F. N.: J. Heterocyclic Chem. *18*, 1047 (1981)
c) Torrini, J., Paradisi, M. P., Romeo, A.: ibid. *18*, 1451 (1981)
243. Shin, C., Ohmatsu, H., Sato, Y., Yoshimura, J.: Chem. Lett. *1981*, 701
244. Shibata, S., Matsushita, H., Kaneko, H., Noguchi, M., Saburi, M., Yoshikawa, S.: ibid. *1981*, 217

245. a) Shibata, S., Matsushita, H., Kaneko, H., Noguchi, M., Saburi, M., Yoshikawa, S.: Heterocycles *16*, 1901 (1981)
    b) Shibata, S., Matsushita, H., Kaneko, H., Noguchi, M., Saburi, M., Yoshikawa, S.: Bull. Chem. Soc. Jpn. *55*, 3546 (1982)
246. a) Shibata, S., Matsushita, H., Noguchi, M., Saburi, M., Yoshikawa, S.: Chem. Lett. *1978*, 1305
    b) Shibata, S., Matsushita, H., Kato, K., Noguchi, M., Saburi, M., Yoshikawa, S.: Bull. Chem. Soc. Jpn. *52*, 2938 (1979)
247. Shibasaki, M., Terashima, S., Yamada, S.: Chem. Pharm. Bull. *24*, 315 (1976)
248. Hiroi, K., Yamada, S.: ibid. *23*, 1103 (1975)
249. a) Belokon, Y. N. et al.: Tetrahedron *33*, 2551 (1977)
    b) Belokon, Y. N. et al.: ibid. *36*, 2327 (1980)
    e) Belokon, Y. N. et al.: Inorg. Chim. Acta *55*, 117 (1981)
250. a) Marks, K., Siemion, I. Z., Sucharda-Sobczyk, A.: Pol. J. Chem. *56* (1983), in press
    b) Siemion, I. Z., Marks, K., Sucharda-Sobczyk, A.: Bull. Acad. Pol. Chim. *1983*, in press
251. a) Koizumi, T., Kobayashi, Y., Amitani, H., Yosshii, E.: J. Org. Chem. *42*, 3459 (1977)
    b) Koizumi, T., Amitani, H., Yoshii, E.: Tetrahedron Lett. *1978*, 3741
    c) Koizumi, T., Amitani, H., Yoshii, E.: Synthesis *1979*, 110
    d) Koizumi, T., Takagi, H., Yoshii, E.: Chem. Lett. *1980*, 1403
    e) Koizumi, T., Yanada, R., Takagi, H., Hirai, H., Yoshii, E.: Tetrahedron Lett. *1981*, 477, 571
    f) Kobayashi, Y., Koizumi, T., Yoshii, E.: Chem. Pharm. Bull. *27*, 1641 (1979)
252. Lavine, T. F.: J. Biol. Chem. *151*, 281 (1943)
253. a) Billington, D. C., Golding, B. T.: Tetrahedron Lett. *1979*, 2937
    b) Billington, D. C., Golding, B. T.: J. Chem. Soc. Perkin I *1982*, 1283
254. a) Bordignon, E., Cattalini, L., Natile, G., Scatturin, A.: J. Chem. Soc. Chem. Commun. *1973*, 878
    b) Natile, G., Bordignon, E., Cattalini, L.: Inorg. Chem. *15*, 246 (1976)
255. a) Ottenheijm, H. C. J., Liskamp, R. M. J., Van Nispen, S. P. J. M., Boots, H. A., Tijhuis, M. W.: J. Org. Chem. *46*, 3273 (1981)
    b) Iwakawa, M., Pinto, B. M., Szarek, W. A.: Can. J. Chem. *56*, 326 (1978)
    c) Nachtergaele, W. A., Anteunis, M. J. O.: Bull. Soc. Chim. Belg. *89*, 749 (1980)
256. Kjaer, A., Malver, O.: Chem. Scripta *20*, 42 (1982)
257. a) Taniguchi, M., Koga, K., Yamada, S.: Tetrahedron *30*, 3547 (1974)
    b) Koga, K., Taniguchi, M., Yamada, S.: Tetrahedron Lett. *1971*, 263
258. Koga, K., Yamada, S., Yoh, M., Mizoguchi, T.: Carbohydr. Res. *36*, C 9 (1974)
259. Szechner, B.: Tetrahedron *37*, 949 (1981)
260. a) Losse, G., Wehrstedt, K.-D.: J. Prakt. Chem. *323*, 869 (1981)
    b) Yamagishi, K. et al.: Chem. Pharm. Bull. *28*, 2783 (1980)
    c) Kutney, J. P. et al.: J. Am. Chem. Soc. *100*, 938 (1978)
    d) Kametani, T., Kanaya, M., Ihara, M.: J. Chem. Soc. Perkin I *1981*, 959
    e) Nakamura, E.: Tetrahedron Lett. *1981*, 663
261. Takano, S., Iwata, H., Ogasawara, K.: Heterocycles *9*, 845 (1978)
262. a) Steglich, W. et al.: Proc. of the 8th European Peptide Symp. Noordwijk 1966, Amsterdam: North-Holland Publ. Co. (1967), p. 67
    b) Teramoto, T.: Tokyo Institute of Polytechnics, pers. commun.
263. Büchi, G., DeShong, P. R., Katsumura, S., Sugimura, Y.: J. Am. Chem. Soc. *101*, 5084 (1979)
264. Nagao, Y., Yagi, M., Ikeda, T., Fujita, E.: Tetrahedron Lett. *1982*, 201
265. a) Terashima, S., Yamada, S., Nara, M.: ibid. *1977*, 1001
    b) Terashima, S., Nara, M., Yamada, S.: ibid. *1978*, 1487
266. a) Terashima, S., Nara, M., Yamada, S.: ibid. *1978*, 3379
    b) Nara, M., Terashima, S., Yamada, S.: Tetrahedron *36*, 3171 (1980)
267. Nara, M., Terashima, S., Yamada, S.: ibid. *36*, 3161 (1980)
268. Mukaiyama, T., Yamashita, H., Asami, M.: Chem. Lett. *1983*, 385
269. a) Yamaguchi, M. et al.: Inorg. Chem. *19*, 2010 (1980)
    b) Yamaguchi, M. et al.: ibid. *20*, 3179 (1981)
    c) Slyudkin, O. P., Kerzhentsev, M. A.: Russ. J. Inorg. Chem. *24*, 1695 (1979)

d) Pasini, A., Casella, L.: J. Inorg. Nucl. Chem. *36*, 2133 (1974)
e) Yamaguchi, M. et al.: Inorg. Chem. *21*, 4138 (1982)
f) Phipps, D. A.: J. Mol. Catal. *5*, 81 (1979)

270. a) Ueoka, R., Murakami, Y.: J. Chem. Soc. Perkin II *1983*, 219
b) Kawaguchi, K., Tanihara, M., Imanishi, Y.: Polymer J. *15*, 97 (1983)
c) Kikuchi, Y., Tanihara, M., Imanishi, Y.: Int. J. Biol. Macromol. *4*, 305 (1982)
d) Tanihara, M., Kikuchi, Y., Imanishi, Y.: ibid. *4*, 297 (1982)
e) Ihara, Y., Hosako, R., Nango, M., Kuroki, N.: J. Chem. Soc. Perkin II *1983*, 5
f) Ogino, K., Tomita, I., Machiya, K., Tagaki, W.: Chem. Lett. *1982*, 1875
g) Ihara, Y., Kunikiyo, N., Kunimasa, T., Nango, M., Kuroki, N.: ibid. *1981*, 667
h) Brown, J. M., Bunton, C. A.: J. Chem. Soc. Chem. Commun. *1974*, 969
i) Brown, J. M., Bunton, C. A.: ibid. *1974*, 971
j) Ihara, Y., Nango, M., Kuroki, N.: J. Org. Chem. *45*, 5009 (1980)
k) Ihara, Y.: J. Chem. Soc. Perkin II *1980*, 1483
l) Brown, J. M., Elliot, R. L., Criggs, C. G., Helmchen, G., Nill, G.: Angew. Chem. *93*, 906 (1981); Angew. Chem. Int. Ed. Engl. *20*, 890 (1981)
m) Murakami, Y., Nakano, A., Yoshimatsu, A., Fukuya, K.: J. Am. Chem. Soc. *103*, 728 (1981), and references cited therein
n) Cho, I., Shin, J.-S.: Makromol. Chem. *183*, 2041 (1982)

Author Index Volumes 101–125

Contents of Vols. 50–100 see Vol. 100
Author and Subject Index Vols. 26–50 see Vol. 50

The volume numbers are printed in italics

GPSR Compliance

The European Union's (EU) General Product Safety Regulation (GPSR) is a set of rules that requires consumer products to be safe and our obligations to ensure this.

If you have any concerns about our products, you can contact us on ProductSafety@springernature.com

In case Publisher is established outside the EU, the EU authorized representative is:

Springer Nature Customer Service Center GmbH
Europaplatz 3
69115 Heidelberg, Germany

Batch number: 10620881

Printed by Printforce, the Netherlands